HAIRY ROOTS

HAIRY ROOTS
Culture and Applications

edited by

Pauline M. Doran

University of New South Wales, Australia

harwood academic publishers
Australia • Canada • China • France • Germany • India • Japan
Luxembourg • Malaysia • The Netherlands • Russia • Singapore
Switzerland • Thailand • United Kingdom

Amsteldijk 166
1st Floor
1079 LH Amsterdam
The Netherlands

British Library Cataloguing in Publication Data

Hairy roots : culture and application
  1. Roots (Botany) 2. Transgenic plants 3. Plant genetic
  transformation
  I. Doran, Pauline M.
  631.5'233

  ISBN 90-5702-117-X

CONTENTS

PREFACE

Genetically transformed or 'hairy' roots are used in a wide variety of research applications, and have received an increasing amount of attention over the past decade. Interest in hairy roots has expanded tremendously from initial work on the molecular biology of agrobacterial infection and investigations into the culture requirements and *in vitro* characteristics of excised hairy roots. Hairy roots are now applied in studies of plant secondary metabolism and its genetic manipulation, as hosts for production of foreign proteins, for plant propagation in agriculture, in environmental research, and for development of new engineering technology for large-scale production of plant chemicals. With the realisation that many of the limitations associated with whole plant systems and unorganised cells can be overcome using hairy roots, hairy root culture has become an active and important research area.

This book on hairy roots has been prepared for two main reasons. First, I thought that the field of hairy root research had progressed sufficiently by 1995 that it was time for someone to try drawing together the various strands to coordinate a review of the major developments that have occurred. Although several books have appeared in recent years covering the broader area of plant tissue culture and its applications, there was none focusing exclusively on hairy roots. Yet I believed that the intensity of interest in this topic in laboratories around the world, as well as the increasingly widespread inclusion of hairy root material in undergraduate and postgraduate curricula, warranted a dedicated volume. The rapidly rising number of hairy root papers appearing in the literature provides ready evidence of the expansion of interest in this form of plant tissue culture. As hairy root culture continues to be a fast moving field of scientific endeavour, this book can only claim to provide a snapshot of the main areas of development at the time of writing. It is to be expected that new topics will continue to emerge.

The second motivation behind this book was to present hairy roots as a subject of interdisciplinary interest. Scientists and engineers from a range of backgrounds have contributed to our understanding of hairy roots. In the past the biological aspects have received somewhat greater prominence, but key technological advances are now being made as chemical engineers examine hairy roots from a process development point of view. By bringing together all these perspectives, with additional input from agricultural scientists and those engaged in ecological and environmental studies, I have aimed to provide an opportunity for wider appreciation of activity in this field, and for cross-disciplinary exchange of ideas, scientific approaches and techniques. If this objective is fulfilled and the book inspires or stimulates further innovative and interesting research on hairy roots, it will be serving a worthwhile purpose. My intentions in this area are assisted by the involvement of authors from many different locations around the world, whose work perhaps would not be otherwise gathered together in a single volume.

The book has been prepared primarily as a resource for scientists and engineers engaged in research. However, in the first chapter, an overview of hairy roots and detailed instructions for experimental procedures are provided, making this section of the book appropriate for preparation of lectures and laboratory class teaching. The remainder of the work provides the research student with a review of recent achievements and a perspective on future developments in the field. There is a distinct emphasis in the book on methods for application and exploitation of hairy roots. Fundamental aspects of the molecular basis of hairy root initiation and *rol* gene action are summarised principally within the opening chapter.

Preparation of this book has relied on contributions from 53 authors and co-authors. I would like to thank all those who have written chapters for their generous and often enthusiastic involvement in this project. The efforts of these authors and their tolerance of my many editorial intrusions are much appreciated.

Pauline M. Doran

CONTRIBUTORS

Bassetti, L
Department of Food Science and Technology
Food and Bioprocess Engineering Group
Wageningen Agricultural University, PO Box 8129
6700 EV Wageningen
The Netherlands

Bhadra, R
Department of Environmental Science and
    Engineering, MS-317
Rice University
Houston TX 77005-1892
USA

Buitelaar, R M
Agrotechnological Research Institute
ATO-DLO, PO Box 17
6700 AA Wageningen
The Netherlands

Carvalho, E B
Department of Chemical Engineering
The Pennsylvania State University
519 Wartik Laboratory
University Park PA 16802-5807
USA

Chang, H N
Bioprocess Engineering Research Center &
    Department of Chemical Engineering
Korea Advanced Institute of Science and Technology
Taejon 305-701
Korea

Christey, M C
Crop & Food Research
Private Bag 4704,
Christchurch
New Zealand

Curtis, W R
Department of Chemical Engineering and
    Biotechnology Institute
The Pennsylvania State University
519 Wartik Laboratory
University Park PA 16802-5807
USA

Doran, P M
Department of Biotechnology
University of New South Wales
Sydney NSW 2052
Australia

Ford-Lloyd, B V
School of Biological Sciences
University of Birmingham
Edgbaston Birmingham B15 2TT
UK

Geerlings, A
Division of Pharmacognosy, Leiden/Amsterdam
Center for Drug Research, Leiden University
Gorlaeus Laboratories, PO Box 9502
2300 RA Leiden
The Netherlands

Green, K D
Advanced Centre for Biochemical Engineering
University College
Torrington Place
London, WC1E 7JE
UK

Hallard, D
Division of Pharmacognosy, Leiden/Amsterdam
Center for Drug Research, Leiden University
Gorlaeus Laboratories, PO Box 9502
2300 RA Leiden
The Netherlands

Hamill, J D
Department of Genetics & Developmental Biology
Monash University
Clayton VIC 3168
Australia

Hjortso, M A
Department of Chemical Engineering
Louisiana State University
Baton Rouge LA 70803
USA

Hoge, J H C
Institute of Molecular Plant Sciences
Leiden University, Clusius Laboratory
PO Box 9505
2300 RA Leiden
The Netherlands

Holihan, S
Department of Chemical Engineering
The Pennsylvania State University
519 Wartik Laboratory
University Park PA 16802-5807
USA

Holmes, P
Fluid And Surface Transport (FAST) Team
School of Chemical Engineering
University of Birmingham
Edgbaston Birmingham B15 2TT
UK

Kanokwaree, K
Department of Biotechnology
University of New South Wales
Sydney NSW 2052
Australia

Kino-oka, M
Department of Chemical Engineering
Faculty of Engineering Science, Osaka University
1-3 Machikaneyama-cho
Toyonaka Osaka 560
Japan

Kobayashi, T
Department of Biotechnology
Faculty of Engineering, Nagoya University
Chikusa-ku Nagoya 464-01
Japan

Kotrba, P
Department of Biochemistry & Microbiology
Faculty of Food and Biochemical Technology
ICT Prague, Technická 3
166 28 Prague 6
Czech Republic

Kwok, K H
Department of Biotechnology
University of New South Wales
Sydney NSW 2052
Australia

Li, S-L
Fermentation Research
1, Tung-Hsing Street
Shu-Lin Centre
Taipei Hsein
Taiwan

Lidgett, A J
Department of Genetics & Developmental Biology
Monash University
Clayton VIC 3168
Australia

Lopes Cardoso, M I
Institute of Molecular Plant Sciences
Leiden University, Clusius Laboratory
PO Box 9505
2300 RA Leiden
The Netherlands

Macek, T E
Institute of Organic Chemistry and Biochemistry
Academy of Sciences of the Czech Republic
Flemingovo n. 2
166 10 Prague 6
Czech Republic

Macková, M
Department of Biochemistry & Microbiology
Faculty of Food and Biochemical Technology
ICT Prague, Technická 3
166 28 Prague 6
Czech Republic

Mahagamasekera, M G P
Department of Biotechnology
University of New South Wales
Sydney NSW 2052
Australia

Mugnier, J
Rhône-Poulenc Agrochimie
69263 Lyon
Cedex 09
France

Parr, A J
Institute of Food Research, Norwich Laboratory
Norwich Research Park
Colney Norwich NR4 7UA
UK

Pearsall, B
Department of Chemical Engineering
The Pennsylvania State University
519 Wartik Laboratory
University Park PA 16802-5807
USA

Rhodes, M J C
Institute of Food Research, Norwich Laboratory
Norwich Research Park
Colney Norwich NR4 7UA
UK

Ruml, T
Department of Biochemistry & Microbiology
Faculty of Food and Biochemical Technology
ICT Prague, Technická 3
166 28 Prague 6
Czech Republic

Shanks, J V
Department of Chemical Engineering, MS 362
Institute of Biosciences and Bioengineering
Rice University
Houston TX 77005-1892
USA

Sim, S J
Bioprocess Engineering Research Center &
   Department of Chemical Engineering
Korea Advanced Institute of Science and Technology
Taejon 305-701
Korea

Singh, G
Department of Chemical Engineering
The Pennsylvania State University
133 Fenske Laboratory
University Park PA 16802
USA

Skácel, F
Department of Analytical Chemistry
Faculty of Chemical Engineering
ICT Prague, Technická 5
166 28 Prague 6
Czech Republic

Subroto, M A
Department of Biotechnology
University of New South Wales
Sydney NSW 2052
Australia

Taya, M
Department of Chemical Engineering
Faculty of Engineering Science
Osaka University
1-3 Machikaneyama-cho
Toyonaka Osaka 560
Japan

Thomas, N H
Fluid And Surface Transport (FAST) Team
School of Chemical Engineering
University of Birmingham
Edgbaston Birmingham B15 2TT
UK

Tone, S
Department of Chemical Engineering
Faculty of Engineering Science
Osaka University
1-3 Machikaneyama-cho
Toyonaka Osaka 560
Japan

Tramper, J
Department of Food Science and Technology
Food and Bioprocess Engineering Group
Wageningen Agricultural University
PO Box 8129
6700 EV Wageningen
The Netherlands

Uozumi, N
Department of Biotechnology
Faculty of Engineering
Nagoya University
Chikusa-ku Nagoya 464-01
Japan

van der Heijden, R
Division of Pharmacognosy, Leiden/Amsterdam
Center for Drug Research, Leiden University
Gorlaeus Laboratories, PO Box 9502
2300 RA Leiden
The Netherlands

Verpoorte, R
Division of Pharmacognosy, Leiden/Amsterdam
Center for Drug Research, Leiden University
Gorlaeus Laboratories, PO Box 9502
2300 RA Leiden
The Netherlands

Walton, N J
Institute of Food Research, Norwich Laboratory
Norwich Research Park
Colney Norwich NR4 7UA
UK

Weathers, P J
Department of Biology and Biotechnology
Worcester Polytechnic Institute
100 Institute Road
Worcester MA 01609
USA

Whipple, M
Genzyme Corporation
500 Soldier's Field Road
Allston, MA 02134
USA

Williams, G R C
Department of Biotechnology
University of New South Wales
Sydney NSW 2052
Australia

Wilson, P D G
Institute of Food Research, Norwich Laboratory
Norwich Research Park
Colney Norwich NR4 7UA
UK

Wongsamuth, R
Department of Biotechnology
University of New South Wales
Sydney NSW 2052
Australia

Wyslouzil, B E
Department of Chemical Engineering
Worcester Polytechnic Institute
100 Institute Road
Worcester MA 01609
USA

Yoshikawa, T
School of Pharmaceutical Sciences
Kitasato University
Minato-ku
Tokyo 108
Japan

Yu, S
Department of Biotechnology
University of New South Wales
Sydney NSW 2052
Australia

Hairy Root Cultures – Opportunities and Key Protocols for Studies in Metabolic Engineering

John D. Hamill and Angela J. Lidgett

Department of Genetics and Developmental Biology, Monash University, Clayton, Melbourne, Victoria 3168, Australia; email: John.Hamill@sci.monash.edu.au

Introduction

In writing this chapter it is our intention to give a brief overview of the molecular biology of hairy root formation and to summarise the value of hairy root cultures as a tissue culture tool for secondary metabolite studies. We have also provided several laboratory protocols which work reproducibly in our hands with a wide range of dicotyledonous species. These may be useful for laboratories wishing to

- establish hairy root cultures,
- prove they are transformed by *Agrobacterium rhizogenes* T-DNA, and
- introduce foreign genes into such cultures and demonstrate that these genes are expressed.

Historical Perspectives

Pioneering work in the early part of this century showed that a soil bacterium was responsible for a neoplastic outgrowth of fine roots at the site of infection. As infected plants had reduced vigour, particularly in nursery stock, the symptoms became known as "Hairy Root Disease". *Agrobacterium rhizogenes* was subsequently identified as being the aetiological agent of the disease (Riker *et al.*, 1930;

Hildebrand, 1934). More than 450 species of many different genera and families are known to be susceptible to infection by *A. rhizogenes* (Porter, 1991) and opines, produced by infected plant tissues, act as a nutritive substrate for agrobacteria in the rhizosphere. A number of separate isolates of *A. rhizogenes* has resulted in strains being classified according to opine type, with commonly used strains being of the agropine, mannopine or cucumopine type. Comparative mapping and DNA:DNA hybridisation experiments indicated that all strains have highly conserved "core" DNA which is essential for hairy root formation (Filetici *et al.*, 1987).

The history of using hairy roots (or transformed roots as they are more accurately called) grown *in vitro* as an experimental tissue is a comparatively short one. Reports from the early 1980s indicated that roots of several species transformed by *Agrobacterium rhizogenes* were capable of quite rapid growth *in vitro* in defined nutrient medium without the requirement for exogenous phytohormones (Tepfer and Tempé, 1981; Willmitzer *et al.*, 1982). The growth characteristics of these root cultures were found to be stable over time and they were capable of producing significant quantities of opine. For example, in a carrot hairy root culture produced by infection with *A. rhizogenes* strain A4, up to 3% of the dry weight was found to be agropine (Tepfer and Tempé, 1981). At about the same time, several groups showed that T-DNA from the Ri plasmid of *A. rhizogenes*, like T-DNA from the Ti plasmid of its close relative *A. tumefaciens*, was integrated into the plant genome resulting in the differentiation and subsequent growth of hairy roots (hence the term "Transformed Roots") (Chilton *et al.*, 1982; White *et al.*, 1982; Willmitzer *et al.*, 1982). Agropine strains such as A4, HRI and LBA9402 (the latter containing the Ri plasmid from strain 1855) are effective in transforming a wide range of plant species, and their Ri plasmids contain two discrete sections of T-DNA referred to as T_L and T_R (Jouanin, 1984; Huffman *et al.*, 1984; De Paolis *et al.*, 1985). Comparative restriction mapping of the Ri plasmids from agropine strains HRI, 1855 and A4 indicated that these plasmids are almost identical to one another (Jouanin, 1984). Southern blotting and DNA: DNA hybridisation experiments showed that while there is some homology between parts of the T-DNA of agropine strains of *A. rhizogenes* and the T-DNA of certain strains of *A. tumefaciens* – notably in the genes encoding auxin biosynthesis and opine synthesis – there is also quite a large section of Ri T-DNA which is unique to *A. rhizogenes* (White and Nester, 1980; Jouanin, 1984). Auxin and opine biosynthesis genes were located on T_R T-DNA while T_L T-DNA was found to contain sequences unique to *A. rhizogenes* (Huffman *et al.*, 1984; Jouanin, 1984; De Paolis *et al.*, 1985). Mannopine strains such as 8196 and 2659 have one section of T-DNA which contains sequences homologous to T_L T-DNA of agropine strains, but they do not contain auxin biosynthesis genes. This explains their reduced capacity to induce hairy roots on tissue with low endogenous levels of auxin such as the basal (i.e. closest to the shoot) side of carrot discs (Cardarelli *et al.*, 1985; Filetici *et al.*, 1987). Agropine strains are capable of inducing roots with equivalent efficiency on either surface of carrot discs. Interestingly, after differentiation, roots formed by mannopine strains are capable of axenic growth rates similar to those of roots obtained by infection with agropine strains (Cardarelli *et al.*, 1985).

The complete sequence of T_L T-DNA from A4 was reported by Slightom *et al.* (1986) and led to the unravelling of the molecular basis for hairy root disease. Work of Spena *et al.* (1987) and Schmülling *et al.* (1988) showed that of the four genes originally recognised by transposon mutagenesis as being important for hairy disease (designated *rol*A, *rol*B, *rol*C and *rol*D, rol = <u>r</u>oot <u>l</u>ocus) (White *et al.*, 1985), three were of particular importance for hairy root formation in tobacco. The three genes *rol*A, *rol*B and *rol*C (ORF 10, 11 and 12 of T_L T-DNA) were

transferred individually and in combination into tobacco using binary vectors in conjunction with disarmed *A. tumefaciens*. Phenotypic analysis of transgenic plants showed that the genes functioned synergistically when they were together and caused plagiotropic roots to form in tobacco which had growth characteristics typical of roots formed by intact Ri T-DNA (Spena *et al.*, 1987; Schmülling *et al.*, 1988). Other genes on T_R T-DNA (auxin biosynthesis genes) and on T_L T-DNA (e.g. ORF 13 and ORF 14) appear to play a supportive role in the formation/ maintenance of hairy roots, possibly by adjusting the hormonal balance of infected plant tissues to levels which are optimal for root differentiation and growth (Offringa *et al.*, 1986; Capone *et al.*, 1989). Overexpression of ORF 13 in tobacco produced a phenotype with some characteristics of plants with elevated cytokinin levels (Hansen *et al.*, 1993). Elegant studies, reported in the early 1990s, appeared to offer an explanation for the action of key *rol* genes, at a biochemical level, in terms of increased rates of conversion of inactive auxin and cytokinin conjugates to active phytohormones (reviewed in Hamill, 1993). However, more recent experimentation has suggested that alterations in hormone perception, and possibly rates of hormone turnover *in vivo*, may be key to understanding the biochemical basis of *rol* gene action *in vivo* (Nilsson *et al.*, 1993a, 1993b; Costantino *et al.*, 1994; Filippini *et al.*, 1994; Maurel *et al.*, 1994; Delbarre *et al.*, 1994; Sandberg *et al.*, 1995). Though their precise mode of action is still unknown, clearly, genes from *A. rhizogenes* have evolved to a high degree of complexity and operate synergistically to cause growth of fast growing, plagiotropic and hormone autotrophic roots of a wide range of plant species following transfer to the plant's genome.

A. rhizogenes as a Gene Delivery System

In the mid–late 1980s there was quite a lot of interest in using *A. rhizogenes* as a vector for plant transformation given the observation that plants could, in many cases, be regenerated from hairy roots and would grow in soil – a phenomenon first reported by Ackermann (1977) (Davey *et al.*, 1987; Morgan *et al.*, 1987; Robaglia *et al.*, 1987; Stougaard *et al.*, 1987). This was especially so following reports that genes in binary vectors derived from *A. tumefaciens* were transferred to the plant genome at high frequency along with the Ri T-DNA (Simpson *et al.*, 1986; Hamill *et al.*, 1987). Plants regenerated from hairy roots often had an altered phenotype designated the T phenotype (Tepfer, 1984), which included wrinkled leaves, phenotypically altered flowers with reduced male fertility, shortened internodes, and a more branched root system than controls. A more extreme version of this phenotype, referred to as T', was also seen in some off-spring of transformed plants and was somewhat unstable as T' plants could revert to a T phenotype (Tepfer, 1984). Nevertheless, for some species, the altered phenotype was not regarded as being a problem and in some plants such as *Medicago sativa* and *Lotus corniculatus* it was not particularly apparent in aerial parts (Sukhapinda *et al.*, 1987; Spano *et al.*, 1987).

It is probably fair to say however that although the use of *A. rhizogenes* can be a relatively easy way to recover transgenic plants, advances made in recent years in transforming crop plants using either disarmed *A. tumefaciens* or effective physical uptake procedures (Klein *et al.*, 1987, 1988; Fromm *et al.*, 1990; Kaeppler *et al.*, 1992) have diminished the importance of *A. rhizogenes* as a vector for generating transformed plants. On the other hand, it remains a very useful tool for generating transgenic roots which can be used for a range of biological studies. One of the more apparent manifestations of interest in roots formed by transformation with *A. rhizogenes* is their application for secondary metabolite biosynthesis and metabolic engineering studies.

Secondary Metabolite Biosynthesis by Tissues Cultured *In Vitro*

Pioneering work of White in the 1930s showed that non-transformed roots of tomato could be cultured indefinitely *in vitro* in a relatively simple salt mixture containing yeast extract (White, 1934). Over the following 20–30 years, a number of important studies in root biology were carried out by several laboratories using non-transformed axenic root cultures as an experimental tissue (reviewed in Butcher and Street, 1964). In the early 1940s, Dawson reported that nicotine was produced by root cultures of tobacco cultured *in vitro*, with most of the alkaloid being found in the medium (Dawson, 1942a). A number of studies in the 1970s and 1980s did utilise non-transformed axenic roots for secondary metabolite studies. Though there have been reports of fast growing, non-transformed root cultures which produced high levels of secondary metabolites in defined medium devoid of phytohormones (e.g. pyrrolizidine alkaloids produced by *Senecio* species: Hartmann and Toppel, 1987; Toppel *et al.*, 1987), in most cases phytohormones need to be added to encourage such root cultures to grow at an adequate rate (e.g. *Duboisia* root cultures: Endo and Yamada, 1985; *Hyoscyamus* root cultures: Hashimoto *et al.*, 1986). This can make it difficult to maintain productive cultures. In addition, the focus of most laboratories in the early–mid 1980s was on production of secondary metabolites by disorganised cell suspensions/callus cultures and the

potential benefits of using root cultures for secondary metabolite studies were often ignored (Kurz and Constabel, 1985; Parr, 1989).

Disorganised cell cultures can offer rapid and reproducible rates of growth *in vitro*. These attributes are vital for experimental purposes and also for the development of industrial processes to produce secondary metabolites by tissues grown *in vitro*. There have been some notable successes in elucidating pathways leading to secondary metabolite biosynthesis using cell suspensions (e.g. berberine synthesis: Zenk *et al.*, 1985; Zenk, 1988). In addition, several commercially competitive processes for secondary metabolite biosynthesis by cell suspensions have been developed, e.g. the red anti-inflammatory dye shikonin produced at high concentrations by cell cultures of *Lithospermum erythrorhizon* (Fujita *et al.*, 1985; Deno *et al.*, 1987a), and the yellow antimicrobial dye berberine produced by cell cultures of *Coptis japonica* (Fujita, 1990). However many reports in the literature have indicated that disorganised cultures often tend to be unstable and unproductive over time with respect to biosynthesis of secondary metabolites. A great deal of energy has often been expended in devising media and growth conditions which promote stable secondary metabolite biosynthesis, with variable but generally limited success (Deus-Neumann and Zenk, 1984; Fowler, 1985; Wink, 1987; Parr, 1989; Drewes and van Staden, 1995a).

Hairy Roots for Secondary Metabolite Studies – Prospects and Limitations

In the mid-1980s several laboratories realised the potential of using roots transformed by *A. rhizogenes* for secondary metabolite biosynthesis studies and reported the biosynthesis of metabolites by hairy roots of a number of species, e.g. tropane alkaloids by hairy roots of *Hyoscyamus muticus* (Flores and Filner, 1985), pyridine alkaloids and betacyanin pigments by hairy roots of *Nicotiana rustica* and *Beta vulgaris*, respectively (Hamill *et al.*, 1986), and tropane alkaloids by hairy roots of *Atropa belladonna* and *Scopolia japonica* (Kamada *et al.*, 1986, and Mano *et al.*, 1986, respectively). Since then a large number of reports has been published on the production of secondary metabolites by transformed root cultures which are summarised in Table 1.1. The main advantage of transforming with *A. rhizogenes* is that the roots are generally easy to grow in defined media such as Gamborg's B5 (Gamborg *et al.*, 1968) or MS (Murashige and Skoog, 1962) without the requirement for external phytohormones. Often the roots grow rapidly with mass doubling times resembling those of disorganised cell suspensions, but unlike the latter they are fully differentiated tissues which tend to produce secondary metabolites characteristic of the species which has been transformed

Table 1.1. Examples of secondary metabolites produced by hairy roots.

Genus	Metabolite	Reference
Ajuga	Hydroxyecydsone	Tanaka and Matsumoto (1993)
Ambrosia	Thiophenes	Flores *et al.* (1988)
Armoracia	Fusicoccin	Babakov *et al.* (1995)
Artemisia	Artemisinin	Qin *et al.* (1994), Weathers *et al.* (1994), Jaziri *et al.* (1995)
Astragalus	Astragalosides	Hirotani *et al.* (1994)
Atropa	Tropane alkaloids	Kamada *et al.* (1986), Jung and Tepfer (1987), Sharp and Doran (1990)
Beta	Betalain pigments	Hamill *et al.* (1986), Taya *et al.* (1992, 1994)
Bidens	Polyacetylenes	Marchant (1988)
Brugmansia	Tropane alkaloid	Giulietti *et al.* (1993)
Campanula	Polyacetylenes	Tada *et al.* (1996)
Carthamus	Thiophenes	Flores *et al.* (1988)
Cassia	Anthraquinones	Asamizu *et al.* (1988)
	Polyketide pigments	Ko *et al.* (1995)
Catharanthus	Indole alkaloids	Parr *et al.* (1988), Toivonen *et al.* (1989), Bhadra *et al.* (1993), Sim *et al.* (1994), Jung *et al.* (1994)
Centranthus	Valepotriates	Gränicher *et al.* (1995b)
Chaenactis	Polyines	Constabel and Towers (1988)
Cinchona	Indole alkaloids	Hamill *et al.* (1989)
Coreopsis	Polyacetylenes	Marchant (1988)
Datura	Tropane alkaloids	Payne *et al.* (1987), Christen *et al.* (1989), Robins *et al.* (1990), Parr *et al.* (1990), Dupraz *et al.* (1994), Rhodes *et al.* (1994)
	Sesquiterpenes	Furze *et al.* (1991)
Daucus	Flavonoids	Bel-Rhlid *et al.* (1993)
	Anthocyanin	Kim *et al.* (1994)
Digitalis	Cardioactive glycosides	Saito *et al.* (1990)
Duboisia	Tropane alkaloid	Deno *et al.* (1987b), Mano *et al.* (1989), Yukimune *et al.* (1994)
Echinacea	Alkamides	Trypsteen *et al.* (1991)
Fragaria	Polyphenol	Motomori *et al.* (1995)
Glycyrrhiza	Glycyrrhizin	Ko *et al.* (1989)
Gynostemma	Saponin	Fei *et al.* (1993)
Hyoscyamus	Tropane alkaloids	Flores and Filner (1985), Parr *et al.* (1990), Doerk-Schmitz *et al.* (1994)
	Piperidone alkaloids	Sauerwein *et al.* (1991)
	Sesquiterpenes	Signs and Flores (1989)
Lactuca	Sesquiterpene lactones	Kisiel *et al.* (1995), Song *et al.* (1996)
Leontopodium	Anthocyanins and essential oils	Hook (1994)
Linum	Lignans	Berlin *et al.* (1988)
Lippia	Sesquiterpenes	Sauerwein *et al.* (1991)
Lithospermum	Naphthoquinone (shikonin)	Shimomura *et al.* (1991), Sim and Chang (1993)
Lobelia	Piperidine alkaloid	Yonemitsu *et al.* (1990)
	Polyacetylenes	Ishimaru *et al.* (1994), Tada *et al.* (1995a), Yamanaka *et al.* (1996)

Table 1.1. Continued.

Genus	Metabolite	Reference
Lotus	Condensed tannins	Carron *et al.* (1994)
Nicotiana	Pyridine alkaloids	Hamill *et al.* (1986), Parr and Hamill (1987), Hamill *et al.* (1990), Green *et al.* (1992), Larsen *et al.* (1993)
	Sesquiterpenoids	Wibberley *et al.* (1994)
Panax	Saponins	Yoshikawa and Furuya (1987), Inomata *et al.* (1993)
Platycodon	Polyacetylenes	Tada *et al.* (1995b)
Podophyllum	Lignans	Berlin *et al.* (1988)
Rauwolfia	Indole alkaloids	Benjamin *et al.*, (1994)
Rubia	Anthraquinone	Sato *et al.* (1991), van der Heijden *et al.* (1994), Kino-oka *et al.* (1994)
Rudbeckia	Thiophenes	Flores *et al.* (1988), Daimon and Mii (1995)
Salvia	Diterpenoid	Hu and Alfermann (1993)
Scoparia	Methoxybenzoxazolinone	Hayashi *et al.* (1994)
Scopolia	Tropane alkaloids	Mano *et al.* (1986), Parr *et al.* (1990), Ahn *et al.* (1993)
*Senecio**	Pyrrolizidine	Toppel *et al.* (1987), Hartmann and Toppel (1987)
Serratula	Ecdysteroid	Delbecque *et al.* (1995)
Sesamum	Naphthoquinone	Ogasawara *et al.* (1993)
Solanum	Steroids	Subroto and Doran (1994), Alvarez *et al.* (1994), Drewes and van Staden (1995b), Ikenaga *et al.* (1995), Yu *et al.* (1996)
Swainsona	Swainsonine	Ermayanti *et al.* (1994)
Tagetes	Thiophenes	Westcott (1988), Croes *et al.* (1989), Buitelaar *et al.* (1993), Talou *et al.* (1994), Jacobs *et al.* (1995)
Trichosanthes	Bryonolic acid	Takeda *et al.* (1994)
Valeriana	Valepotriates	Gränicher *et al.* (1994)
	Iridoid diester	Gränicher *et al.* (1995a)
Withania	Withanolides	Banerjee *et al.* (1994)

*In this case, fast growing root cultures were established in medium devoid of phytohormones without being transformed with *A. rhizogenes*. This serves to remind us that it is the fact that fully differentiated roots are being cultured, and not transformation by Ri T-DNA *per se*, which accounts for the large number of reports of secondary metabolite formation by hairy roots as indicated in Table 1.1.

(reviewed in Signs and Flores, 1990, and Hamill and Rhodes, 1993; Table 1.1).

Whilst hairy roots have much to offer for secondary metabolism studies, one possible limitation which should be noted is that they may not produce compounds which are produced in aerial tissues of plants. For example, Parr *et al.* (1988) found that vindoline and the derivative anti-leukaemic drug vinblastine were produced at very low levels in hairy roots of *Catharanthus roseus*. Previous studies have shown that these compounds are found in leaves of shoot cultures grown *in vitro* (Endo *et al.*, 1987). Similarly, Nguyen *et al.* (1992) established hairy roots from several *Psoralea* species which contain significant quantities of the furanocoumarins, psoralen and angelicin, in both aerial tissues and roots of plants grown in soil. No furanocoumarins were detected in any of these hairy root cultures, suggesting that synthesis *in planta* may be exclusively in aerial tissues (Nguyen *et al.*, 1992). In other studies, secondary metabolites have been found in hairy roots even though the main site of metabolite deposition *in planta* may be in aerial tissues. For example quinine, found at high levels in bark of *Cinchona* species, is synthesised at appreciable levels in hairy roots of *Cinchona ledgeriana* (Hamill *et al.*, 1989). Pyridine and tropane alkaloids are also synthesised in roots but not in shoots of species from several solanaceous genera (Dawson, 1942b; Waller and Nowacki, 1978; Saito *et al.*, 1989; Hashimoto *et al.*, 1991; Subroto *et al.*, 1996a, 1996b). Thus, unless it is known that roots cannot synthesise a particular metabolite which is of interest, it is well worth trying to establish a hairy root culture of a producing species to assess its biosynthetic capacity.

Examples of Difficult or Recalcitrant Species

The *Cinchona ledgeriana* hairy roots referred to above are also noteworthy because this species was very difficult to establish *in vitro* as a hairy root culture. A stable axenic hairy root culture of *C. ledgeriana* was established only when appropriate growth conditions were determined by an empirical approach (Hamill *et al.*, 1989). It is impossible to assess with accuracy how many experiments have been carried out with the aim of producing hairy root cultures but which have failed. In this laboratory, we have attempted to produce transformed roots of *Castanospermum australe*, an Australian subtropical tree species which produces in its seeds large quantities of castanospermine which may have clinical value for treatment of viral infections. So far we have obtained roots occasionally on seedlings after infection with agropine strains of *A. rhizogenes*, but these have failed to grow *in vitro* in standard media. We have also induced roots to form on seedlings of several temperate eucalypt species by inoculation with agropine strains of *A. rhizogenes* (Pelosi, unpublished), but again these have failed to grow *in vitro* in standard media. This is interesting as there has been a report of a vigorous hairy root culture of *E. grandis* (a subtropical eucalypt species) after transformation with *A. rhizogenes* strain LBA9402 using standard media and growth conditions (MacRae and van Staden, 1993). In a report by Mugnier (1988), he noted that roots were formed on *Amaranthus retroflexus*, *Sedum acris* L., *Coleus blumei* Benth, *Capsella bursa-pastoris*, *Thlaspi arvensis*, several *Brassica* species, *Artemisia absinthium*, *Achillea millefolium*, *Matricaria chamomilla* L., *Senecio vulgaris* L., *Hibiscus esculentus* L., *Lysimachia vulgaris* L., *Borrago officinalis* L., *Silene inflata* L., *Rumex obtusifollus* L. and *R. acetosella* L., and several species in the genus *Euphorbia* following inoculation with *A. rhizogenes* A4. However, poor development prevented hairy root cultures of these species from being established. Handa (1991) noted that roots formed on plants of *Dianthus chinensis*, *Brassica campestris*, *Prunus incisa*, *Lupinus polyphyllus*, *Curcubita pepo*, *Lagenaria siceraria* and *Solanum melongena* following infection with an agropine (A4) or two mikimopine (1724 and A13) strains, but that it was not possible to establish them as root cultures *in vitro*. Mugnier (1988) also noted a lack of success in inducing roots to form in species belonging to the Ranunculaceae and Papaveraceae after

infection with *A. rhizogenes* A4, and we too have failed to induce roots on several species of *Papaver* using LBA9402, 15834 (agropine) and 8196 (mannopine) strains of *A. rhizogenes* (Hamill, unpublished). Recently, Williams and Ellis (1993) did manage to transform *Papaver somniferum* with *A. rhizogenes*, though the culture was difficult to maintain *in vitro* as transformed roots and grew in hormone-free medium as a cell suspension which did not produce morphinan alkaloids in a stable manner. It is possible that the use of different isolates or genetically engineered strains of *A. rhizogenes* with altered combinations of Ri genes, in conjunction with better tissue culture media, may enable transformed roots to be induced and cultivated from "challenging" species such as those noted above.

Hairy root cultures have also not been reported for any monocotyledonous species. Infection of 16 monocotyledonous species with *A. rhizogenes* produced no symptoms (De Cleene and De Ley, 1981), though tumour induction was reported in onion after inoculation with several agropine or mannopine strains of *A. rhizogenes* (Dommisse *et al.*, 1990). One strain, HRI, produced tumours from which roots differentiated; however, agropine or mannopine was not reported to have been produced by these tumorous tissues (Dommisse *et al.*, 1990). *A. rhizogenes* was also shown to be effective in promoting agroinfection of cereals (using it as a vector to deliver an infective viral genome cloned in a binary vector: Boulton *et al.*, 1989), and recent reports indicate that the close relative of *A. rhizogenes*, *A. tumefaciens*, can be used successfully to transform rice (Hiei *et al.*, 1994). Thus it may be more correct to state that the genes carried on Ri T-DNA of *A. rhizogenes* do not usually induce roots to form on monocots. The reason for this difference between monocots and dicots is of fundamental interest for root biology studies but is largely a matter of speculation at present and thus will not be considered further here.

Stability of Metabolite Production by Hairy Roots

Stability in synthesis, together with reproducibility between production runs, are essential prerequisites for any commercial production process and are also highly desirable for experimental purposes. If handled properly, hairy roots can offer this property for secondary metabolite studies. There are reports of some cytological variability in hairy root cultures, including variation in ploidy, chromosome number and chromosome structure (Banerjee-Chattopadhay *et al.*, 1985; Hänisch ten Cate *et al.*, 1987; Ramsay and Kumar, 1990; Ermayanti *et al.*, 1993). However, in general transformed roots show a high degree of chromosomal stability which is expected of tissues in which a functional meristem is maintained (Aird *et al.*, 1988a). Disruption of the differentiated state can be induced by addition of phytohormones to promote growth of undifferentiated cells at the expense of differentiated roots (Flores and Filner, 1985; Aird *et al.*, 1988b). It can also occur if root cultures are not maintained carefully, e.g. applying excessive mechanical forces during subculturing, or subculturing roots into fresh medium at a high density. Loss of differentiated root tissue is likely to be accompanied by severely reduced levels of secondary metabolite biosynthesis or accumulation (Flores and Filner, 1985; Rhodes *et al.*, 1989; Robins *et al.*, 1991; Hamill, unpublished).

Several reports note that hairy roots, properly maintained and subcultured at regular intervals, remain more or less stable in their growth and/or secondary metabolite productivity characteristics for a number of years (Mugnier, 1988; Tepfer, 1989; Signs and Flores, 1990). In a study carried out by Maldonado-Mendoza *et al.* (1993), it was found that the capacity of some hairy root lines of *Datura stramonium* to synthesise tropane alkaloids remained at the same level (~1% dry weight) for 5 years, representing about 75 subcultures. In recent experiments, we found that the hairy root culture of *N. rustica* which we first reported to produce nicotine at about 300 μg g^{-1} fresh weight (Hamill *et al.*, 1986) continues to produce nicotine at similar levels more than a decade later (Lidgett and Hamill, unpublished). This particular culture has been maintained by a number of personnel and has been passaged *in vitro* approximately at monthly intervals since it was established in 1985.

Prospects for Alteration of Secondary Metabolite Levels by Direct Manipulation

As has been noted previously, foreign genes ligated into binary vectors can be easily introduced into *A. rhizogenes* and efficiently incorporated into the plant genome at the same time as *A. rhizogenes* T-DNA transformation (Simpson *et al.*, 1986; Hamill *et al.*, 1987). Alternatively, if a transformation procedure using disarmed *A. tumefaciens* or a DNA physical uptake method can be used to generate transgenic plants, these can be transformed with wild-type *A. rhizogenes* to produce hairy roots which also contain and express the foreign gene (e.g. Hashimoto *et al.*, 1993). Fusion of coding sequences to strong constitutive promoters, or inducible promoters, can be used to assess the effects of overexpressing particular genes on the capacity of hairy roots to synthesise or accumulate metabolites. Expression of genes in an antisense orientation to downregulate a particular enzymic step is also quite straightforward, technically speaking. These procedures allow an iterative approach to be taken to assess the effects of specific manipulations in path-

ways on secondary metabolite levels in hairy roots as suggested by Bailey (1991).

Such experimentation requires the availability of genes encoding enzymic steps relevant to the secondary metabolic pathway of interest. Numerous genes which encode enzymes controlling key reactions at the interface between primary and secondary metabolism in plants have been isolated from a range of organisms. In addition, full length plant cDNAs have been isolated which encode enzymes in a number of secondary metabolic pathways, indicating a growing interest in the molecular control of secondary metabolism in plants. For example, Kutchan *et al.* (1988) cloned the gene for strictosidine synthase which condenses tryptamine and secologanin, the first committed enzymic step in the secondary metabolic pathway leading to indole alkaloid biosynthesis. In the area of pyridine and tropane alkaloid metabolism, cDNAs have been cloned for putrescine methyl transferase from *N. tabacum* (Hibi *et al.*, 1994), hyoscyamine 6β-hydroxylase from *H. niger* (Matsuda *et al.*, 1991) and tropinone reductase from *D. stramonium* (Nakajima *et al.*, 1993). The cDNA for the berberine bridge enzyme was isolated from *Eschscholtzia californica* by Dittrich and Kutchan (1991). This enzyme catalyses a key step in the synthesis of benzophenanthridine alkaloids. Recently the same laboratory isolated the gene for berbamunine synthase from *Berberis stolofinera*, a plant cytochrome P450 enzyme which catalyses the synthesis of the bisbenzylisoquinoline alkaloid berbamunine (Kraus and Kutchan, 1995).

To date, relatively few experiments have been carried out which attempt to alter levels of secondary metabolites specifically in hairy roots by direct genetic manipulation, though a larger number has been undertaken with intact transgenic plants (reviewed in Hashimoto and Yamada, 1994; Kutchan, 1995). Several reports have described experiments designed to increase the availability of substrate for a secondary metabolite pathway *in vivo* by increasing levels of enzyme at the interface between primary and secondary metabolism in hairy roots. Nessler (1994) has referred to this approach as "pushing" the pathway by overexpressing genes encoding its early enzymes. Thus, Hamill *et al.* (1990) overexpressed the yeast ornithine decarboxylase (yODC) gene in hairy roots of *N. rustica* and recovered some lines with an increase in measurable yODC specific activity. Analysis of transgenic tissues expressing yODC, compared to tissues lacking the yODC gene, indicated an elevation in their putrescine content and also an increase in specific activity of putrescine methyl transferase which catalyses the first committed step in the pyridine alkaloid biosynthetic pathway (Hamill *et al.*, 1990). This was correlated with a modest (up to 2-fold) but nevertheless significant elevation in nicotine levels of manipulated lines. In addition, overexpression of the yODC gene in hairy root cultures of a low alkaloid variety of *N. tabacum* resulted in a 3–4-fold increase in nicotine content of some lines relative to controls (Lidgett and Hamill, unpublished). Overexpression of tryptophan decarboxylase (TDC) from *Catharanthus roseus* in hairy roots of *Peganum harmala* resulted in a substantial increase in measurable TDC activity in several lines (Berlin *et al.*, 1993). This was correlated with a 10–20-fold increase in serotonin levels, reaching 1.5–2% of dry weight in some hairy root lines. Yields of serotonin could be enhanced even further (up to 3–5% of dry weight) by feeding L-tryptophan to hairy root lines overexpressing the *C. roseus* TDC gene (Berlin *et al.*, 1993). Overexpression of a bacterial gene encoding lysine decarboxylase (LDC) in hairy roots of *N. tabacum* led to an LDC activity of up to 1 pkat mg^{-1} protein in transgenic lines (LDC was not detectable in controls). This was accompanied by up to a 10-fold increase in cadaverine content in some lines compared to controls (Fecker *et al.*, 1993). Levels of anabasine, which is derived from cadaverine, were elevated 3-fold in several root lines expressing the bacterial LCD gene (Fecker *et al.*, 1993) compared to controls. Feeding of lysine to these root cultures led to a substantial further increase in levels of cadaverine and anabasine.

Overexpression of genes encoding enzymes late in a metabolic pathway has also been attempted in hairy roots — a strategy referred to by Nessler (1994) as "pulling" the pathway. For example, the enzyme hyoscyamine 6β-hydroxylase (H6H) catalyses the conversion of hyoscyamine to scopolamine, which is significantly more valuable than hyoscyamine, both commercially and medicinally. Constitutive expression of the gene encoding H6H from *H. niger* in hairy roots of *Atropa belladonna*, which produces predominantly hyoscyamine and relatively little scopolamine, resulted in the root tissues of *A. belladonna* converting most of their hyoscyamine into scopolamine (Hashimoto *et al.*, 1993).

These experiments represent a modest beginning towards the development of more productive hairy root cultures as a result of direct gene transfer. Increasingly, the availability of genes encoding important steps in secondary metabolism will lead to a better understanding of the mechanisms regulating these pathways at a molecular level and may also enable the isolation of key regulatory genes. As has been alluded to in several recent reviews on this general topic (Nessler, 1994; Hashimoto and Yamada, 1994; Kutchan, 1995), the availability of such regulatory genes will facilitate the successful manipulation of complex secondary metabolic pathways in plants and hairy root tissues cultured *in vitro*.

Manipulating the expression of regulatory genes may be most effective if carried out in conjunction with experiments designed to alter expression of genes at the interface between primary and secondary metabolism. Of course, one possible outcome of such manipulations may be that the supply of metabolites from primary metabolism may become a limiting factor governing secondary metabolite biosynthesis. In such cases, working with hairy roots rather than intact plants could be advantageous as it may be possible to feed precursors to such genetically manipulated root tissues with the aim of harvesting high levels of secondary metabolites. If such chemical precursors were cheap, the production of valuable secondary metabolites by hairy roots cultured in bioreactors may well be feasible on a commercial scale.

Key Protocols for Introducing Foreign Genes into Hairy Roots

As the intended readership of this book includes scientists other than practising plant molecular biologists, we have included several laboratory protocols which are in routine use in this laboratory and in undergraduate practical classes taught at Monash University. They represent, for the most part, slight modifications of protocols which have been published elsewhere and further details can be found by referral to quoted texts, technical manuals in molecular biology (e.g. Sambrook *et al.*, 1989), or technical literature produced by companies producing molecular biology products. Generally we avoid the use of commercially available kits due to their usually expensive nature and sometimes lack of detail on their components. Exceptions are those we have found to be particularly useful or cost-effective. In our hands the methodology as described below is applicable to hairy roots of a wide range of species. We have not included protocols for secondary metabolite analysis as this represents a diverse array of methodology. Such procedures can be accessed by reference to the papers cited in Table 1.1. We assume practitioners are starting with *E. coli* containing a binary vector compatible with *A. rhizogenes*, e.g. pBin19 (Bevan, 1984) or derivatives, with the gene of interest fused to appropriate regulatory sequences. Full details on standard molecular biology protocols for gene cloning, ligation and manipulation of DNA *in vitro* are contained in appropriate technical manuals such as Sambrook *et al.* (1989). Texts by Old and Primrose (1994), Brown (1990, 1992) or Nicholl (1994) are recommended for those readers who are not molecular biologists but who wish to gain an appreciation of the methodology and relevant background theory.

1. Commonly Used Solutions

Typically, procedures in molecular biology utilise a number of concentrated stock solutions which are used when required to prepare working solutions. Commonly used stocks are listed here. High quality sterile water (double glass distilled or Milli-Q purified to 18 MΩ CM purity or better if possible) is used to make all solutions and dilutions which, unless stated otherwise, are sterilised by autoclaving at 121 psi for 20 min.

- 1 M Tris HCl, pH 7.5 (1 L)
  Dissolve 121 g Trizma base ($C_4H_{11}NO_3$) in 800 mL H_2O. Adjust pH to 7.5 with concentrated HCl. Make to volume with H_2O. Store at room temperature.
- 0.5 M Na_2 EDTA, pH 8.0 (1 L)
  Dissolve 186 g Na_2 EDTA ($[CH_2N(CH_2COOH)CH_2COONa]_2 \cdot 2H_2O$) in ~500 mL H_2O. Adjust pH to 8.0 with 10 M NaOH (solution will not dissolve until pH is adjusted). Make to volume with H_2O. Store at room temperature.
- 10 $\times$ TBE stock buffer (1 L)
  Dissolve 108 g Trizma base, 55 g boric acid (H_3BO_3) and 9.3 g Na_2 EDTA in 1 L H_2O. pH should be ~8.3. Store in darkness at room temperature.
- 20 $\times$ SSC solution (1 L)
  Dissolve 175.3 g NaCl and 88.2 g trisodium citrate ($Na_3C_6H_5O_7 \cdot 2H_2O$) in 800 mL H_2O. Adjust pH to 7.0 with 1 M NaOH and make to 1 L with H_2O. Store at room temperature.
- 20 $\times$ SSPE solution (1 L)
  Dissolve 175.3 g NaCl and 31.2 g $NaH_2PO_4 \cdot H_2O$ in 800 mL H_2O. Adjust pH to 7.7 with 1 M NaOH. Add 40 mL of 0.5 M Na_2 EDTA (pH 8.0). Make to volume with H_2O and store at room temperature.
- 3 M K acetate, pH 4.8 (1 L)
  Dissolve 294 g potassium acetate (CH_3COOK) in about 200 mL H_2O (heating the solution will assist it to dissolve). Add about 400 mL glacial acetic acid. Cool to room temperature. Check pH is 4.8, adding more glacial acetic acid if necessary. Make to volume with H_2O and store at room temperature after autoclaving. The resulting solution is 3 M with respect to K^+ and 5 M with respect to acetate.
- 3 M Na acetate, pH 6.0 (1 L)
  Dissolve 246 g sodium acetate (CH_3COONa) in ~400 mL H_2O. Adjust pH to 6.0 with glacial acetic acid. Make to volume and dispense into 10 mL aliquots. Store at room temperature.
- TE (1 L)
  Contains 10 mM Tris HCl (pH 7.5) and 1 mM EDTA. Prepare by adding 10 mL 1 M Tris HCl

(pH 7.5) and 2 mL 0.5 M Na_2 EDTA (pH 8.0) to 988 mL H_2O. Dispense into 10 mL aliquots and store at 4°C.
- Phenol solution
  Melt 500 g high quality phenol crystals at 65°C in a waterbath placed in a fumehood. Add 100 mL TE and stir for 15 min. Allow the layers to separate and remove and discard the supernatant. Repeat the extractions until the pH of the supernatant is 7.5–8.0. Freeze the phenol solution in 50 mL aliquots and store at –20°C. When using the solution, store in darkness at 4°C in a glass bottle under a 1–2 cm layer of TE. Discard if the pH drops below 7 or the solution becomes pink. (Normally, a phenol working solution is stable for 1–2 months if stored in darkness at 4°C.)
- Chloroform/IAA
  Combine 24 parts chloroform with 1 part isoamyl alcohol.
- Phenol/chloroform
  Combine equal volumes of phenol solution and chloroform/IAA. Mix and allow to settle. Store at 4°C in darkness. (The phenol/chloroform is the lower solution. The upper layer is aqueous and should be maintained as a 1–2 cm layer to reduce the rate of phenol oxidation.)
- Blue dye loading buffer (100 mL)
  Dissolve 15 g Ficoll type 400 in 80 mL H_2O. Add 0.25 g bromophenol blue and stir gently to dissolve. Make to 100 mL with H_2O and dispense into 5 mL aliquots.
- 10% SDS (1 L)
  Dissolve 100 g sodium dodecyl sulphate (SDS) in ~800 mL H_2O with gentle stirring and make to volume. Do not autoclave. If SDS precipitates, heat to 65°C with gentle shaking to redissolve.
- 1 M Na phosphate buffer, pH 7.0
  Dissolve 35.49 g dihydrogen orthophosphate (Na_2HPO_4) in 500 mL H_2O. Titrate to pH 7.0 with orthophosphoric acid (H_3PO_4).
- Salmon/herring sperm DNA (100 mL)
  Add 0.5 g salmon or herring testes DNA to 100 mL H_2O and microwave with gentle shaking to dissolve. Pass the solution through a fine gauge needle (23 G) several times to shear the DNA. Store at –20°C in 1 mL aliquots.
- EtBr stock solution
  Dissolve ethidium bromide powder (e.g. Sigma) in H_2O at a concentration of 10 mg mL^{-1}. Store in darkness at 4°C. (Handle with care!)
- 10 mg mL^{-1} RNase A (10 mL)
  Dissolve 100 mg RNase A in 9.75 mL H_2O. Add 100 μL 1 M Tris HCl (pH 7.5) and 150 μL 1 M NaCl. Place in a boiling waterbath for 15 min to remove any contaminating DNase activity. Allow the solution to cool to room temperature

and store at –20°C in 100 µL aliquots. Note: do not use in equipment which is also used for RNA isolation/analysis, as RNase activity is difficult to eradicate from contaminated equipment and can degrade RNA in Northern hybridisation experiments.

- 10 M NH$_4$ acetate (100 mL)
  Dissolve 77 g ammonium acetate (CH$_3$COONH$_4$) in 100 mL H$_2$O. Sterilise by filtration and store in 5 mL aliquots at room temperature.

- 5 × MOPS RNA buffer (1 L)
  Dissolve 20.6 g of 3-(N-morpholino) propane-sulfonic acid (MOPS; C$_7$H$_{15}$NO$_4$S) in 700 mL H$_2$O. Add 13.3 mL of 3 M Na acetate. Adjust pH to 7.0 with 1 M NaOH. Add 10 mL of 0.5 M Na$_2$ EDTA (pH 8.0) and make to volume. Store at room temperature in darkness. After autoclaving, the buffer will have a light yellow colouration. If a dark yellow colour develops, it should be discarded.

- 50 × Denhardt's solution (100 mL)
  Dissolve 1 g BSA fraction V, 1 g Ficoll type 400 and 1 g soluble polyvinylpyrrolidone (PVP) in 100 mL H$_2$O. Do not autoclave. Store in 5 mL aliquots at –20°C.

- YMB medium for *A. rhizogenes* (1 L)
  Dissolve 0.5 g K$_2$HPO$_4$, 0.2 g MgSO$_4 \cdot$7H$_2$O, 0.1 g NaCl, 0.4 g yeast extract and 10 g mannitol in 800 mL H$_2$O. Adjust pH to 7.0 with 1 M NaOH. Make to volume and dispense into 10 mL aliquots. Store at 4°C after autoclaving.

- MGL medium for *A. rhizogenes* (1 L)
  Dissolve 5 g tryptone, 2.5 g yeast extract, 5 g mannitol, 1 g L-glutamic acid, 0.25 g KH$_2$PO$_4$ and 0.1 g NaCl in 800 mL H$_2$O. Add 10 µL of a 100 µg mL^{-1} aqueous stock solution of biotin, adjust pH to 7.0 with 1 M NaOH, and make to volume with H$_2$O. Dispense into 10 mL aliquots and store at 4°C after autoclaving.

- LB medium (1 L)
  Dissolve 10 g tryptone, 5 g yeast extract and 10 g NaCl in 800 mL H$_2$O. Adjust to pH 7.5 with 1 M NaOH, and make to volume with H$_2$O. Dispense into 10 mL aliquots and store at 4°C after autoclaving.

For agar-solidified media, add 15 g L^{-1} agar (e.g. Bacto Difco) to liquid media as above and dissolve by heating before autoclaving.

Agarose gels for separation of DNA and RNA by electrophoresis are made by combining 1 g agarose per 100 mL 1 × TBE buffer, heating (microwave or boiling water bath) until the agarose dissolves, and then adding 3 µL EtBr stock solution per 100 mL of molten agarose. The solution is cooled to ~50°C and the gels poured. No further staining of gels to detect DNA is necessary.

Gels and electrophoresis buffer containing EtBr must be disposed of properly as EtBr is a mutagen. Plasticware and glassware used for molecular biology work should be packed with gloved hands, sterilised by autoclaving or dry heat as appropriate, and stored in a dust-free environment. Particular care should be taken when working with RNA to avoid contamination with RNases present on skin or dust. Separate solutions, equipment and working areas for RNA work are recommended. Many solutions used in molecular biology research are potentially toxic and some are suspected carcinogens. Due caution should therefore be exercised in handling reagents which should be disposed of in a safe manner.

2. Miniprep Binary Vector Plasmid DNA Isolation From *E. coli* Using an Alkali Lysis Procedure (adapted from Sambrook *et al.*, 1989)

This method can be used for low copy and high copy plasmids.

Procedure

1. Grow an overnight culture of *E. coli* containing the plasmid of interest at 37°C from a single colony inoculation in LB medium containing antibiotic to ensure the binary vector is maintained. Transfer 4 × 1.5 mL aliquots to appropriately labelled microcentrifuge tubes. Store the remainder of the culture at 4°C until needed.

2. Centrifuge cells at 14,000 rpm (~12,000 g) for 30 s at room temperature. Discard **all** supernatant (remove using a micropipette tip) and resuspend each pellet in 100 µL of ice-cold plasmid isolation buffer. (Plasmid isolation buffer contains 50 mM glucose, 25 mM Tris HCl (pH 8.0) and 10 mM Na$_2$ EDTA (pH 8.0)). Use a sterile toothpick to loosen the pellets if necessary and vortex briefly to disperse the cells.

3. Add 20 µL of proteinase K solution (10 mg mL^{-1} dissolved in TE, incubated at 37°C for 1 h and stored at –20°C in aliquots) to each tube, and incubate at 37°C for 30 min.

4. Add 20 µL of cell lysis buffer to each tube. (Cell lysis buffer contains 200 mM NaOH and 1% SDS.) Mix gently and store on ice for 5 min.

5. Add 150 µL of 3 M K acetate (pH 4.8) to each tube and invert sharply several times. A precipitate should be visible. Vortex briefly and return tubes to ice for 5 min.

6. Centrifuge at 14,000 rpm at room temperature for 5 min and transfer supernatant to tubes containing 500 µL of phenol/chloroform. Vortex for 30 s and centrifuge tubes at 14,000 rpm for 10 min.

7. Remove ~400 µL of supernatant (taking care not to disturb the debris at the interface) and transfer to a fresh microcentrifuge tube containing 500 µL chloroform/IAA and vortex for 30 s. Centrifuge tubes at 14,000 rpm for 10 min. Remove 350 µL of supernatant to a fresh microcentrifuge tube.

8. Add 35 µL of 3 M Na acetate (pH 6.0) and 800 µL of ice-cold 100% ethanol to each tube of aqueous crude DNA solution. Mix by inverting tubes sharply. Leave for 5 min at room temperature and centrifuge for 15 min at 14,000 rpm. Carefully remove and discard the supernatant without disturbing the pellet.

9. Add 100 µL of water to each microcentrifuge tube, re-cap and invert several times to dissolve the pellet. Pulse spin in a microcentrifuge and combine all samples into one microcentrifuge tube (400 µL in total). Add 40 µL of 3 M Na acetate (pH 6.0), mix gently and add 900 µL of cold 100% ethanol. Re-cap tubes and invert several times. Leave at room temperature for 5 min and centrifuge for 15 min at 15,000 rpm. Remove and discard the supernatant without disturbing the pellet. Rinse the pellet with 100 µL of 70% ethanol and remove as much liquid as possible.

10. Leave the pellets to dry for 5–10 min at 37°C, or dry for 1–2 min under vacuum (do not dry completely under vacuum as the DNA may be difficult to redissolve). Redissolve pellets in 20 µL of sterile distilled H$_2$O. Store at –20°C if not used straight away.

3. Digestion of Miniprep DNA With Restriction Enzymes

Plasmid DNA isolated by the miniprep method should be analysed to check that the plasmid contains the insert. Normally this is done with restriction enzymes which will allow the insert to be released from the plasmid.

Procedure

1. In a sterile microcentrifuge tube, combine 10 µL of plasmid miniprep DNA isolated as in Section 2, 3 µL of 10 × appropriate buffer, 1 µL of restriction enzyme containing 2–5 units µL^{-1}, 1 µL of 20 mM spermidine, and 15 µL of H$_2$O. (Spermidine is optional and can aid the digestion of crude DNA by restriction enzymes; store a filter-sterilised solution (20 mM) in aliquots at –20°C.)

2. Mix the contents of the tube gently, spin for 10 s and incubate at 37°C for 2–3 h.

3. Add 1 µL of 10 mg mL^{-1} RNase A and incubate at 37°C for 3–5 min.

4. Add 5 µL of blue dye loading buffer, mix the contents of the tube gently and load 20 µL onto a 1% agarose TBE gel. Electrophorese at 60 mA for 2–3 h or 10 mA overnight as appropriate. Store the remainder of the restriction digest reaction at –20°C to re-run sample if required.

5. A fragment containing the insert at the appropriate size should be visible when viewed under short-wave UV illumination. The binary vector should also be visible.

6. Cells containing these plasmids should then be used to prepare an overnight culture to use in a triparental mating procedure to transfer the plasmid to *A. rhizogenes* or to isolate pure plasmid DNA by a large-scale version of the alkaline miniprep procedure.

4. Introduction of Binary Vector Containing Foreign Genes into Hairy Roots via *A. rhizogenes*

A number of *A. rhizogenes* strains have been used to produce hairy roots, including mannopine-producing, cucumopine-producing and agropine-producing strains. We tend to favour the use of agropine strains as they usually have a high level of efficiency with respect to hairy root formation. The following protocols have been determined for the agropine strain LBA9402.

Binary vectors containing foreign genes which are compatible with *A. rhizogenes*, e.g. pBin19 (Bevan, 1984) and common derivatives which contain the *uidA* (GUS) gene such as pBI121 (Jefferson *et al.*, 1987), can be easily transferred from *E. coli* to *A. rhizogenes* via a triparental mating procedure. Electroporation is also an effective means of delivery into *A. rhizogenes*.

4.1. Triparental mating

Procedure

1. On Day 1, inoculate 10 mL of YMB medium with a colony of *A. rhizogenes* and culture for two nights (~40 h) at 25–28°C with shaking.

2. On Day 2, inoculate separate 10 mL aliquots of LB medium with *E. coli* containing:
   (a) helper plasmid pRK2013
   (b) a binary vector containing the foreign gene. Add antibiotic (kanamycin) for (a) and usually kanamycin for (b) at 50 µg mL^{-1} to ensure the plasmids are retained. Incubate overnight at 37°C with shaking.

3. On Day 3, combine 100 µL of each *E. coli* suspension and mix together with 200 µL of *A. rhizogenes* suspension. Plate the mixture on conjugation medium comprised of 50% LB agar and 50% YMB agar. Incubate at 25–28°C overnight.

4. On Day 4, take a loopful of bacteria from the triparental mating plate and spread it thinly on selection medium consisting of YMB agar medium containing 100 µg mL^{-1} kanamycin and 100 µg mL^{-1} rifampicin. (The kanamycin selects against parental *A. rhizogenes*, whilst rifampicin selects against parental *E. coli*.) Incubate at 25–28°C for several days.

5. When discrete colonies are discernible (3–5 d), restreak on YMB agar + kan + rif medium. These should be *A. rhizogenes* containing the binary vector. Repeat the restreaking procedure to ensure no parental bacterial cells are maintained. Isolation of total DNA as noted in Section 5 and probing with the foreign gene and a T-DNA marker as outlined in Section 10 is recommended to ensure that no mutant *E. coli* or *A. rhizogenes* parents have grown on the selection medium. (This can happen but is unusual in our experience.) Polymerase chain reaction (PCR) amplification directly from fresh colonies as outlined in Section 12 is also a useful and rapid procedure which can be used for this purpose.

4.2. *Electroporation of DNA into A. rhizogenes*
(adapted from Mattanovich *et al.*, 1989)

If an electroporator is available, e.g. Biorad Gene Pulser, this is an effective procedure which is more rapid than the triparental mating procedure and eliminates concerns about recovery of parental *E. coli*. As noted in Section 4.1, Step 5, it is advisable to carry out a DNA preparation and Southern hybridisation to check that the binary vector carrying the foreign DNA is present in a structurally intact manner in the *A. rhizogenes* cells, rather than an antibiotic resistant mutant of *A. rhizogenes* having been recovered.

Procedure

1. Select an *A. rhizogenes* colony from a freshly streaked YMB agar plate and check by replica plating that it is sensitive to kanamycin. Inoculate 10 mL of YMB medium and grow for two nights (~40 h) at 25°C with shaking.

2. Inoculate 1 mL of the *A. rhizogenes* broth into 100 mL of YMB medium and incubate at 25–28°C with shaking until OD$_{600}$ is approx. 0.5. Chill the culture on ice and harvest the cells by centrifugation at 5000 rpm for 10 min.

3. Wash the cells 2–3 times in 10 mL of 1 mM HEPES (pH 7.0) and once in 10 mL of 10% glycerol with centrifugation at 5000 rpm for 10 min between washings. Resuspend the final pellet in 1 mL of 10% glycerol.

4. Divide into 100 µL aliquots, snap freeze in liquid N$_2$ and store at –70°C until needed.

5. For electroporation, thaw an aliquot of cells on ice and add plasmid (10–200 ng in 10 µL H$_2$O).

6. Place 40 µL in an electroporation cuvette with electrode distance of 0.2 cm (Biorad). Apply one pulse of 2.5 kV with capacitance set at 25 µF (Biorad Gene Pulser).

7. Immediately transfer cells into 0.5 mL of YMB liquid medium and incubate at 25–28°C for 2–3 h.

8. Make several serial dilutions of electroporated *A. rhizogenes* cells and plate on YMB agar medium containing antibiotic (e.g. kanamycin for pBin19 derivatives at 50–75 µg mL^{-1}). Incubate at 25°C for 2–3 d.

9. Colonies recovered are restreaked and used for DNA extraction to confirm the presence of binary vector T-DNA as noted in Section 4.1, Step 5.

5. **Total DNA Isolation From *A. rhizogenes***
(adapted from the method of Slusarenko, 1990)

This procedure usually allows isolation of DNA which is easily digested by restriction enzymes. Cells are grown in MGL medium containing appropriate antibiotics for 1–2 d at 25–28°C with shaking. This leads to a reduction in polysaccharide levels which can interfere with DNA isolation and which can be a problem if cells are grown in YMB.

Procedure

1. Grow *A. rhizogenes* cells in MGL medium containing appropriate antibiotics for 1–2 d at 25–28°C with shaking. Pellet the cells from 5 mL broth by centrifugation at 5000 rpm for 5 min.

2. Resuspend the pellet in 900 µL of lysis buffer. (Lysis buffer contains 125 µg mL^{-1} proteinase K, 8.3 mg mL^{-1} lysozyme and 1.25% v/v Sarcosyl.) Mix and incubate the solution at 37°C for 1 h.

3. Centrifuge at 14,000 rpm for 15 min and transfer supernatants to new microcentrifuge tubes. Add 0.5 volume of 7.5 M NH$_4$ acetate, mix and leave on ice for 20–30 min.

4. Centrifuge at 14,000 rpm for 15 min, remove the supernatant into a fresh microcentrifuge tube and repeat the centrifugation. A small pellet of precipitated protein should be visible. Remove the supernatant to a fresh microcentrifuge tube.

5. Add an equal volume of isopropanol, mix gently and leave on ice for 30 min. Strands of DNA should be visible.

6. Remove DNA with a sterile plastic micropipette by twirling around to recover strands. (Centrifugation at this stage can result in large amounts of polysaccharides contaminating the sample.) Dissolve the DNA strands in a total of 50 µL of H$_2$O.

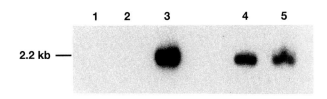

Figure 1.1. Detection of the *uidA* (GUS) coding sequence in *Agrobacterium*. DNA was extracted from agrobacteria as described in Section 5. 10 µL of crude DNA was digested to completion with *Eco*RI and *Bam*HI restriction enzymes, transferred to a nylon (Hybond N⁺) membrane and probed using α^{32}P-labelled GUS gene sequences as a probe as described in Section 10. Tracks 1 and 2 contain DNA extracted from control agrobacteria not containing the binary vector. Tracks 3, 4 and 5 contain DNA extracted from agrobacteria containing the binary vector pBI121 (Jefferson *et al.*, 1987). The presence of a 2.2 kb fragment in tracks 3–5 indicates that the binary vector was present and the construct was in a structurally intact form.

7. Digest 10 µL of the crude DNA to completion with appropriate restriction enzymes and electrophorese the digest on an agarose gel. Transfer the DNA to a membrane and probe for the presence of the gene in the binary vector as described in Section 10. An example of a Southern blot of total DNA extracted from agrobacteria containing the binary vector pBI121 and probed with the GUS gene is shown in Figure 1.1.

6. Transformation of Leaf Tissue or Explants With *A. rhizogenes* Carrying a Binary Vector

6.1. Infection of large leaf explants/stem tissues (e.g. tobacco leaf)

Where it is possible to use tissue of several centimetres in length, we tend to use this method due to its ease of execution.

Procedure

1. Surface sterilise a leaf or stem tissue with a 10% dilution of commercial bleach (e.g. Domestos) for 20–30 min and rinse thoroughly in sterile water 3–4 times. Remove damaged tissues and trim so that the explant will fit inside an appropriate culture pot (e.g. Figure 1.2a).
2. Using the tip of a needle attached to a syringe containing agrobacteria (e.g. Figure 1.2b) or a sterile scalpel, make a number of small cuts and incisions in the petiole and major veins or in the stem tissue. Inoculate 1–2 drops of fresh (2-d-old) undiluted *A. rhizogenes* culture into the wounded portion of tissue. (Remove excess liquid as this may lead to excessive growth of agrobacteria on the surface of the medium.)

3. Carefully place the tissue into a culture jar containing MS or B5 medium + 2% sucrose solidified with 1.8 g L⁻¹ Phytagel (Sigma) or 8 g L⁻¹ agar (e.g. Bacto Difco) (Figure 1.2c). Make sure the infected portion of the explant does not come in contact with the agar surface or the sides of the jar, or condensation will carry agrobacteria to the surface of the agar where they will proliferate.
4. Re-cap the culture jar and incubate for 2–3 weeks at 25°C (light or darkness). Usually, roots will differentiate around the wound spot and grow to about 5–10 mm on the leaf or stem explant (Figure 1.2d).
5. Remove the leaf/stem tissue around the roots leaving a 5–10 mm diameter of parent tissue (Figure 1.2e), and place the wound spot with roots into liquid medium. Culture the tissue in MS or B5 medium (or ¹/₂-strength medium for some species, e.g. *C. roseus*) with 2–3% w/v sucrose (pH 6.0) containing ampicillin (Na salt, Sigma) or cefotaxime (e.g. Claforan, Roussel Pharmaceuticals) at 500 µg mL⁻¹ to inhibit growth of agrobacteria. The inclusion of kanamycin at 25–50 µg mL⁻¹ will ensure the selection of hairy roots containing Ri T-DNA and binary vector T-DNA. Normally, we culture rooted explants in 20–25 mL medium on an orbital shaker operated at 50–60 rpm, and select a single healthy growing root tip from the edge of the culture to establish a clonal hairy root culture (Figure 1.2f). Ampicillin or cefotaxime should be maintained at 500 µg mL⁻¹ for 2–3 subcultures to prevent agrobacterial contamination. Thereafter ampicillin can often be omitted, though we usually maintain cultures in medium containing 125 µg mL⁻¹ ampicillin or cefotaxime to ensure asepsis.

6.2. Infection of small explants or small-leaved species with A. rhizogenes carrying a binary vector

Procedure

1. Surface sterilise tissues (e.g. potato tubers, seedlings/small-leafed species) by immersion in a 10% w/v solution of commercial bleach (e.g. Domestos) for 15–30 min and rinse thoroughly in sterile water (3–4 times).
2. Remove damaged tissues with a sharp scalpel (also serves to create freshly wounded tissues for agrobacterial infection and transformation). Take a fresh (2-d-old) culture of *A. rhizogenes*, dilute 1:50 in YMB medium, and immerse the tissue explant in the diluted agrobacteria for 1–3 min. Remove the explant and blot off excess liquid on sterile filter paper.
3. Incubate the tissue explant for 2–3 d at 25°C in low light/darkness on MS medium + 2% w/v

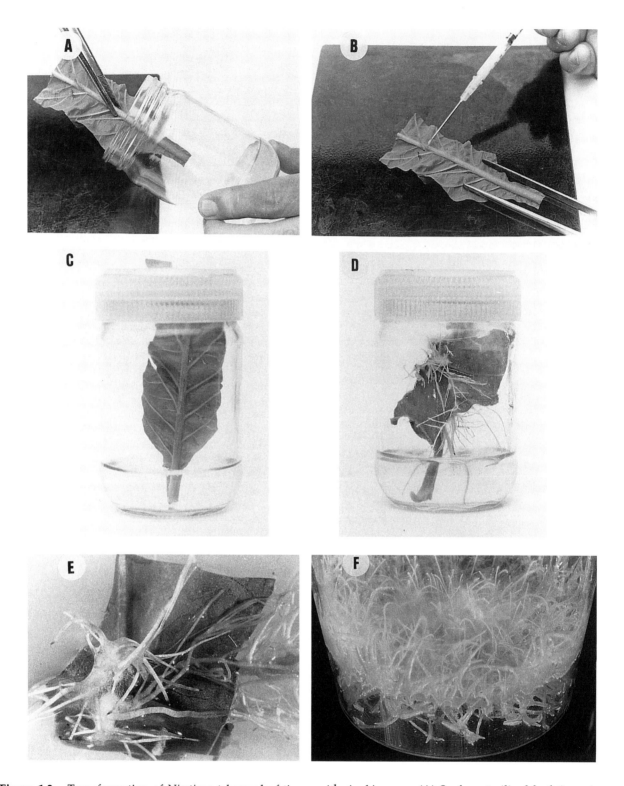

Figure 1.2. Transformation of *Nicotiana tabacum* leaf tissue with *A. rhizogenes*. (A) Surface-sterilised leaf tissue is trimmed to fit inside a culture vessel. (B) Mid-rib and major veins are wounded and 1–2 drops (~30–50 µL) of a suspension of agrobacteria is applied to each wound spot. (C) Inoculated plant tissue is placed upright in the culture vessel and cultured at ~25°C. (D) Hairy roots differentiate from wounded tissues 2–3 weeks after inoculation. (E) Leaf explants containing hairy roots are cultured in liquid medium. (F) A hairy root culture which has been established from a single root tip.

sucrose containing 1 mg L^{-1} 6-benzylamino-purine (BAP) (cytokinin stimulates cell division in many species) and solidified with 1.8 g L^{-1} Phytagel (Sigma) or 8 g L^{-1} agar (e.g. Bacto Difco).

4. Transfer explants to MS medium + 2% sucrose solidified with Phytagel or agar also containing 500 µg mL^{-1} ampicillin (Na salt, Sigma) or 500 µg mL^{-1} cefotaxime (e.g. Claforan, Roussel Pharmaceuticals). Repeat the transfer to fresh medium at 2–4 d intervals. Root differentiation should be apparent within 15–20 d and will usually originate mainly from vascular tissue on leaf and stem tissue explants. Culture of root tissue is as described in Section 6.1, Step 5.

7. Isolation of DNA From Hairy Roots

7.1. Isolation of high quality DNA

The following method adapted from Martin et al. (1985) is slightly more labour intensive than some published protocols but we find that it gives good quality DNA of high molecular weight which can be digested easily with most restriction enzymes.

Procedure

1. Add 10 mL extraction buffer (containing 0.1 M Na_2 EDTA (pH 8.0) and 0.1 M sodium diethyl-thiocarbonate (DIECA) in $3 \times SSC$) and 2 mL 10% SDS to a clean, sterile 50-mL Oakridge tube or equivalent and heat to 65°C. (Note: extraction buffer should not be autoclaved. It is stable for 1 year or more at 4°C. The DIECA will form a brownish sediment but this will not interfere with DNA extraction if it is not disturbed.)

2. Place 1–5 g plant material in a cold mortar or coffee grinder (cleaned with 10% SDS, then with 100% ethanol, and allowed to dry). Grind the tissue in liquid N_2 until it becomes a fine powder. If grinding more than one sample, wipe the mortar and pestle or coffee grinder thoroughly between use with 100% ethanol, air dry, and then reuse.

3. Transfer the powder with a pre-cooled spatula to the buffer in the tube, avoiding any that has thawed out. Periodically mix the powder with the buffer before it thaws. (This becomes more difficult the larger the sample weight.) Incubate the capped tube in a 37°C waterbath for a few minutes with gentle shaking to ensure the SDS is redissolved.

4. In a fumehood, add 10 mL phenol/chloroform to the ground tissue in buffer, re-cap the tube, shake well, and centrifuge at 10,000 rpm at 5°C for 15 min.

5. Remove about 14 mL of supernatant into clean tubes. Add 2 volumes of ethanol to this solution and cool to –20°C. (Tubes can be left overnight at –20°C if desired.)

6. Centrifuge at 10,000 rpm at 4°C for 20 min. Remove the supernatant and leave the tubes to drain for a few minutes.

7. Dissolve the crude DNA pellet in 9 mL TE. Add 9 g caesium chloride ($CsCl_2$) and 500 µL of EtBr stock solution.

8. Using a Pasteur pipette, carefully transfer the solution into an ultracentrifuge tube (e.g. an opaque Beckman quickseal centrifuge tube, filling almost up to the bottom of the spout). Balance the tubes (a 1 g mL^{-1} $CsCl_2$ solution should be available for balancing) and seal them.

9. Set the temperature of an ultracentrifuge to 18°C and spin at 45,000 rpm for 16 h and then 40,000 rpm for 4–5 h, or at 40,000 rpm for about 60 h (or over a weekend). Set the deceleration at the slowest rate possible to avoid disturbing the gradient. Examine the tubes under long-wave UV light — a fluorescent band of DNA should be clearly visible.

10. Push a needle into the top of the tube to admit air. Under UV light, puncture the side of the tube ~0.5 cm underneath the bottom of the band with an 18 or 19 G needle attached to a 10 mL disposable syringe. Draw out the fluorescent DNA layer by moving the needle back and forth underneath the DNA and place the DNA solution (normally ~2 mL) in a clean test tube.

11. In a fumehood, add 6–8 mL butanol saturated with H_2O to tubes containing the DNA/$CsCl_2$/EtBr solution. Shake vigorously and allow the solution to settle until layers are clearly defined. Remove the pink butanol layer with a Pasteur pipette and extract twice more with 10 mL H_2O-saturated butanol until the DNA solution is colourless. Remove the butanol layer completely. The volume of the DNA–$CsCl_2$ solution should be ~2 mL.

12. Make up to 5 mL with sterile H_2O. Transfer to 30-mL glass Corex tubes, add 0.5 mL 3 M Na acetate (pH 6.0) and 11 mL 100% ethanol, and cool at –20°C for at least 1 h. Centrifuge at 10,000 rpm for 20 min.

13. Remove the supernatant completely and dissolve the pellet in 0.5 mL sterile H_2O. Transfer to a 1.5-mL microcentrifuge tube.

14. Add 50 µL of 3 M Na acetate (pH 6.0) and 1 mL of 100% ethanol. Cool at –20°C for 2–3 h or at –70°C for 15 min.

15. Centrifuge at 14,000 rpm for 15 min. Remove the supernatant and rinse the pellet carefully with 200 µL of 70% ethanol. Remove the supernatant and allow the pellet to dry almost completely (but not totally!) either by inversion on tissue paper at room temperature for 20–30 min, or under vacuum for 2–3 min. (The DNA

pellet should appear slightly damp and will be translucent.)

16. Dissolve the pellet in 100–200 μL TE and store at 4°C.

17. Quantify levels of DNA using a UV spectrophotometer set at 230, 260 and 280 nm. (A good DNA preparation will have a 230:260:280 ratio of ~1:2:1.) An OD_{260} value of 1 indicates a DNA concentration of 50 μg mL^{-1}.

18. Pure DNA can be stored in a stable condition at –20°C at a concentration of ~1 mg mL^{-1} for at least 2 years.

7.2. Miniprep procedure for isolating DNA from hairy roots

Sometimes it is desirable or necessary to extract DNA rapidly from many samples or from small amounts of tissue (1 g fresh weight or less), and the protocol described above may not be appropriate. There is a variety of miniprep procedures for recovering DNA from plant tissues and some trial-and-error experimentation may be needed to find the best method (both in terms of quantity and quality) for extraction of DNA for any particular species. The following method adapted from that described by Pich and Schubert (1993) is capable of producing 50–100 μg of high quality DNA per g fresh weight root tissue from hairy roots of a number of plant species.

Procedure

1. Place 0.5–1 g of tissue in a mortar and add liquid N_2. Grind the tissue to a very fine powder adding liquid N_2 as required. Transfer the frozen tissue to 3 mL of DNA extraction buffer (extraction buffer contains 500 mM NaCl, 50 mM Tris HCl (pH 8.0), 50 mM Na_2 EDTA (pH 8.0) and 1% v/v β-mercaptoethanol added just prior to use) in a labelled centrifuge tube. Ground tissue should not be allowed to thaw **until** it is in direct contact with extraction buffer.

2. Add 1.3 mL of cold 20% polyvinylpyrrolidone (PVP) solution (store stock at –20°C), mix, and cool the tube on ice.

3. Add 800 μL of 10% SDS, mix, then incubate the tube at 65°C for 10 min with occasional gentle mixing.

4. Add 500 μL of 3 M K acetate (pH 4.8) and mix. Place on ice for 15–30 min.

5. Centrifuge the tube at 13,000 rpm for 10 min at 4°C.

6. Carefully collect the supernatant and transfer to a clean centrifuge tube, estimating the volume while doing so.

7. Add 0.5 volume of ice-cold isopropanol to the supernatant solution, cover with parafilm and mix by inversion. Incubate on ice for about 5–10 min.

8. Centrifuge the tube again at 13,000 rpm for 10 min at 4°C. A faint pellet should be visible. Carefully remove all supernatant and discard it. Redissolve the crude DNA pellet in 500 μL of TE and transfer to a microcentrifuge tube.

9. Add 500 μL of phenol/chloroform and vortex for 30 s. Centrifuge the sample at 14,000 rpm for 5 min.

10. Remove the top aqueous layer from the tube (without disturbing the interface) and place in a clean microcentrifuge tube containing 500 μL chloroform/IAA. Mix by vortexing for 30 s and centrifuge at 14,000 rpm for 5 min. Remove the upper layer to a new microcentrifuge tube without disturbing the interface.

11. Add 50 μL of 3 M Na acetate (pH 6.0) and 1 mL of 100% ethanol to the tube. Re-cap and mix by inverting the tube sharply several times. A precipitate of DNA may be seen. Place at –20°C for 1–2 h (or –70°C for 15–20 min), and then centrifuge the tube at 14,000 rpm for 20 min.

12. Remove and discard the supernatant without disturbing the pellet and rinse with 100 μL of 70% ethanol. Centrifuge for 5 min at 14,000 rpm. Remove all the supernatant without disturbing the pellet and allow to become almost dry at 37°C for 5–10 min or under vacuum for 1–2 min.

13. Redissolve the DNA in a total volume of 30 μL of TE. Remove 2 μL to a separate tube and add 3 μL of TE and 3 μL of blue dye loading buffer. Run the 8 μL on a 1% TBE agarose gel overnight at 20 mA, using 5 and 10 μL of uncut λ DNA at 30 ng μL^{-1} as a standard to assist in estimating the quantity and quality of extracted DNA.

14. The remaining DNA should be stored at –20°C until needed for DNA analysis by Southern hybridisation or polymerase chain reaction (PCR) techniques.

8. RNA Extraction From Hairy Roots

The following method adapted from Verwoerd *et al.* (1989) works well with small amounts of tissue (0.1–0.5 g fresh weight). It can be scaled up for larger RNA preparations.

Procedure

1. Tissue should be harvested, rinsed with sterile H_2O, blotted dry, wrapped in aluminium foil in 0.1–0.5 g aliquots and frozen in liquid N_2. It can be stored at –70°C until required (for at least 6 months).

2. Add 750 μL TLES buffer (containing 100 mM Tris HCl (pH 8.0), 100 mM LiCl, 10 mM EDTA, 1% SDS) and 750 μL phenol solution to screw-topped sterile plastic centrifuge tubes (e.g. Oakridge) and incubate at 60°C in a waterbath in a fume cupboard.

3. Grind the root tissue to a very fine powder in liquid N_2 using a clean mortar and pestle (wipe with tissue dampened with 10% SDS and then with 100% ethanol between extractions).

4. Add ground root powder to the hot TLES/phenol mixture and mix by vortexing for 30 s. (It is important that the powder does not thaw until it comes in contact with the TLES/phenol mixture.)

5. Add 750 µL chloroform/IAA and mix by vortexing for 30 s.

6. Transfer all of the extract to 2×1.5-mL microcentrifuge tubes and centrifuge at 14,000 rpm for 5 min.

7. Remove the supernatant to separate microcentrifuge tubes and add an equal volume of 4 M LiCl to each tube. Store at 0–4°C overnight to allow RNA to precipitate from solution.

8. Centrifuge samples for 30 min at 14,000 rpm at 4°C. Remove the supernatant (contains DNA which can be recovered as a crude DNA preparation if desired by addition of 2 volumes of ethanol) and resuspend the RNA pellet in 250 µL H_2O. Overnight storage at 4°C may assist the RNA to dissolve fully.

9. Add 25 µL 3 M Na acetate (pH 6.0) and 550 µL 100% ethanol to reprecipitate the RNA. Store at –20°C for 3 h/overnight, or at –70°C for 30–60 min.

10. Centrifuge at 14,000 rpm for 30 min. Remove and discard the supernatant, rinse the pellet with 70% ethanol and dry under vacuum for 2–3 min (pellet should appear slightly damp). Redissolve in 50–100 µL H_2O. Use 5 µL for RNA quantification using a UV/visible spectrophotometer set at 230, 260 and 280 nm. Good quality RNA should have a 230:260:280 ratio of approximately 1:2:1. An OD_{260} of 1 indicates an RNA concentration of 40 µg mL^{-1}.

Note: the RNA extracted using this procedure will remain stable for periods in excess of one year if stored at –70°C. If repeated use is envisaged, it is advisable to freeze in small aliquots to avoid repeated freeze–thawing of the sample which can lead to degradation of RNA.

9. Freeze Method for Isolating DNA Fragments From Agarose to be Used as Probes in Southern and Northern Hybridisations

We have found Pharmacia NA agarose (low sulphate) to give more reliable results using this technique.

Procedure

1. Digest the plasmid containing the DNA insert to be used as a probe with appropriate restriction enzymes and run a 1% agarose gel in $1 \times$ TBE buffer. Visualise on a clean UV transilluminator and cut out the appropriate band with a sterile scalpel. Trim off all excess agarose and transfer to a microcentrifuge tube.

2. Crush the agarose block until it is fully homogenised.

3. Add 500 µL of phenol solution and mix by vortexing for 30 s.

4. Freeze at –70°C for 15 min or at –20°C for 3 h. Centrifuge at 14,000 rpm for 15 min.

5. Remove the aqueous (top) layer to a new tube. Re-extract the phenol layer by adding 200 µL TE, mix by vortexing, and centrifuge at 14,000 rpm for 10 min. Remove 150 µL of the aqueous layer and add to the previous sample.

6. Extract the aqueous layer twice with chloroform/IAA, removing the upper aqueous layer to a clean microcentrifuge tube each time. Following extraction and removal of all chloroform/IAA, precipitate the DNA by adding a $^1/_{10}$ volume of 3 M Na acetate (pH 6.0) and 2 volumes of 100% ethanol to the aqueous layer. Incubate overnight at –20°C, or for 30–60 min at –70°C.

7. Centrifuge at 14,000 rpm for 30 min. Remove the supernatant and rinse the pellet with 70% ethanol. Remove the supernatant. A small pellet should be visible. Dry under vacuum for 1 min or at 37°C for 5–10 min. Redissolve the pellet in 10 µL H_2O and use 1 µL to check for yield and integrity by running on a 1% agarose minigel. The yield can be estimated by comparison of the fluorescence intensity with a known amount of standard.

10. Southern Blotting and DNA:DNA Hybridisation to Confirm That Foreign Genes Are Present

For *A. rhizogenes* DNA, 1 µg or less total DNA is adequate for Southern blotting and will allow a signal to be produced on an autoradiograph after overnight exposure. For plant genomic Southerns, 3–5 µg DNA is adequate for small genomes (e.g. *Arabidopsis, Beta vulgaris*), while 10–20 µg DNA may be necessary to allow detection of a single copy of a foreign gene in the genome of medium–large genomes (e.g. *N. tabacum*). It may be necessary to expose the filter to X-ray film for several days in order to obtain an adequate signal.

10.1. Setting up the blot

Procedure

1. To a sterile microcentrifuge tube, add DNA solution + H_2O to a volume of 23 µL; 3 µL restriction enzyme at 5–10 units µL^{-1}; 3 µL of $10 \times$ appropriate buffer; and 1 µL of 20 mM spermi-

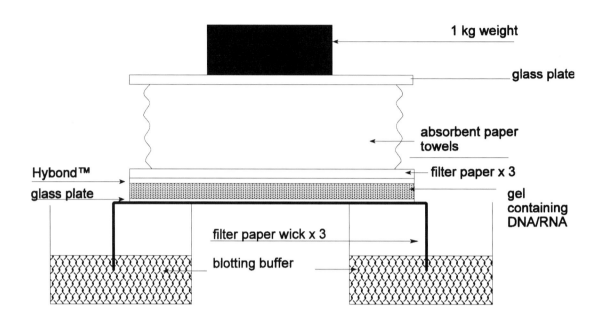

A typical capillary blotting apparatus

Figure 1.3. Diagrammatic representation of a Southern/Northern blot to transfer DNA/RNA to a membrane for DNA/RNA hybridisation.

dine from frozen stock (optional: often aids digestion of "difficult" or miniprep DNA).

2. Incubate at an appropriate temperature for 3–4 h. Remove 1 μL after 2–3 h and run on a minigel (1% agarose at 50 mA for 1 h) to check for digestion. A smear should be apparent. If satisfied that digestion is complete, add 5 μL of blue dye loading buffer to the main digest, mix, and load into wells of a large 1% agarose/EtBr gel.

3. Electrophorese the samples and appropriate markers (e.g. BstEII digested λ) overnight (12–16 h) at 25–30 mA (until the blue marker dye has migrated 15–20 cm down the gel). Marker bands of 4 kb should have moved about 10 cm down the gel.

4. Photograph the gel beside a graduated ruler and treat with acid wash (0.25 M HCl) for 20 min, followed by denaturing solution (1.5 M NaCl + 0.5 M NaOH) for 30 min, followed by neutralising solution (1.5 M NaCl + 1 M Tris HCl) for 30 min.

5. Transfer DNA from the gel to nylon or nitrocellulose membranes (we find Hybond N+ or Hybond C+ from Amersham to be satisfactory) overnight using 20 × SSC as transfer buffer (see Figure 1.3).

6. Remove the membrane noting its orientation (often it is useful to remove a corner of the membrane to assist with orientation) and rinse

gently in 3 × SSC for 30 s. Allow the membrane to air dry on blotting or chromatography paper.

7. To immobilise the DNA onto the membrane, the following protocols are appropriate for membranes made by Amersham:
   (a) for Hybond C+, bake at 80°C under vacuum for 1 h;
   (b) for Hybond N+, place on filter paper dampened with 0.4 M NaOH for 15 min followed by treatment with 2 × SSC for 30 s.

Following immobilisation of DNA, the membrane is allowed to dry at room temperature and can be stored in a dust-free environment until needed.

10.2. Preparation of radiolabelled probe

For DNA hybridisation, we routinely use DNA probes labelled according to the random primer method of Feinberg and Vogelstein (1984). Though it is easy to make a labelling buffer from individual components (see Hamill and Parr (1994) for recipe), we now usually use a commercial labelling kit (e.g. Gigaprime: Bresatec, Australia) due to its cost-effectiveness with respect to efficiency and saving of time. Usually 100–150 ng of DNA (estimated by comparison with standards) is used to prepare a probe and we use α^{32}P-dATP as the label. Labelled DNA is separated efficiently and quickly from unincorporated nucleotides by passage through a small Sephadex G50 column (Sambrook *et al.*, 1989). This is easily prepared by autoclaving Sephadex G50 (Pharmacia, medium grade) and pouring it into a 2-mL Pasteur pipette with sterile glass or polypropylene wool placed in the neck of the pipette to retain the Sephadex. The labelled DNA mixture is placed on top of the column and eluted with TE. Fractions are collected, each containing 8 drops, and most of the labelled DNA is usually eluted in Fractions 3 and 4. This can easily be checked by a Geiger counter and verified by Cherenkoff counting using a scintillation counter.

10.3.   DNA hybridisation

Procedure

1. Make 50 mL of a hybridisation solution containing $5 \times$ SSPE, $5 \times$ Denhardt's solution and 0.5% SDS. Heat to 65°C.
2. Denature salmon/herring sperm DNA by boiling an aliquot from the –20°C stock for 10 min. Add to the hybridisation solution at a final concentration of 20 µg mL^{-1}.
3. Soak the membrane containing DNA in 50 mL of the above solution in a flat-bottomed plastic container such as a lunch box (or in 10 mL of buffer in a cylindrical hybridisation tube) at 65°C for 3–4 h. Remove the prehybridisation solution and add fresh hybridisation solution made as above to just cover the membrane: ~40 mL if a 10 cm × 20 cm lunch box is used; ~8–10 mL if a hybridisation tube is used.
4. Boil the DNA probe for 5–10 min and add to the hybridisation solution and membrane at a final concentration of ~20 ng mL^{-1}. Incubate overnight at 65°C with gentle shaking if a lunch box is used, or 2–3 h rotation at 65°C if a hybridisation tube is used. Ensure that the membrane does not dry due to evaporation of the hybridisation solution as strong background problems will occur on the resulting autoradiograph.
5. Rinse the membrane once with low stringency solution ($2 \times$ SSC, 0.5% SDS) at room temperature, and wash once in 100 mL of low stringency solution for 30 min at 65°C. Wash twice more with high stringency solution ($0.1 \times$ SSC, 0.5% SDS) at 65°C for 30 min. Wrap the membrane in plastic food wrap to ensure it remains damp, and expose to X-ray film at –70°C in a cassette with an intensifier screen. Normally, single-copy foreign genes can be visualised in the tobacco genome within 2–3 d.

An example of a Southern blot using DNA extracted from hairy roots is shown in Figure 1.4. This blot was prepared to confirm the presence of the yeast ornithine decarboxylase yODC gene in DNA extracted from tobacco hairy roots.

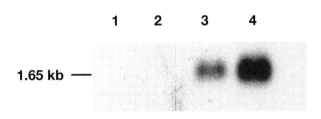

Figure 1.4. Detection of a foreign gene in DNA extracted from hairy roots. Hairy roots were formed in tobacco by infection with *A. rhizogenes* containing a binary vector containing the yeast ornithine decarboxylase (yODC) gene fused to a derivative of the CaMV35S promoter (Hamill *et al.*, 1990) as described in Section 6.1. Tracks 1 and 2 contain DNA from tissues infected with wild-type *A. rhizogenes*. Tracks 3 and 4 contain DNA from tissues infected with *A. rhizogenes* containing the yODC gene. DNA was extracted as in Section 7.1 and digested with *Xba*I prior to electrophoresis and Southern transfer as described in Section 10. A band of 1.65 kb in tracks 3 and 4 indicates that the yODC gene is present in these lines and is structurally intact.

11.   Northern Blot Analysis to Confirm That the Foreign Gene is Transcribed in Hairy Roots

11.1.   Preparation of RNA and running of the gel

Procedure

1. Combine 4.5 µL of RNA at 4.4 µg mL^{-1} with 2 µL of $5 \times$ MOPS RNA buffer, 3 µL formaldehyde and 10 µL formamide.
2. Heat samples to 65°C for 15 min, chill on ice for 2 min, and add 4 µL of blue dye loading buffer. Mix, pulse spin, and store on ice until needed.
3. Combine 4.5 g of agarose with 60 mL of $5 \times$ MOPS RNA buffer and 225 mL H$_2$O. Microwave or place in a boiling waterbath to dissolve. Cool to 50°C.
4. Add 16.2 mL formaldehyde and 20 µL of EtBr stock solution and mix gently.
5. Pour the gel in a fume cupboard and allow it to set.
6. Run the gel overnight at 25–30 mA in $1 \times$ MOPS RNA buffer until the blue dye has migrated about 10–12 cm. Photograph the gel beside a graduated ruler.

11.2.   Setting up a Northern blot

Procedure

1. Soak the gel twice in $10 \times$ SSC for 20 min with gentle shaking to remove the formaldehyde.
2. Carefully flip the gel over so that transfer will be through the bottom to the membrane. This also has the advantage that the orientation of the samples on the gel/blot will be the same as in the photograph.

3. Set up the Northern blot in the same way as for a Southern blot (Section 10.1) but using $10 \times$ SSC as transfer buffer (see also Figure 1.3). After the transfer is complete, rinse the membrane in $2 \times$ SSC and allow it to air dry.
4. Fix RNA samples to the membrane, either:
   (a) by baking membranes at 80°C for 1 h under vacuum if a nitrocellulose membrane such as Hybond C$^+$ is used; or
   (b) by alkali fixation via contact with a filter paper pad soaked in 0.05 M NaOH for 5 min followed by rinsing in $2 \times$ SSPE for 30 s if a positively charged nylon membrane such as Hybond N$^+$ is used.

After immobilisation of RNA, allow the membrane to air dry and store in a dust-free environment until required.

11.3. Probe preparation and hybridisation

Both RNA and DNA probes can be used to detect specific mRNA transcripts. A common problem in using some probes is that they detect rRNA bands due to non-specific hybridisation. We have found that the protocol below using RNase digestion will usually remove non-specific hybridisation with RNA probes. However, if membranes have been treated with RNase, they may not produce reliable results if reprobed with other RNA probes. The DNA probe procedure noted below using commercially available hybridisation solution also usually avoids detection of rRNA bands. It also has the advantage of being very rapid and allows membranes to be reprobed with fresh DNA probes.

11.3.1. Probe preparation (RNA probe)

The gene to be used as a probe should be available in a plasmid vector containing bacterial promoters such that antisense transcript can be produced by use of appropriate polymerase (e.g. pGEM series or pBluescript). T7 polymerase is cheaper than Sp6 polymerase and is at least as effective in producing a suitable RNA probe.

Procedure

1. Make up 50 mL of a hybridisation solution containing 50% formamide, 0.25 M Na phosphate buffer (pH 7.0), 0.25 M NaCl, 1 mM Na$_2$ EDTA (pH 8.0), $5 \times$ Denhardt's solution, denatured salmon/herring sperm DNA at 20 µg mL^{-1}, and 7% w/v SDS. Soak the membrane containing RNA in this solution for 3 h at 47°C with gentle shaking.
2. In a microcentrifuge tube, combine 4 µL of $5 \times$ transcription buffer (usually supplied with RNA polymerase), 2 µL of 100 mM dithiothreitol (DTT), 4 µL of cold rTP mix (this consists of rATP, rCTP and rGTP each at 2.5 mM and rUTP at 0.0625 mM), 0.5 µL RNase inhibitor at 40 units mL^{-1}, and 3 µL of linearised plasmid containing the gene of interest dissolved in H$_2$O at 0.5–1 µg mL^{-1}. Note: do not prepare the reaction below ~18°C as spermidine in the transcription buffer may cause precipitation of DNA. Usually we prepare the reaction using a water bath set at 25°C.
3. Add 1 µL of T7 or Sp6 polymerase as appropriate (at ~19 units µL^{-1}) and 5 µL of α^{32}P-rUTP.
4. Incubate the reaction at 37°C for 1.5–2 h and then add 2 µL of RQ1 (RNase-free DNase).
5. Incubate at 37°C for 15–20 min and then purify the radiolabelled probe by passing through a Sephadex G50 (medium grade) column made in a 2-mL Pasteur pipette (as noted in Section 10.2).
6. Combine the fractions containing the probe and heat to 65°C for 10 min. Add to the hybridisation mix and incubate the membrane overnight at 47–50°C in a heat-tolerant plastic container with gentle shaking. Ensure that the box has a tight lid to avoid evaporation.

11.3.2. Processing the filter after hybridisation

Note that it is essential to keep the membrane damp at all times.

Procedure

1. After hybridisation is complete, rinse the filter twice at room temperature using $2 \times$ SSC, 0.1% SDS with 15–20 min per rinse.
2. Rinse the filter twice in 100 mL of 25 mM Na phosphate buffer (pH 7.0), 1 mM EDTA, 0.1% SDS, at 50°C for 15–20 min each time.
3. Expose the filter to X-ray film with an intensifier screen at –70°C for 2–3 h or overnight.
4. If rRNA bands are visible, RNase treatment will be necessary which removes unbound/trapped riboprobe from rRNA bands. Add RNase A at 1 µg mL^{-1} final concentration to 50 mL of $3 \times$ SSC and incubate with the membrane at 37°C for 15–30 min.
5. Rinse the membrane twice with 100 mL of 25 mM Na phosphate buffer (pH 7.0), 1 mM EDTA, 0.1% SDS at 37°C.
6. Cover the membrane with plastic food wrap and expose to an X-ray film at –70°C with intensifier screen. Signal from strong transcripts is usually visible after 1–2 d of exposure. Weaker transcripts may require 7–10 d of exposure to X-ray film to produce a clear image.

An example of RNA extracted from hairy root tissue of tobacco and probed for expression of the yeast ornithine decarboxylase yODC gene using an RNA probe is shown in Figure 1.5. Figure 1.5A shows an

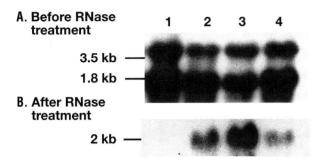

A. Before RNase treatment 1 2 3 4

3.5 kb —
1.8 kb —

B. After RNase treatment

2 kb —

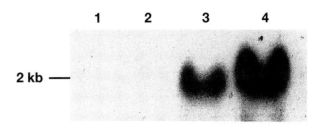

1 2 3 4

2 kb —

Figure 1.5. Detection of foreign gene expression in hairy roots using a RNA probe. RNA was extracted from the plant tissues noted in Figure 1.4 using the methods described in Section 8, and probed to detect the yODC transcript using an RNA probe prepared as described in Section 11.3.1. Track 1 contains RNA from control tissues infected with wild-type *A. rhizogenes*. Tracks 2, 3 and 4 contain RNA from hairy root tissues expressing the yODC gene. (A) Nonspecific interaction between the RNA riboprobe and rRNA bands. (B) The same filter as described in (A) following digestion with RNase A. Specific detection of the yODC transcript (2 kb) in tracks 2–4 is clearly demonstrated.

image of the filter before RNase treatment indicating the interaction of the probe with rRNA bands. Figure 1.5B shows an image of the filter after treatment with RNase A as noted above to reveal the transcript from the yeast ODC gene.

11.3.3. DNA probe to detect specific RNA transcripts

With the advent of commercially available hybridisation solutions (e.g. Express Hyb from Stratagene), we have found Northern hybridisation to be effective and rapid using DNA probes without problems of non-specific hybridisation between the probe and rRNA bands. Whilst these hybridisation solutions are expensive, only a small amount is required and the time-saving is significant. The following protocol works well in our hands using a temperature-controlled oven in conjunction with rotating hybridisation tubes.

Procedure

1. Label 100 ng of DNA to be used as a probe with α³²P-dATP and purify as described in Section 10.2.
2. Place the membrane prepared as in Section 11.2 in a hybridisation tube with the RNA-containing side facing away from the tube wall. Heat the hybridisation solution noted above (e.g. Express Hyb) to 65°C for 10–15 min, mix gently to ensure any precipitates are dissolved, and add 5–7 mL to the membrane in the hybridisation tube. Incubate with gentle rotation at 65°C for 1 h.

Figure 1.6. Detection of foreign gene expression in hairy roots using a DNA probe. RNA was extracted from the tobacco hairy roots noted in Figure 1.4 and probed to detect the yODC transcript using a DNA probe prepared as described in Section 11.3.3. Tracks 1 and 2 contain RNA from control tissues infected with wild type *A. rhizogenes*. Tracks 3 and 4 contain RNA from hairy root tissues expressing the yODC gene. The yODC transcript (2 kb) in tracks 3 and 4 is clearly visible.

3. Boil the DNA probe for 5 min. Remove the solution from the hybridisation tube containing the membrane and add 3 mL of fresh hybridisation solution and the boiled DNA probe. Incubate with gentle rotation for 30–60 min at 65°C.
4. Remove the probe solution, rinse the filter in 2 × SSC, then rinse once with 2 × SSC and 0.05% SDS at room temperature for 20 min. Rinse twice with 0.1 × SSC, 0.1% SDS at 50°C (high stringency RNA wash) for 15 min per rinse.
5. Cover the membrane in plastic food wrap and expose to X-ray film at –70°C with intensifier screen for 1–2 d or longer if required.

Note: it is important to ensure that the membranes never dry out during any of the hybridisation or washing steps. It is possible to strip membranes hybridised with a DNA probe and reuse them. This process can be repeated several times. To do this, boil a solution of 0.5% SDS and pour onto the membrane in a flat-bottomed container. Place on a shaker (20–30 rpm) and allow it to cool to room temperature. If the previous hybridisation signal was strong, it may be necessary to repeat this process, checking by exposure to X-ray film to ensure that all the probe is removed.

An example of a Northern blot using RNA extracted from tobacco hairy roots expressing the yeast ornithine decarboxylase yODC gene is shown in Figure 1.6. A DNA probe was used as noted above.

12. Detection of Gene Sequences by the Polymerase Chain Reaction (PCR) Technique

PCR techniques are routinely used in almost all areas of modern biology where detection of specific DNA sequences is required. We find PCR a useful technique to check whether specific sequences are present in DNA from transformed roots or in *A. rhizogenes* cells used to transform plant material (Hamill *et al.*, 1991). The technique is a powerful one and is sensitive to contamination of samples with plasmid DNA, so it is advisable to check that the constructs are correct or that DNA has integrated into the plant genome using procedures outlined in Sections 2, 3 and 7–10. We usually use oligonucleotides of 21 bases in length, chosen to have an (A+T)% of approximately 50%. To calculate T_m (T_m is the temperature at which 50% of the helical structure of DNA is lost), $2(A+T) + 4(G+C)$ is a reasonably accurate estimate for oligonucleotides of this length. Usually, a hybridisation temperature of ~5–10°C below T_m is utilised and often 55–60°C is adequate to enable specific amplification of target sequences. Pure or crude plant DNA can be utilised for PCR with input quantities ranging from 20–200 ng. Bacterial colonies (both *E. coli* and *A. rhizogenes*) can be used directly by dispersing a fresh colony in 50 µL H₂O and using 10 µL directly after boiling to denature the DNA and release some DNA from the cells.

The following protocol works well and will allow detection of a 1.2 kb fragment from the GUS gene in *E. coli*/*A. rhizogenes* carrying the binary vector pBI121, and also in DNA from transformed roots. Suitable primers are:

(a)   (Forward)
5'GGTGGGAAAGCGCGTTACAAG3'
(b)   (Reverse) 5'GTTTACGCGTTGCTTCCGCCA3'
(Sequences derived from (a) positions 400–420 and (b) positions 1599–1579 from within the GUS gene: Jefferson *et al.*, 1986.)

Procedure

1.   A double strength solution should be prepared containing 400 µM dATP, 400 µM dTTP, 400 µM dGTP, 400 µM dCTP, 0.02% w/v gelatin, 3 mM MgCl₂, 100 mM KCl, 20 mM Tris HCl (pH 8.3), and 140 ng (~20 pmole) of each primer. Place 25 µL in a 0.5-mL microcentrifuge tube and store on ice until needed. Note: many suppliers of thermostable polymerase such as Taq polymerase supply a concentrated buffer to which nucleotides, oligonucleotide primers, DNA and enzyme need only be added.
2.   Combine DNA/bacterial cells and water to a volume of 24 µL. Boil for 3–4 min, pulse spin to settle the contents in the tube, and place on ice for 2–3 min.
3.   Set a thermocycler to 72°C and incubate the double strength reaction mixture from Step 1 for 3–4 min.
4.   Add the DNA solution from Step 2, incubate for a further 2–3 min, and then add 1 µL of Taq polymerase containing 1 unit µL⁻¹.

5.   Add 50 µL (~ 2 drops) of mineral oil to each tube and thermocycle 25–35 times as follows:
1 min at 92°C (denaturing temperature)
2 min at 55°C (oligonucleotide hybridisation temperature)
2 min at 72°C (DNA extension temperature).
Allow the samples to cool to room temperature before analysis by electrophoresis.
6.   After the PCR reaction is complete, combine 20 µL of each PCR sample with 4 µL of blue dye loading buffer and electrophorese the samples on a 0.8% agarose gel at 40 mA for 2–3 h. The remainder of each sample can be stored at 4°C or –20°C as appropriate in the event that it is required. A discrete band (1.2 kb in the example noted above) should be visible. Southern blotting/DNA:DNA hybridisation procedures can be used to confirm that the amplified bands do represent the target DNA sequences.

Acknowledgments

We are grateful to Mrs J. Elliston for preparation of the manuscript and to Nicole De Rycke and other members of this laboratory for helpful comments.

References

Ackermann C (1977). Pflanzen aus *Agrobacterium rhizogenes*-tumoren an *Nicotiana tabacum*. *Plant Sci Lett* 8: 23–30.

Ahn JC, BG Jung, YW Paek, YJ Kim, KM Ko, SJ Hwang and B Hwang (1993). Production of tropane alkaloids by hairy root cultures of *Scopolia parviflora*. *Korean J Bot* 36: 225–231.

Aird ELH, JD Hamill and MJC Rhodes (1988a). Cytogenetic analysis of hairy root cultures from a number of plant species transformed by *Agrobacterium rhizogenes*. *Plant Cell Tiss Organ Cult* 15: 47–57.

Aird ELH, JD Hamill, RJ Robins and MJC Rhodes (1988b). Chromosome stability in transformed hairy root cultures and the properties of variant lines of *Nicotiana rustica* hairy roots. In *Manipulating Secondary Metabolism in Culture*, edited by RJ Robins and MJC Rhodes, pp. 137–144. Cambridge University Press, Cambridge.

Alvarez MA, JR Talou, NB Paniego and AM Giulietti (1994). Solasodine production in transformed organ cultures (roots and shoots) of *Solanum eleagnifolium* Cav. *Biotechnol Lett* 16: 393–396.

Asamizu T, K Akiyama and I Yasuda (1988). Anthraquinones production by hairy root culture in *Cassia obtusifolia*. *Yakagaku Zasshi* 108: 1215–1218.

Babakov AV, LM Bartova, IL Dridze, AN Maisuryan, GU Margulis, RR Oganian, VD Voblikova and GS Muromtsev (1995). Culture of transformed horseradish roots as a source of fusicoccin-like ligands. *J Plant Growth Reg* 14: 163–167.

Bailey JE (1991). Toward a science of metabolic engineering. *Science* 252: 1668–1675.

Banerjee S, AA Naqui, S Mandal and PS Ahuja (1994). Transformation of *Withania somnifera* (L) Dunal by *Agrobacterium rhizogenes* — infectivity and phytochemical studies. *Phytotherapy Res* 8: 452–455.

Banerjee-Chattopadhyay S, AM Schwemmin and DJ Schwemmin (1985). A study of karyotypes and their alterations in cultured and *Agrobacterium* transformed roots of *Lycopersicon peruvianum*. *Theor Appl Genet* 71: 258–262.

Bel-Rhlid R, S Chabot, Y Piche and R Chenevert (1993). Isolation and identification of flavonoids from Ri T-DNA transformed roots (*Daucus carota*) and their significance in vesicular-arbuscular mycorrhiza. *Phytochemistry* 33: 1369–1371.

Benjamin BD, G Roja and MR Heble (1994). Alkaloid synthesis by root cultures of *Rauwolfia serpentina* transformed by *Agrobacterium rhizogenes*. *Phytochemistry* 35: 381–383.

Berlin J, N Bedorf, C Mollenschott, V Wray, F Sasse and G Höfle (1988). On the podophyllotoxins of root cultures of *Linum flavum*. *Planta Med* 54: 204–206.

Berlin J, C Ruegenhagen, P Dietze, LF Fecker, OJM Goddijn and JHC Hoge (1993). Increased production of serotonin by suspension and root cultures of *Peganum harmala* transformed with a tryptophan decarboxylase cDNA clone from *Catharanthus roseus*. *Transgenic Res* 2: 336–344.

Bevan M (1984). Binary *Agrobacterium* vectors for plant transformation. *Nuc Acid Res* 12: 8711–8721.

Bhadra R, S Vani and JV Shanks (1993). Production of indole alkaloids by selected hairy root lines of *Catharanthus roseus*. *Biotechnol Bioeng* 41: 581–592.

Boulton MI, WG Bucholz, MS Marks, PG Markham and JW Davies (1989). Specificity of *Agrobacterium*-mediated delivery of maize streak virus DNA to members of the Graminaceae. *Plant Mol Biol* 12: 31–40.

Brown TA (1990). *Gene Cloning: An Introduction*. Chapman and Hall, London.

Brown TA (1992). *Genetics: A Molecular Approach*. Chapman and Hall, London.

Buitelaar RM, EJTM Leenen, G Geurtsen, Æ De Groot and J Tramper (1993). Effects of the addition of XAD-7 and of elicitor treatment on growth, thiophene production and excretion of hairy roots of *Tagetes patula*. *Enzyme Microb Technol* 15: 670–676.

Butcher DN and HE Street (1964). Excised root culture. *Bot Rev* 30: 513–586.

Capone I, L Spano, M Cardarelli, D Bellincampi, A Petit and P Costantino (1989). Induction and growth properties of carrot roots with different complements of *Agrobacterium rhizogenes* T-DNA. *Plant Mol Biol* 13: 43–52.

Cardarelli M, L Spano, A De Paolis, ML Mauro, G Vitali and P Costantino (1985). Identification of the genetic locus responsible for non-polar root induction by *Agrobacterium rhizogenes* 1855. *Plant Mol Biol* 5: 385–391.

Carron TR, MP Robbins and P Morris (1994). Genetic modification of condensed tannin biosynthesis in *Lotus corniculatus*: 1. Heterologous and antisense dihydroflavonol reductase down-regulates tannin accumulation in "hairy root" cultures. *Theor Appl Genet* 87: 1006–1015.

Chilton M-D, DA Tepfer, A Petit, C David, F Casse-Delbart and J Tempé (1982). *Agrobacterium rhizogenes* inserts T-DNA into the genomes of the host plant root cells. *Nature* 295: 432–434.

Christen P, MF Roberts, JD Phillipson and WC Evans (1989). High yield production of tropane alkaloids by hairy-root cultures of a *Datura candida* hybrid. *Plant Cell Rep* 8: 75–77.

Constabel CP and GHN Towers (1988). Thiarubrine accumulation in hairy root cultures of *Chaenactis douglasii*. *J Plant Physiol* 133: 67–72.

Costantino P, I Capone, M Cardarelli, A De Paolis, ML Mauro and M Trovato (1994). Bacterial plant oncogenes; the *rol* gene's saga. *Genetica* 94: 203–211.

Croes AF, AJR Van den Berg, M Bosveld, H Breteler and GJ Wullems (1989). Thiophene accumulation in relation to morphology in roots of *Tagetes patula*: effects of auxin and transformation by *Agrobacterium*. *Planta* 179: 43–50.

Daimon H and M Mii (1995). Plant regeneration and thiophene production in hairy root cultures of *Rudbeckia hirta* L used as an antagonistic plant to nematodes. *Jap J Crop Sci* 64: 650–655.

Davey MR, BJ Mulligan KMA Gartland, E Peel, AW Sargent and AJ Morgan (1987). Transformation of *Solanum* and *Nicotiana* species using an Ri plasmid vector. *J Exp Bot* 38: 1507–1516.

Dawson RF (1942a). Nicotine synthesis in excised tobacco roots. *Am J Bot* 29: 813–815.

Dawson RF (1942b). The localization of the nicotine synthetic mechanism in the tobacco plant. *Science* 94: 396–397.

De Cleene M and J De Ley (1981). The host range of infectious hairy root. *Bot Rev* 47: 147–194.

Delbarre A, P Muller, V Imhoff, H Barbier-Brygoo, C Maurel, N Leblanc, C Perrot-Rechenmann and J Guern (1994). The *rol*B gene of *Agrobacterium rhizogenes* does not increase the auxin sensitivity of tobacco protoplasts by modifying the intracellular auxin concentration. *Plant Physiol* 105: 563–569.

Delbecque JP, P Beydon, L Chapuis and MF Corio-Costet (1995). *In vitro* incorporation of radiolabelled cholesterol and mevalonic acid into ecdysteroid by hairy root cultures of a plant, *Serratula tinctoria*. *Eur J Entomol* 92: 301–307.

Deno H, C Suga, T Morimoto and Y Fujita (1987a). Production of shikonin derivatives by cell suspension cultures of *Lithospermum erythrorhizon*. VI. Production of shikonin derivatives by a two-layer culture containing an organic solvent. *Plant Cell Rep* 6: 197–199.

Deno H, T Yamagata, T Emoto, T Yoshioka, Y Yamada and Y Fujita (1987b). Scopolamine production by root cultures of *Duboisia myoporoides*. II. Establishment of a hairy root culture by infection with *Agrobacterium rhizogenes*. *J Plant Physiol* 131: 315–323.

De Paolis A, ML Mauro, M Pomponi, M Cardarelli, L Spano and P Costantino (1985). Localization of agropine-synthesizing functions in the T_R-region of the root inducing plasmid of *Agrobacterium rhizogenes* 1855. *Plasmid* 13: 1–7.

Deus-Neumann B and MH Zenk (1984). Instability of indole alkaloid production in *Catharanthus roseus* cell suspension cultures. *Planta Med* 50: 427–431.

Dittrich H and TM Kutchan (1991). Molecular cloning, expression and induction of berberine bridge enzyme, an enzyme essential to the formation of benzophenanthridine alkaloids in the response of plants to pathogenic attack. *Proc Nat Acad Sci USA* 88: 9969–9973.

Doerk-Schmitz K, L Witte and AW Alfermann (1994). Tropane alkaloid patterns in plants and hairy roots of *Hyoscyamus albus*. *Phytochemistry* 35: 107–110.

Dommisse EM, DWM Leung, ML Shaw and AJ Conner (1990). Onion is a monocotyledonous host for *Agrobacterium*. *Plant Sci* 69: 249–257.

Drewes FE and J van Staden (1995a). Attempts to produce solasodine in callus and suspension cultures of *Solanum mauritianum* Scop. *Plant Growth Regul* 17: 21–25.

Drewes FE and J van Staden (1995b). Initiation of and solasodine production in hairy root cultures of *Solanum mauritianum* Scop. *Plant Growth Regul* 17: 27–31.

Dupraz JM, P Christen and I Kapetanidis (1994). Tropane alkaloids in transformed roots of *Datura quercifolia*. *Planta Med* 60: 158–162.

Endo T and Y Yamada (1985). Alkaloid production in cultured roots of three species of *Duboisia*. *Phytochemistry* 24: 1233–1236.

Endo T, A Goodbody and M Misawa (1987). Alkaloid production in root and shoot cultures of *Catharanthus roseus*. *Planta Med* 53: 479–482.

Ermayanti TM, JA McComb and PA O'Brien (1993). Cytological analysis of seedling roots, transformed root cultures and roots regenerated from callus of *Swainsona galegifolia* (Andr.) R. Br. *J Exp Bot* 44: 375–380.

Ermayanti TM, JA McComb and PA O'Brien (1994). Growth and swainsonine production of *Swainsona galegifolia* (Andr.) R. Br. untransformed and transformed root cultures. *J Exp Bot* 45: 633–639.

Fecker LF, C Rügenhagen and J Berlin (1993). Increased production of cadaverine and anabasine in hairy root cultures of *Nicotiana tabacum* expressing a bacterial lysine decarboxylase gene. *Plant Mol Biol* 23: 11–21.

Fei HM, KF Mei, X Shen, YM Ye, ZP Lin and LH Peng (1993). Transformation of *Gynostemma pentaphyllum* by *Agrobacterium rhizogenes* saponin production in hairy root cultures. *Acta Bot Sinica* 35: 626–631.

Feinberg AP and BV Vogelstein (1984). A technique for radiolabelling DNA restriction endonuclease fragments to high specific activity. *Anal Biochem* 137: 266–267.

Filetici P, L Spano and P Costantino (1987). Conserved region in the T-DNA of different *Agrobacterium rhizogenes* root-inducing plasmids. *Plant Mol Biol* 9: 19–26.

Filippini F, F Lo Schiavo, M Terzi, P Costantino and M Trovato (1994). The plant oncogene *rol*B alters binding of auxin to plant cell membranes. *Plant Cell Physiol* 35: 767–771.

Flores HE and P Filner (1985). Metabolic relationships of putrescine, GABA and alkaloids in cell and root cultures of Solanaceae. In *Primary and Secondary Metabolism of Plant Cell Cultures*, edited by KH Newmann, W Barz and E Reinhard, pp. 174–186. Springer-Verlag, Heidelberg.

Flores HE, JJ Pickard and MW Hoy (1988). Production of polyacetylenes and thiophenes in heterotrophic and photosynthetic root cultures of Asteraceae. In *Chemistry and Biology of Naturally Occurring Acetylenes and Related Compounds (NOARC)*, edited by J Lam, H Breheler, T Arnason and L Hansen. *Bioactive Molecules* 7: 233–254.

Fowler MW (1985). Plant cell culture and natural product synthesis: an academic dream or a commercial possibility? *BioEssays* 3: 172–175.

Fromm ME, F Morrish, C Armstrong, R Williams, J Thomas and TM Klein (1990). Inheritance and expression of chimeric genes in the progeny of transgenic maize plants. *Bio/Technol* 8: 833–839.

Fujita Y (1990). The production of industrial compounds. In *Plant Tissue Culture: Applications and Limitations*, edited by SS Bhojwani, pp. 259–275. Elsevier, Amsterdam.

Fujita Y, S Takahashi and Y Yamada (1985). Selection of cell lines with high productivity of shikonin derivatives by protoplast culture of *Lithospermum erythrorhizon* cells. *Agric Biol Chem* 49: 1755–1759.

Furze JM, MJC Rhodes, AJ Parr, RJ Robins, IM Whitehead and DR Threlfall (1991). Abiotic factors elicit sesquiterpenoid phytoalexin production but not alkaloid production in transformed root cultures of *Datura stramonium*. *Plant Cell Rep* 10: 111–114.

Gamborg OL, RA Miller and K Ojima (1968). Nutrient requirements of suspension cultures of soy bean root cells. *Exp Cell Res* 50: 151–158.

Giulietti AM, AJ Parr and MJC Rhodes (1993). Tropane alkaloid production in transformed root cultures of *Brugmansia candida*. *Planta Med* 59: 428–431.

Gränicher F, P Christen and P Vuagnat (1994). Rapid high performance liquid chromatographic quantification of valepotriates in hairy root cultures of *Valeriana officinalis* L var. *sambucifolia*. *Phytochem Anal* 5: 297–301.

Gränicher F, P Christen, P Kamalaprija and U Burger (1995a). An iridoid diester from *Valeriana officinalis* var. *sambucifolia* hairy roots. *Phytochemistry* 38: 103–105.

Gränicher F, P Christen and I Kapetanidis (1995b). Production of valepotriates by hairy root cultures of *Centranthus ruber* DC. *Plant Cell Rep* 14: 294–298.

Green KD, NH Thomas and JA Callow (1992). Product enhancement and recovery from transformed root cultures of *Nicotiana glauca*. *Biotechnol Bioeng* 39: 195–202.

Hamill JD (1993). Alterations in auxin and cytokinin metabolism of higher plants due to expression of specific genes from pathogenic bacteria: a review. *Aust J Plant Physiol* 20: 405–423.

Hamill JD and MJC Rhodes (1993). Manipulating secondary metabolism in culture. In *Biosynthesis and Manipulation of Plant Products*, edited by D Grierson, pp. 178–209. Blackie, London.

Hamill JD and AJ Parr (1994). Methods for production of alkaloids in root cultures and analysis of products. In *Modern Methods of Plant Analysis*, edited by HF Linskens and JF Jackson, pp. 191–214, vol. 15. Springer-Verlag, Berlin.

Hamill JD, AJ Parr, RJ Robins and MJC Rhodes (1986). Secondary product formation by cultures of *Beta vulgaris* and *Nicotiana rustica* transformed with *Agrobacterium rhizogenes*. *Plant Cell Rep* 5: 111–114.

Hamill JD, A Prescott and C Martin (1987). Assessment of the efficiency of cotransformation of the T-DNA of disarmed binary vectors derived from *Agrobacterium tumefaciens* and the T-DNA of *A. rhizogenes*. *Plant Mol Biol* 9: 573–584.

Hamill JD, RJ Robins and MJC Rhodes (1989). Alkaloid production by transformed root cultures of *Cinchona ledgeriana*. *Planta Med* 55: 354–357.

Hamill JD, RJ Robins, AJ Parr, DM Evans, JM Furze and MJC Rhodes (1990). Over-expressing a yeast ornithine decarboxylase gene in transgenic roots of *Nicotiana rustica* can lead to enhanced nicotine accumulation. *Plant Mol Biol* 15: 27–38.

Hamill JD, S Rounsley, A Spencer, G Todd and MJC Rhodes (1991). The use of the polymerase chain reaction in plant transformation studies. *Plant Cell Rep* 10: 221–224.

Handa T (1991). Establishment of hairy root lines by inoculation with *Agrobacterium rhizogenes*. *Bull RIAR Ishkiawa Agr Coll* 2: 13–18.

Hänisch ten Cate CH, K Sree Ramulu, P Dijkhuis and B de Groot (1987). Genetic stability of cultured hairy roots by *Agrobacterium rhizogenes* on tubers of potato cv Bintje. *Plant Sci* 49: 217–222.

Hansen G, J Vaubert, JN Héron, J Clerot, J Tempé and J Breven (1993). Phenotypic effects of overexpression of *Agrobacterium rhizogenes* T-DNA ORF13 in transgenic tobacco plants are mediated by diffusible factor(s). *Plant J* 4: 581–585.

Hartmann T and G Toppel (1987). Senecionine *N*-oxide, the primary product of pyrrolizidine alkaloid biosynthesis in root cultures of *Senecio vulgaris*. *Phytochemistry* 26: 1639–1643.

Hashimoto T and Y Yamada (1994). Alkaloid biogenesis: molecular aspects. *Ann Rev Plant Physiol Plant Mol Biol* 45: 257–285.

Hashimoto T, Y Yukimura and Y Yamada (1986). Tropane alkaloid production in *Hyoscyamus* root cultures. *J Plant Physiol* 124: 61–75.

Hashimoto T, A Hayashi, Y Amano, J Kohno, H Iwanari, S Usuda and Y Yamada (1991). Hyoscyamine 6β-hydroxylase, an enzyme involved in tropane alkaloid biosynthesis, is localised at the pericycle of the root. *J Biol Chem* 266: 4648–4653.

Hashimoto T, D-J Yun and Y Yamada (1993). Production of tropane alkaloids in genetically engineered root cultures. *Phytochemistry* 32: 713–718.

Hayashi T, K Gotoh, K Ohnishi, K Okamura and T Asamizu (1994). 6-Methoxy-2-benzoxazolinone in *Scoparia dulcis* and its production by cultured tissues. *Phytochemistry* 37: 1611–1614.

Hibi N, S Higashiguchi, T Hashimoto and Y Yamada (1994). Gene expression in tobacco low-nicotine mutants. *Plant Cell* 6: 723–735.

Hiei Y, S Ohta, T Komari and T Kumashiro (1994). Efficient transformation of rice (*Oryza sativa* L.) mediated by *Agrobacterium* and sequence analysis of the boundaries of the T-DNA. *Plant J* 6: 271–282.

Hildebrand E (1934). Life history of the hairy-root organism in relation to its pathogenesis on nursery apple trees. *J Agric Res* 48: 857–885.

Hirotani M, Y Zhou, H Lui and T Furuya (1994). Astragalosides from hairy root cultures of *Astragalus membranaceus*. *Phytochemistry* 36: 665–670.

Hook I (1994). Secondary metabolites in hairy root cultures of *Leontopodium alpinum* Cass (edelweiss). *Plant Cell Tiss Organ Cult* 38: 321–326.

Hu Z-B and AW Alfermann (1993). Diterpenoid production in hairy root cultures of *Salvia miltiorrhiza*. *Phytochemistry* 32: 699–703.

Huffman GA, FF White, MP Gordon and EW Nester (1984). Hairy root inducing plasmid: physical map and homology to tumor-inducing plasmids. *J Bacteriol* 157: 269–276.

Ikenaga T, T Oyama and T Muranaka (1995). Growth and steroidal saponin production in hairy root cultures of *Solanum aculeatissimum*. *Plant Cell Rep* 14: 413–417.

Inomata S, M Yokoyama, Y Gozu, T Shimizu and M Yanagi (1993). Growth pattern and ginsenoside production of *Agrobacterium*-transformed *Panax ginseng* roots. *Plant Cell Rep* 12: 681–686.

Ishimaru K, H Arakawa, M Yamanaka and K Shimomura (1994). Polyacetylene in *Lobelia sessilifolia* hairy roots. *Phytochemistry* 35: 365–369.

Jacobs JJ, RR Aaroo, EA De-Koning, AJ Klunder, AF Croes and GJ Wullems (1995). Isolation and characterization of mutants of thiophene synthesis in *Tagetes erecta*. *Plant Physiol* 107: 807–814.

Jaziri M, K Shimomura, K Yoshimatsu, ML Fauconnier, M Marlier and J Homes (1995). Establishment of normal and transformed root cultures of *Artemisia annua* L for artemisinin production. *J Plant Physiol* 145: 175–177.

Jefferson RA, SM Burgess and D Hirsh (1986). β-Glucuronidase from *Escherichia coli* as a gene fusion marker. *Proc Nat Acad Sci USA* 83: 8447–8451.

Jefferson RA, TA Kavanagh and MW Bevan (1987). GUS fusion: β-glucuronidase as a sensitive and versatile gene fusion marker in higher plants. *EMBO J* 6: 3301–3307.

Jouanin L (1984). Restriction map of an agropine-type Ri plasmid and its homologies with Ti plasmids. *Plasmid* 12: 91–102.

Jung G and D Tepfer (1987). Use of genetic trans-formation by the Ri T-DNA of *Agrobacterium rhizogenes* to stimulate biomass and tropane alkaloid production in *Atropa belladonna* and *Calystegia sepium* roots grown *in vitro*. *Plant Sci* 50: 145–151.

Jung KH, SS Kwak, CY Choi and JR Liu (1994). Development of two stage culture process by optimisation of inorganic salts for improving catharanthine production in hairy root cultures of *Catharanthus roseus*. *J Ferment Bioeng* 77: 57–61.

Kaeppler HF, DA Somers, HW Rines and AF Cockburn (1992). Silicon carbide fiber-mediated stable transformation of plant cells. *Theor Appl Genet* 84: 560–566.

Kamada H, N Okamura, M Satake, M Harada and K Shimomura (1986). Alkaloid production by hairy root cultures in *Atropa belladonna*. *Plant Cell Rep* 5: 239–242.

Kim CH, SW Lee and IS Chung (1994). Hairy root culture of *Daucus carota* for anthocyanin production in a fluidised-bed bioreactor. *Agric Chem Biotechnol* 37: 237–242.

Kino-oka M, K Mine, M Taya, S Tone and T Ichi (1994). Production and release of anthraquinone pigments by hairy roots of madder (*Rubia tinctorum* L.) under improved culture conditions. *J Ferment Bioeng* 77: 103–106.

Kisiel W, A Stojakowska, J Malarz and S Kohlmunzer (1995). Sesquiterpene lactones in *Agrobacterium rhizogenes*-transformed hairy root culture of *Lactuca virosa*. *Phytochemistry* 40: 1139–1140.

Klein TM, ED Wolf, R Wu and JC Sandford (1987). High velocity microprojectiles for delivery of nucleic acids into living cells. *Nature* 327: 70–73.

Klein TM, L Kornstein, JC Sandford, ME Fromm and P Maliga (1988). Stable genetic transformation of intact *Nicotiana* cells by the particle bombardment process. *Proc Nat Acad Sci USA* 85: 8502–8505.

Ko KS, H Noguchi, Y Ebizuka and U Sankawa (1989). Oligoside production by hairy root cultures transformed by Ri plasmids. *Chem Pharm Bull* 37: 249–252.

Ko KS, Y Ebizuka, H Noguchi and U Sankawa (1995). Production of polyketide pigments in hairy root cultures of *Cassia* plants. *Chem Pharm Bull* (Tokyo) 43: 274–278.

Kraus PFX and TM Kutchan (1995). Molecular cloning and heterologous expression of a cDNA encoding berbamunine synthase, a *c-o* phenol-coupling cytochrome P-450 from the higher plant *Berberis stolonifera*. *Proc Nat Acad Sci USA* 92: 2071–2075.

Kurz WGW and F Constabel (1985). Aspects affecting biosynthesis of secondary metabolites in plant cell cultures. *CRC Crit Rev Biotechnol* 2: 105–118.

Kutchan TM (1995). Alkaloid biosynthesis — the basis for metabolic engineering of medicinal plants. *Plant Cell* 7: 1059–1070.

Kutchan TM, N Hampp, F Lottspeich, K Beyreuther and MH Zenk (1988). The cDNA clone for strictosidine synthase from *Rauvolfia serpentina*: DNA sequence determination and expression in *Escherichia coli*. *FEBS Lett* 257: 40–44.

Larsen WA, JT Hsu, HE Flores and AE Humphrey (1993). A study of nicotine release from tobacco hairy roots by transient technique. *Biotechnol Techniq* 7: 557–562.

MacRae S and J van Staden (1993). *Agrobacterium rhizogenes*-mediated transformation to improve rooting ability of eucalypts. *Tree Physiol* 12: 411–418.

Maldonado-Mendoza IE, T Ayora-Talavera and VM Loyola-Vargas (1993). Establishment of hairy root cultures of *Datura stramonium*. *Plant Cell Tiss Organ Cult* 33: 321–329.

Mano Y, S Nabeshima, C Matsui and H Ohkawa (1986). Production of tropane alkaloids by hairy root cultures of *Scopolia japonica*. *Agric Biol Chem* 50: 2715–2722.

Mano Y, H Ohkawa and Y Yamada (1989). Production of tropane alkaloids by hairy root cultures of *Duboisia leichhardtii* transformed by *Agrobacterium rhizogenes*. *Plant Sci* 59: 191–201.

Marchant YY (1988). *Agrobacterium rhizogenes*-transformed root cultures for the study of polyacetylene metabolism and biosynthesis. In: *Chemistry and Biology of Naturally Occurring Acetylenes and Related Compounds (NOARC) Bioactive Molecules* 7: 217–231.

Martin C, R Carpenter, H Sommer, H Saedler and ES Coen (1985). Molecular analysis of instability in flower pigmentation of *Antirrhinum majus*, following isolation of the pallida locus by transposon tagging. *EMBO J* 4: 1625–1630.

Matsuda J, S Okabe, T Hashimoto and Y Yamada (1991). Molecular cloning of hyoscyamine 6β-hydroxylase, a 2-oxoglutarate-dependent dioxygenase from cultured roots of *Hyoscyamus niger*. *J Biol Chem* 266: 9460–9464.

Mattanovich D, F Rüker, A Machado da Câmara, M Laimer, F Regner, H Steinkellner, G Himmler and H Katinger (1989). Efficient transformation of *Agrobacterium* spp. by electroporation. *Nuc Acid Res* 17: 6747.

Maurel C, N Leblanc, H Barbier-Brygoo, C Perrot-Rechenmann, M Bouvier-Durand and J Guern (1994). Alterations of auxin perception in *rol*B transformed tobacco protoplasts. *Plant Physiol* 105: 1209–1215.

Morgan A, P Cox, D Turner, E Peel, M Davey, K Gartland and B Mulligan (1987). Transformation of tomato using an Ri plasmid vector. *Plant Sci* 49: 37–49.

Motomori Y, K Shimomura, K Mori, H Kunitake, T Nakashima, M Tanaka, S Miyazaki and K Ishimaru (1995). Polyphenol production in hairy root cultures of *Fragaria × ananassa*. *Phytochemistry* 40: 1425–1428.

Mugnier, J (1988). Establishment of new axenic root lines by inoculation with *Agrobacterium rhizogenes*. *Plant Cell Rep* 7: 9–12.

Murashige, T and F Skoog (1962). A revised medium for rapid growth and bioassays with tobacco tissue cultures. *Physiol Plant* 15: 473–479.

Nakajima K, T Hashimoto and Y Yamada (1993). Two tropinone reductases with different stereo-specificities are short-chain dehydrogenases evolved from a common ancestor. *Proc Nat Acad Sci USA* 90: 9591–9595.

Nessler CL (1994). Metabolic engineering of plant secondary products. *Transgenic Res* 3: 109–115.

Nguyen C, F Bourgaud, P Forlot and A Guckert (1992). Establishment of hairy root cultures of *Psoralea* species. *Plant Cell Rep* 11: 424–427.

Nicholl DST (1994). *An Introduction to Genetic Engineering*. Cambridge University Press, Cambridge.

Nilsson O, A Crozier, T Schmülling, G Sandberg and O Olsson (1993a). Indole-3-acetic acid homeostasis in transgenic tobacco plants expressing the *Agrobacterium rhizogenes rol*B gene. *Plant J* 3: 681–689.

Nilsson O, T Moritz, N Imbault, G Sandberg and O Olsson (1993b). Hormonal characterization of tobacco plants expressing the *rol*C gene of *Agrobacterium rhizogenes* T-DNA. *Plant Physiol* 102: 363–371.

Offringa A, L Melchers, G Regensburg-Tuink, P Costantino, R Schilperoot and P Hooykaas (1986). Complementation of *Agrobacterium tumefaciens* tumour-inducing *aux* mutants by genes from the T_R-region of the Ri plasmid of *Agrobacterium rhizogenes*. *Proc Nat Acad Sci USA* 83: 6934–6939.

Ogasawara T, K Chiba and M Tada (1993). Production in high-yield of a naphthoquinone by a hairy root culture of *Sesamum indicum*. *Phytochemistry* 33: 1095–1098.

Old RW and SB Primrose (1994). *Principles of Gene Manipulation: An Introduction to Genetic Engineering*, 5th ed. Blackwell Scientific, Oxford.

Parr AJ (1989). The production of secondary metabolites by plant cell cultures. *J Biotechnol* 10: 1–26.

Parr AJ and JD Hamill (1987). Relationship between *Agrobacterium rhizogenes* transformed hairy roots and intact uninfected *Nicotiana* plants. *Phytochemistry* 26: 3241–3245.

Parr AJ, ACJ Peerless, JD Hamill, NJ Walton, RJ Robins and MJC Rhodes (1988). Alkaloid production by transformed root cultures of *Catharanthus roseus*. *Plant Cell Rep* 7: 309–312.

Parr AJ, J Payne, J Eagles, BT Chapman, RJ Robins and MJC Rhodes (1990). Variation in tropane alkaloid accumulation within the Solanaceae and strategies for its exploitation. *Phytochemistry* 29: 2545–2550.

Payne J, JD Hamill, RJ Robins and MJC Rhodes (1987). Production of hyoscyamine by hairy root cultures of *Datura stramonium. Planta Med.* 53: 474–478.

Pich U and I Schubert (1993). Midiprep method for isolation of DNA from plants with a high content of polyphenolics. *Nuc Acid Res* 21: 3328.

Porter J (1991). Host range and implications of plant infection by *Agrobacterium rhizogenes. Crit Rev Plant Sci* 10: 387–421.

Qin MB, GZ Li, Y Yun, HC Ye and GF Li (1994). Induction of hairy root from *Artemisia annua* with *Agrobacterium rhizogenes* and its culture *in vitro. Acta Bot Sinica* 36 (Suppl.), 165–170.

Ramsay G and A Kumar (1990). Transformation of *Vicia faba* cotyledon and stem tissues by *Agrobacterium rhizogenes*: infectivity and cytological studies. *J Exp Bot* 41: 841–847.

Rhodes MJC, RJ Robins, ELH Aird, J Payne, AJ Parr and NJ Walton (1989). Regulation of secondary metabolism in transformed root cultures. In *Primary and Secondary Metabolism in Transformed Root Cultures*, edited by WGW Kurz, pp. 58–72. Springer-Verlag, Berlin.

Rhodes MJC, AJ Parr, A Giulietti and ELH Aird (1994). Influence of exogenous hormones on the growth and secondary metabolite formation in transformed root cultures. *Plant Cell Tiss Organ Cult* 38: 143–151.

Riker A, W Banfield, W Wright, G Keitt and H Sagen (1930). Studies on infectious hairy root of nursery apple trees. *J Agr Res* 41: 887–912.

Robaglia C, F Vilaine, V Pautot, F Raimond, J Amselem, L Jouanin, F Casse-Delbart and M Tepfer (1987). Expression vectors based on the *Agrobacterium rhizogenes* Ri plasmid transformation system. *Biochimie* 69: 231–237.

Robins RJ, AJ Parr, J Payne, NJ Walton and MJC Rhodes (1990). Factors regulating tropane-alkaloid production in a transformed root culture of a *Datura candida × D. aurea* hybrid. *Planta* 181: 414–422.

Robins RJ, EG Bent and MJC Rhodes (1991). Studies on the biosynthesis of tropane alkaloids by *Datura stramonium* L. transformed root cultures. 3. The relationship between morphological integrity and alkaloid biosynthesis. *Planta* 185: 385–390.

Saito K, I Murakoshi, D Inze and M Van Montagu (1989). Biotransformation of nicotine alkaloids by tobacco shooty teratomas induced by a Ti plasmid mutant. *Plant Cell Rep* 7: 607–610.

Saito K, M Yamazaki, K Shimomura, K Yoshimatsu and T Murakoshi (1990). Genetic transformation of foxglove (*Digitalis purpurea*) by chimeric foreign genes and production of cardioactive glycosides. *Plant Cell Rep* 9: 121–124.

Sambrook J, EF Fritsch and T Maniatis (1989). *Molecular Cloning: A Laboratory Manual.* 2nd ed. Cold Spring Harbor Laboratory Press, New York.

Sandberg G, F Sitbon, S Eklöf, A Edlund, C Åstot, J Blackwell, T Moritz, B Sundberg and O Olsson (1995). Regulation of auxin and cytokinin turnover in plants. *Abstr 15th Internat Conf Plant Growth Substances.* Minneapolis, Minnesota, USA, 15.

Sato K, T Yamazaki, E Okuyama, K Yoshihira and K Shimomura (1991). Anthraquinone production by transformed root cultures of *Rubia tinctorum*: influence of phytohormones and sucrose concentration. *Phytochemistry* 30: 1507–1510.

Sauerwein M, K Ishimaru and K Shimomura (1991). A piperidone alkaloid from *Hyoscyamus albus* roots transformed with *Agrobacterium rhizogenes. Phytochemistry* 30: 2977–2978.

Schmülling T, J Schell and A Spena (1988). Single genes from *Agrobacterium rhizogenes* influence plant development. *EMBO J* 7: 2621–2629.

Sharp JM and PM Doran (1990). Characteristics of growth and tropane alkaloid synthesis in *Atropa belladonna* roots transformed by *Agrobacterium rhizogenes. J Biotechnol* 16: 171–186.

Shimomura K, H Sudo, H Saga and H Kamada (1991). Shikonin production and secretion by hairy root cultures of *Lithospermum erythrorhizon. Plant Cell Rep* 10: 282–285.

Signs MW and HE Flores (1989). Elicitation of sesquiterpene phytoalexin biosynthesis in transformed root cultures of *Hyoscyamus muticus. Plant Physiol* 89 (Suppl.), 135.

Signs MW and HE Flores (1990). The biosynthetic potential of plant roots. *Bioassays* 12, 7–13.

Sim SJ and HN Chang (1993). Increased shikonin production by hairy roots of *Lithospermum erythrorhizon* in two phase bubble column reactor. *Biotechnol Lett* 15: 145–150.

Sim SJ, HN Chang, JR Liu and KH Jung (1994). Production and secretion of indole alkaloids in hairy root cultures of *Catharanthus roseus*: effects of *in situ* adsorption, fungal elicitation and permeabilization. *J Ferment Bioeng* 78: 229–234.

Simpson RB, A Spielmann, L Margossian and TD McKnight (1986). A disarmed binary vector from *Agrobacterium tumefaciens* functions in *Agrobacterium rhizogenes. Plant Mol Biol* 6: 403–415.

Slightom JL, M Durand-Tardif, L Jouanin and D Tepfer (1986). Nucleotide sequence analysis of *Agrobacterium rhizogenes* agropine type plasmid: identification of open reading frames. *J Biol Chem* 261: 108–121.

Slusarenko AJ (1990). A rapid miniprep for the isolation of total DNA from *Agrobacterium tumefaciens. Plant Mol Biol Reporter* 8: 249–252.

Song Q, ML Gomezbarrios, EL Hopper, MA Hjortso and NH Fischer (1996). Biosynthetic studies of lactucin derivatives in hairy root cultures of *Lactuca floridana. Phytochemistry* 40: 1659–1665.

Spena A, T Schmülling, C Koncz and J Schell (1987). Independent and synergistic activity of rolA, B and C loci in stimulating abnormal growth in plants. *EMBO J* 6: 3891–3899.

Spano L, D Mariotti, M Pezzoti, F Damiani and S Arcioni (1987). Hairy root transformation in alfalfa (*Medicago sativa* L.). *Theor Appl Genet.* 73: 523–530.

Stougaard J, D Abildsten and KA Marcker (1987). The *Agrobacterium rhizogenes* pRi TL-TDNA segment as a gene vector system for transformation of plants. *Mol Gen Genet* 207: 251–255.

Subroto MA and PM Doran (1994). Production of steroidal alkaloids by hairy roots of *Solanum aviculare* and the effect of gibberellic acid. *Plant Cell Tiss Organ Cult* 38: 93–102.

Subroto MA, JD Hamill and PM Doran (1996a). Development of shooty teratomas from several solanaceous plants: growth kinetics, stoichiometry and alkaloid production. *J Biotechnol* 45: 45–57.

Subroto MA, KH Kwok, JD Hamill and PM Doran (1996b). Coculture of genetically transformed roots and shoots for synthesis, translocation and biotransformation of secondary metabolites. *Biotechnol Bioeng* 49: 481–494.

Sukhapinda K, R Spivey and EA Shakin (1987). Ri plasmid as a helper for introducing vector DNA into alfalfa plants. *Plant Mol Biol* 8: 209–216.

Tada H, K Shimomura and K Ishimaru (1995a). Polyacetylenes in hairy root cultures of *Lobelia chinensis* lour. *J Plant Physiol* 146: 199–202.

Tada H, K Shimomura and K Ishimaru (1995b). Polyacetylenes in *Platycodon grandiflorum* hairy root and campanulaceous plants. *J Plant Physiol* 145: 7–10.

Tada H, T Nakashima, H Kuntake, K Mori, M Tanaka and K Ishimaru (1996). Polyacetylenes in hairy root cultures of *Campanula medium* L. *J Plant Physiol* 147: 617–619.

Takeda T, T Kondo, H Mizukami and Y Ogihara (1994). Bryonolic acid production in hairy roots of *Trichosanthes kirilowii* MAX. var. japonica KITAM. transformed with *Agrobacterium rhizogenes* and its cytotoxic activity. *Chem Pharm Bull* (Tokyo) 42: 730–732.

Talou JR, O Cascone and AM Giulietti (1994). Content of thiophenes in transformed root cultures of Argentinian species of *Tagetes*. *Planta Med* 60: 260–262.

Tanaka N and T Matsumoto (1993). Regenerants from *Ajuga* hairy roots with high productivity of 20-hydroxyecdysone. *Plant Cell Rep* 13: 87–90.

Taya M, K Mine, M Kino-oka, S Tone and T Ichi (1992). Production and release of pigments by culture of transformed hairy root of red beet. *J Ferment Bioeng* 73: 31–36.

Taya M, K Yakura, M Kino-oka and S Tone (1994). Influence of medium constituents on enhancement of pigment production by batch culture of red beet hairy roots. *J Ferment Bioeng* 77: 215–217.

Tepfer D (1984). Transformation of several species of higher plants by *Agrobacterium rhizogenes*: sexual transmission of the transformed genotype and phenotype. *Cell* 37: 959–967.

Tepfer D (1989). Ri T-DNA from *Agrobacterium rhizogenes*: a source of genes having applications in rhizosphere biology and plant development, ecology and evolution. In *Plant–Microbe Interactions — Molecular and Genetic Perspectives*, edited by T Kosuge and EW Nester, pp. 294–342, vol. 3. McGraw-Hill, New York.

Tepfer DA and J Tempé (1981). Production d'agropine par des racines formés sous l'action d'*Agrobacterium rhizogenes* souche A₄. *CR Acad Sci Paris* Ser III 292: 153–156.

Toivonen L, J Balsevich and WGW Kurz (1989). Indole alkaloid production by hairy root cultures of *Catharanthus roseus*. *Plant Cell Tiss Organ Cult* 18: 79–93.

Toppel G, L Witte, B Riebesehl, KV Borstel and T Hartmann (1987). Alkaloid patterns and biosynthetic capacity of root cultures from some pyrrolizidine alkaloid producing *Senecio* species. *Plant Cell Rep* 6: 466–469.

Trypsteen M, M Van-Lijsebettens, R van Severen and M van Montagu (1991). *Agrobacterium rhizogenes*-mediated transformation of *Echinacea purpurea*. *Plant Cell Rep* 10: 85–89.

Van der Heijden R, R Verpoorte, SS Hoekstra and JHC Hoge (1994). Nordamnacanthal, a major anthraquinone from an *Agrobacterium rhizogenes* induced root culture of *Rubia tinctorum*. *Plant Physiol Biochem* 32: 399–404.

Verwoerd TC, BMM Dekker and A Hoekema (1989). A small-scale procedure for the rapid isolation of plant RNAs. *Nuc Acid Res* 17: 2362.

Waller GR and EK Nowacki (1978). *Alkaloid Biology and Metabolism in Plants*. Plenum, New York.

Weathers PJ, RD Cheetham, E Follansbee and K Teoh (1994). Artemisinin production by transformed roots of *Artemisia annua*. *Biotechnol Lett* 16: 1281–1286.

Westcott R (1988). Thiophene production from 'Tagetes' hairy roots. In *Manipulating Secondary Metabolism in Culture*, edited by RJ Robins and MJC Rhodes, pp. 233–237. Cambridge University Press, Cambridge.

White F and EW Nester (1980). Relationship of plasmids responsible for hairy root and crown gall tumourigenicity. *J Bacteriol* 141: 1134–1141.

White F, G Chidossi, M Gordon and E Nester (1982). Tumour induction by *Agrobacterium rhizogenes* involves the transfer of plasmid DNA to the plant genome. *Proc Nat Acad Sci USA* 79: 3193–3197.

White F, B Taylor, G Huffman, M Gordon and E Nester (1985). Molecular and genetic analysis of the transferred DNA regions of the root inducing plasmid of *Agrobacterium rhizogenes*. *J Bacteriol* 164: 33–44.

White P (1934). Potentially unlimited growth of excised tomato root tips in a liquid medium. *Plant Physiol* 9: 585–600.

Wibberley MS, JR Lenton and SJ Neill (1994). Sesquiterpenoid phytoalexins produced by hairy roots of *Nicotiana tabacum*. *Phytochemistry* 37: 349–351.

Williams RD and BE Ellis (1993). Alkaloids from *Agrobacterium rhizogenes* transformed *Papaver somniferum* cultures. *Phytochemistry* 32: 719–723.

Willmitzer L, J Sanchez-Serrano, E Buschfeld and J Schell (1982). DNA from *Agrobacterium rhizogenes* is transferred to and expressed in axenic hairy root tissues. *Mol Gen Genet* 186: 16–22.

Wink M (1987). Why do lupin cultures fail to produce alkaloids in large quantities? *Plant Cell Tiss Organ Cult* 8: 103–111.

Yamanaka M, K Ishibashi, K Shimomura and K Ishimaru (1996). Polyacetylene glucosides in hairy root cultures of *Lobelia cardinalis*. *Phytochemistry* 41: 183–185.

Yonemitsu H, K Shimomura, M Satake, S Mochida, M Tanaka, T Endo and A Kaji (1990). Lobeline production by hairy root culture of *Lobelia inflata* L. *Plant Cell Rep* 9: 307–310.

Yoshikawa T and T Furuya (1987). Saponin production by cultures of *Panax ginseng* transformed with *Agrobacterium rhizogenes*. *Plant Cell Rep* 6: 449–453.

Yu S, KH Kwok and PM Doran (1996). Effect of sucrose, exogenous product concentration and other culture conditions on growth and steroidal alkaloid production by *Solanum aviculare* hairy roots. *Enzyme Microb Technol* 18: 238–243.

Yukimune Y, Y Hara and Y Yamada (1994). Tropane alkaloid production in root cultures of *Duboisia myoporoides* obtained by repeated selection. *Biosci Biotechnol Biochem* 58: 1443–1446.

Zenk MH (1988). Biosynthesis of alkaloids using plant cell cultures. In *Recent Advances in Phytochemistry: Plant Nitrogen Metabolism*, edited by JE Poulton, JT Romeo and EE Conn, 23: 429–457.

Zenk MH, M Bueffer, M Amann, B Deus-Neumann and N Nagakura (1985). Benzylisoquinoline biosynthesis by cultivated plant cells and isolated enzymes. *J Nat Prod (Lloydia)* 48: 725–738.

Studies of Secondary Metabolic Pathways in Transformed Roots

Michael J.C. Rhodes, Adrian J. Parr and
Nicholas J. Walton

*Institute of Food Research, Norwich Laboratory, Norwich Research Park, Colney, Norwich,
NR4 7UA, UK; email: nicholas.walton@bbsrc.ac.uk*

Introduction

Transformed ("hairy") roots, developed following integration and expression of genes from *Agrobacterium rhizogenes*, have proved to be valuable model systems for the study of the secondary metabolism of root tissue. The transferred genes involved in hairy root formation include genes leading to the production in the plant of specific bacterial metabolites — the opines — and genes which encode enzymes that promote the hydrolysis of conjugated forms of the plant hormones (Michael and Spena, 1995; see Hamill and Lidgett, Chapter 1 of this volume).

The properties of transformed lines generally reflect those of normal roots but there are some differences. Transformed roots are morphologically different from normal plant roots in that they are much more highly branched. During growth they generate many new lateral meristematic growing points and this can lead to faster overall growth rates than for normal roots. Transformed root cultures have proved valuable in studying secondary metabolic pathways since they generally reflect the *in planta* operation of these pathways both in their route and enzymology. However, roots in culture lack the ability to transport their metabolites to other plant organs. For instance, in *Nicotiana* plants, nicotine is synthesised in the roots and transported to other parts of the plant where it accumulates. In *N. rustica* root cultures, the accumulation of nicotine is somewhat higher than in normal roots. A similar situation exists in *Datura* with hyoscyamine which is also synthesised in the roots and, *in planta*, transported to the aerial parts. In some cases the culture medium may act as a sink for transported product. In *Nicotiana* root cultures, nicotine is rapidly lost to the medium and essentially all the nicotine produced can be mobilised if the medium is frequently changed (Rhodes *et al.*, 1986). However in *Datura* where *in planta* alkaloid transport also occurs, transformed root cultures release very little hyoscyamine into the medium. The reasons for the differences in behaviour between these two relatively closely related species are unclear. The unique features of transformed roots, i.e. their altered hormone metabolism, transport properties and production of opines, give grounds to expect that the regulation of some secondary pathways in transformed and normal roots may prove to be different. Indeed there are a few cases where the accumulation of novel metabolites occurs in transformed roots. For instance, acetyltropine is a major metabolite in transformed roots of *Datura* while it is barely detectable in plant roots (Parr *et al.*, 1990).

In this chapter, we intend to give a perspective on studies with transformed roots which derives from our own work. Root cultures have proved ideal for feeding experiments since relatively high quantities of roots (typically 7–10 g fresh weight in 50 mL) can be fed under sterile conditions with precursors or inhibitors and the effects on the pathway evaluated. Cultures of young root material have proved useful for enzyme extraction and purification, in part because their levels of inhibitory substances such as phenolics appear generally low. Furthermore, since transformed root formation and stability involve the integration of foreign DNA, such roots are amenable to use in genetic manipulation experiments to engineer secondary metabolic pathways (see Hamill and Lidgett, Chapter 1 of this volume). Experiments with reporter genes such as β-glucuronidase (GUS) encoded on plasmids introduced into *A. rhizogenes* strains such as LBA9402 show that the frequency of co-integration of the T-DNA and the reporter gene into the plant genome is relatively high. Thus metabolic genes housed in such plasmids would also be expected to be co-transformed into the plant cell and expressed if supplied behind suitable promoters. There are now several examples of the use of such techniques to study and alter secondary pathways, and some will be described in this chapter.

Precursor-Feeding Studies

Feeding experiments can be of two main kinds. In the first, labelled precursors are provided to elucidate a physiologically-significant biosynthetic pathway or mechanism. In the second, relatively large quantities of metabolic precursors or precursor analogues are supplied to establish whether the normal biosynthetic pathway can be perturbed to produce increased amounts of products, to produce uncharacteristic products, or to be inhibited.

Both types of experiment have been performed with hairy root cultures. As indicated, the ability to bathe cultures under sterile and defined conditions to achieve rapid and reproducible rates of growth makes hairy roots much more attractive for feeding experiments than excised root tissue or intact plants and, as shown below, much higher levels of incorporation can often be obtained.

Studies With Isotopically-Labelled Precursors

Two recent examples are described which indicate the particular suitability of hairy root cultures for studies using isotopically-labelled precursors, together with GC/MS and NMR analysis of resulting metabolites.

The tropane alkaloid, hyoscyamine, is an ester of tropine with (S)-tropic acid. Tropine is derived from the N-methylpyrrolinium ion, itself a metabolite of ornithine and/or arginine (Figure 2.1). (S)-Tropic acid has been known for several years to arise as a result of a side-chain rearrangement of L-phenylalanine (Leete, 1984, 1987). The most obvious possibility was that free 3-phenyllactic acid derived from phenylalanine could be rearranged directly to form free tropic acid which was subsequently activated (most probably as the CoA thioester) and esterified to produce hyoscyamine. Experiments with hairy roots of *Datura stramonium* and a *Brugmansia candida* × *aurea* hybrid analysed by GC/MS and ^{13}C NMR suggested however that this did not occur, since tropic acid did not dilute the incorporation of (RS)-3-phenyl [1,3-$^{13}C_2$] lactic acid into hyoscyamine and its phenyllactoyl counterpart, littorine (Robins *et al.*, 1994c). Specific incorporations of around 40% were obtained in this work, compared with 0.27% reported for the incorporation of the same precursor into intact *D. stramonium* plants, indicating the clear advantage of hairy roots for this type of study. It was then shown using hairy root cultures of *D. stramonium* that labelled littorine (as (RS)-(3-phenyl [1,3-$^{13}C_2$] lactoyl) [*methyl*-$^2H_3$] tropine) was efficiently converted into hyoscyamine without loss of label, thereby demonstrating the direct rearrangement of the phenyllactoyl moiety to the tropoyl moiety in alkaloidal form, esterified with tropine (Robins *et al.*, 1994b). More recently, the mechanism of this process has been investigated by GC/MS, using ^{13}C- and ^{2}H-labelled phenyllactic acids and littorine and examining the incorporation of label not only into hyoscyamine but also into the minor alkaloidal products, 3α-phenylacetoxytropane, 3α-phenylacetoxy-6β,7β-epoxytropane and 3α-(2'-hydroxyacetoxy) tropane (Robins *et al.*, 1995). A novel putative mechanism has been proposed, in which the minor alkaloids are side-products of a rearrangement process leading principally to hyoscyamine (Figure 2.2). The enzymology remains entirely uncharacterised, but it is suggested that the mechanism might be free-radical in nature and involve a cytochrome P450, as described for the flavonone-isoflavonone isomerisation (Hakamatsuka *et al.*, 1991).

In a novel study, hairy root cultures of *Datura stramonium* and *Nicotiana tabacum* grown in the presence of [^{15}N] ammonium sulphate or [^{15}N] potassium nitrate have been investigated by *in vivo* ^{15}N NMR (Ford *et al.*, 1994). In *D. stramonium* cultures grown with each substrate, peaks could be assigned *en bloc* to nitrogenous primary metabolites, including the amide N of glutamine, the α-amino N atoms of glutamate and glutamine, the side-chain N atoms of γ-aminobutyric acid, ornithine and lysine and the

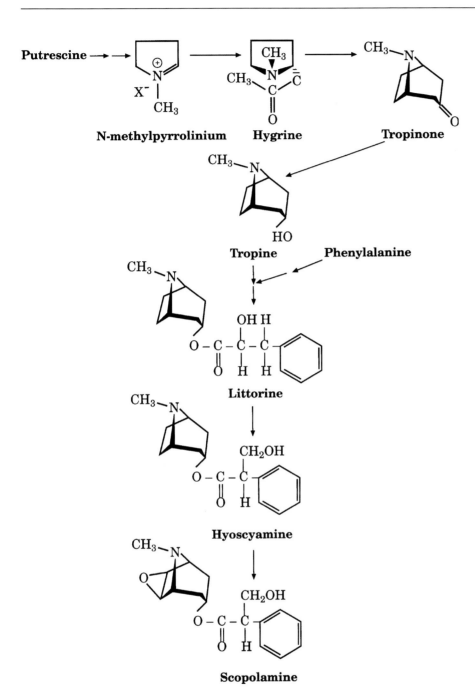

Figure 2.1. The biosynthetic pathway of tropane alkaloids.

ω,ω' and δ N atoms of arginine. It was also possible, however, to assign peaks to other nitrogenous metabolites, including alkaloids; thus peaks could be assigned to the amide N of hydroxycinnamoyl putrescine, to the ring and amide N atoms of agropine, to hyoscyamine, and to the pyridyl and pyrrolidyl N atoms of nicotine. The chemical shift of the pyrrolidyl N atom of nicotine is pH-dependent and it could be concluded that the bulk of the intracellular nicotine was vacuolar at a pH of ca. 5.4. For these experiments, about 2 g fresh weight of hairy roots was used in 20 mL of buffered solution, using an airlift system for oxygenation delivering O_2 at a rate of up to 60 mL min^{-1}. *In vivo* ^{31}P NMR spectra were recorded to monitor the cytoplasmic pH (ca. 7.6) and to confirm adequate oxygenation of the roots. The principal limitation of the technique was sensitivity; assuming 70% ^{15}N-labelling, only metabolites present at concentrations in excess of 1 μmol g^{-1} fresh weight were detectable, ruling out the detection of most intermediary metabolites.

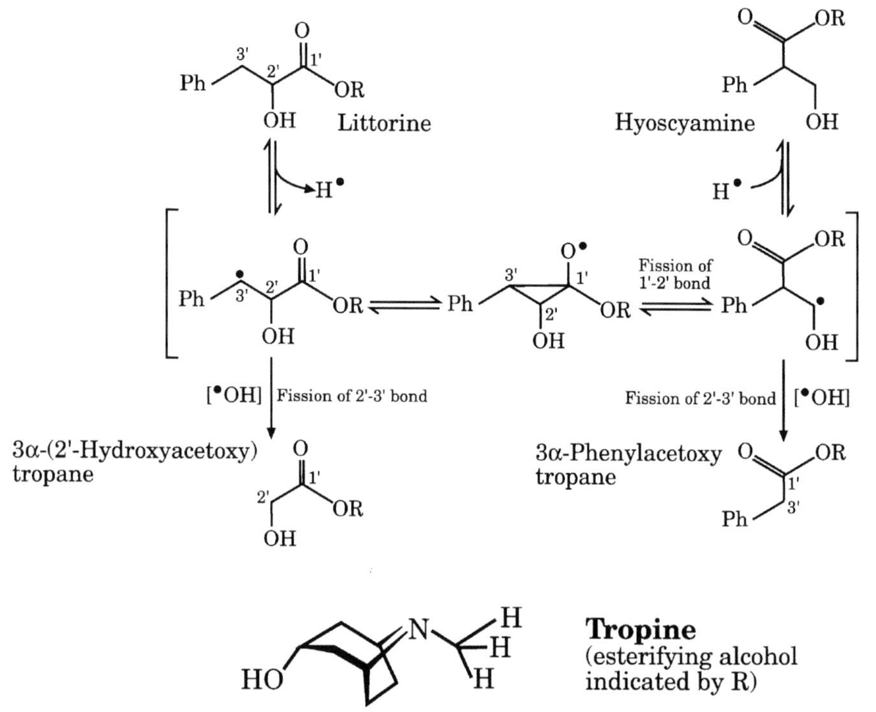

Figure 2.2. The proposed mechanism of rearrangement of littorine to hyoscyamine.

Studies With Metabolite Analogues

Feeding metabolic precursors to hairy roots in attempts to increase the formation of end-products has in general produced unremarkable results. For example, a modest stimulation of nicotine production in hairy root cultures of *Nicotiana rustica* was observed in response to exogenous putrescine or agmatine (Walton *et al.*, 1988). In contrast, feeding analogues of normal metabolites can sometimes produce substantial effects. Thus, cadaverine, the higher homologue of putrescine and the decarboxylation product of lysine, inhibits the formation of nicotine when supplied to *Nicotiana* hairy root cultures, and markedly stimulates the production of the piperidine alkaloid, anabasine, so that this becomes the major alkaloidal product (Figure 2.3) (Walton *et al.*, 1988; Walton and Belshaw, 1988).

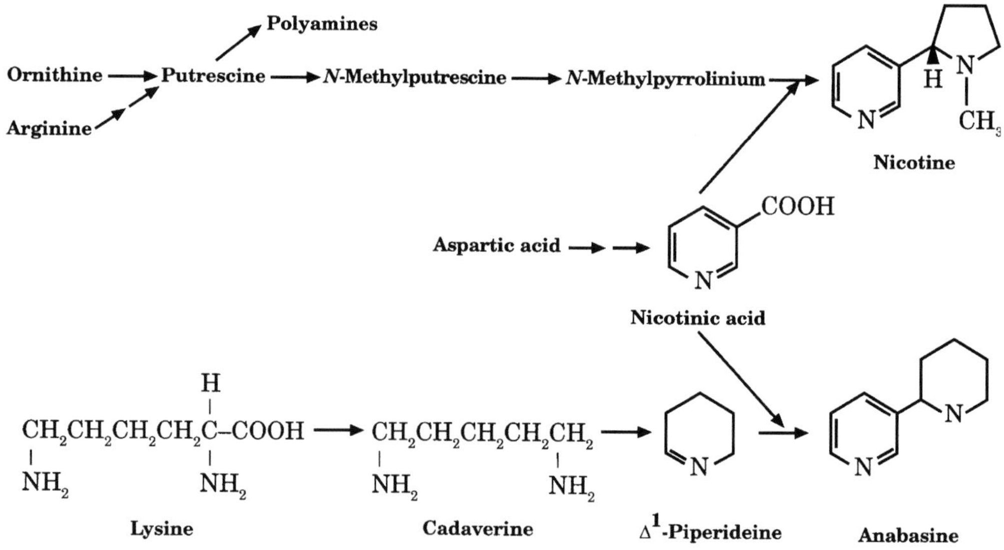

Figure 2.3. The pathways of nicotine and anabasine biosynthesis.

Anabasine is the principal alkaloid in intact plants of *N. glauca* and is optically inactive (Leete, 1979). Using ^{2}H NMR spectroscopy, it was demonstrated that labelled anabasine generated from (S)- and (R)-[1-^{2}H] cadaverine fed to *Nicotiana* hairy root cultures was also racemic, suggesting that anabasine formation in response to exogenous cadaverine occurs by the same mechanism as that normally operating in plants of *N. glauca* (Watson *et al.*, 1990). Cadaverine at high concentrations is a good substrate for *N*-methylputrescine oxidase (MPO; Walton and McLauchlan, 1990) and (in *Datura*) a competitive inhibitor of putrescine *N*-methyltransferase (PMT; $K_i = 0.04$ mM), the immediately-preceding enzyme in the putrescine branch of the nicotine biosynthetic pathway (Walton *et al.*, 1994), so the effects of exogenous cadaverine in stimulating anabasine biosynthesis at the expense of that of nicotine can be interpreted at an enzymic level. Nicotine, however, is an optically-active alkaloid (Wigle *et al.*, 1982) and the final stages of anabasine biosynthesis may not precisely parallel those of nicotine formation.

Following this work, it has been demonstrated by Fecker *et al.* (1992) that the expression of a bacterial gene for lysine decarboxylase (LDC) can, in an analogous fashion, lead to hairy root cultures synthesising elevated levels of cadaverine and thence anabasine.

N-Ethylputrescine is a higher homologue of *N*-methylputrescine. When fed at a concentration of 1 mM to hairy root cultures of *N. rustica*, it was converted to *N*-ethylnornicotine with a yield of 6% (Boswell *et al.*, 1993). Only the *S*-isomer was produced, corresponding to *S*-nicotine itself. *N*-Methylcadaverine, the structural isomer of *N*-ethylputrescine, is also actively metabolised to a novel alkaloid — putatively *N*-methylanabasine — by *Nicotiana* hairy root cultures (C. McClintock, D.J. Robins, R.J. Robins and N.J. Walton, unpublished data). It will be interesting to know whether this also is the *S*-isomer, consistent with the optical activity shown by nicotine rather than the racemic mixture exhibited by anabasine.

Similar experiments have been performed with hairy root cultures of a *Brugmansia candida* × *aurea* hybrid (H.D. Boswell, C. McClintock, D.J. Robins, J. Eagles, A.J. Parr, R.J. Robins and N.J. Walton, unpublished data), alkaloidal extracts of the roots being analysed by GC and by GC/MS for the production of novel alkaloids. As with *Nicotiana*, the most extensive metabolism was found to occur with *N*-ethylputrescine which gave rise to a range of analogues of the normal alkaloids, including analogues of littorine and/or hyoscyamine (resolution incomplete) and — to a smaller extent —

scopolamine. Similar results were obtained with *N*-fluoroethylputrescine, although in this case no scopolamine analogue was detected. The metabolism of higher *N*-alkylputrescines (*N*-*n*-propyl-, *N*-*iso*-propyl-) and of the corresponding cadaverines was less extensive, being restricted to the formation of analogues of hygrine, tropinone, tropine and related compounds; higher *N*-alkylcadaverines (*N*-*n*-propyl-, *N*-*n*-butyl-) were apparently not metabolised at all. These results seem to indicate that the earlier part of the tropane pathway as far as tropine may be comparatively non-specific, but that greater selectivity exists in the later enzymes responsible for aromatic esterification and for epoxidation to scopolamine.

The enzymological implications of these studies have not yet been investigated in great detail. The only analogue which has been studied appreciably in this context is 8-thiabicyclo[3.2.1]octan-3-one (TBON), a sulphur analogue of tropinone. This has been shown to be a substrate for one of the isoenzymes of tropinone reductase (TRI; see below), and is reduced stereospecifically to the corresponding analogue of tropine, designated α-TBOL (Portsteffen *et al.*, 1994). It is not a substrate for the other isoenzyme, TRII. TBON was metabolised by hairy root cultures of *Datura stramonium* to α-TBOL and to its 3-*O*-acetyl ester, but there was no esterification with phenyllactic acid or tropic acid.

These experiments have demonstrated some potential for hairy roots to be used to generate novel secondary products from chemically-synthesised precursors. Such products may prove to be useful in their own right, but the studies are significant also in exploring the plasticity of pathways and in enabling aspects of the enzymology of pathways to be revealed in the absence of detailed characterisation. Some analogues such as TBON have potential as selective enzyme substrates or inhibitors. Finally, as with the effects of cadaverine feeding in *Nicotiana*, these experiments can occasionally be useful pointers to the likely outcome of genetic manipulation studies.

Enzyme Isolation

The advantage of hairy root cultures for enzymological studies lies principally and obviously in the ease of generating relatively abundant quantities of sterile, rapidly-growing tissue under defined conditions. There is no reason to believe that hairy roots confer any overriding advantage, although the proportion of meristematic tissue may be higher and the phenolic content lower than in normal plant roots, leading to an improved yield of enzyme activity in some cases.

Several papers have been published describing the purification of enzymes of alkaloid biosynthesis from hairy roots of solanaceous species (McLauchlan *et al.*, 1993; Portsteffen *et al.*, 1994; Robins *et al.*, 1994a; Walton *et al.*, 1994). In general, high resolution ion-exchange and other standard chromatographic techniques have been used. Affinity chromatography has been successful in the resolution and purification from hairy roots of *D. stramonium* of two tropinone reductase enzymes (TRI, EC 1.1.1.206 and TRII, EC 1.1.1.236) which produce, respectively, tropine and pseudotropine from tropinone (Portsteffen *et al.*, 1994). The isoenzyme TRI was found to be bound by triazine dye matrices, whereas TRII could be bound by 2′5′-ADP linked to Sepharose 4B (Pharmacia); TRI was purified 108-fold, whilst TRII was purified 3410-fold, to homogeneity. Affinity chromatography, in this case using a matrix of L-ornithine carboxyl-bound to ω-aminohexyl Sepharose 4B (Pharmacia), has also been a key step in the ca. 700-fold purification of putrescine *N*-methyltransferase (PMT). The selection of this matrix followed a survey of the structural requirements for substrate activity of a range of putrescine and cadaverine analogues and derivatives (Walton *et al.*, 1994).

The purification to homogeneity of *N*-methylputrescine oxidase (MPO) from transformed roots of *Nicotiana tabacum* L. SC58 by McLauchlan *et al.* (1993) was achieved by standard approaches using a final hydrophobic-interaction chromatography step. In this case, the value of hairy root cultures was shown in a study of the cross-reactivity of polyclonal antibodies raised to the purified enzyme. The antibodies were tested against crude extracts of a range of hairy root cultures of both solanaceous and non-solanaceous species. It was demonstrated that in a range of extracts, including those of both solanaceous and non-solanaceous plants, the antibodies recognised a polypeptide of ca. 53 kDa molecular weight, corresponding to the subunit molecular weight of MPO. This indicated the presence of a protein with antigenic similarity to MPO even in hairy roots of plants not known for the production of putrescine-derived alkaloids (*Vinca minor*, *Beta vulgaris*). There was no cross-reactivity, however, with diamine oxidase extracted from seedlings (roots removed) of peas (*Pisum sativum*). This suggested that MPO might be representative of a class of root-specific diamine oxidases having other functions unrelated to alkaloid biosynthesis.

The relative stability and, in many cases, rapid growth of hairy roots in culture suggests that they will continue to be a favoured material for plant enzymologists interested in the biosynthesis of root-derived compounds, and in comparative studies of biosynthesis across a range of species.

Genetic Manipulation of Secondary Metabolic Pathways

There is increasing interest in the genetic manipulation of the biosynthetic pathways for secondary metabolites in plants, in efforts to influence both levels of production and patterns of product accumulation (see, e.g. Stitt and Sonnewald, 1995; articles in Ellis *et al.*, 1994; and also Hamill and Lidgett, Chapter 1 of this volume). Our own work with hairy root cultures has been concerned principally with attempts to manipulate the levels of (1) putrescine-derived alkaloids, and (2) ascorbic acid. In the first case, the approach has been to try to increase the biosynthetic rate; in the second, an attempt has been made to decrease the rate of degradation.

Manipulating Biosynthesis

The pathway of tropane alkaloid formation in *Datura* and of nicotine alkaloids in *Nicotiana* (Figures 2.1 and 2.3) is closely related to the production of primary metabolites, the polyamines (putrescine, spermine and spermidine). The common intermediate in these two pathways is putrescine, which in plants can arise by two routes: one via arginine and arginine decarboxylase (ADC), and the second via ornithine and ornithine decarboxylase (ODC). Preliminary experiments showed that modest increases in nicotine production in transformed roots of *N. rustica* could be effected by feeding putrescine at 1–5 mM. In an initial experiment, the ODC gene cloned from yeast was transformed into hairy root cultures of a highly-inbred line of *N. rustica* using a vector in which the yeast ODC gene was placed under the control of the cauliflower mosaic virus 35S promoter with duplicated upstream enhancer sequences to increase its strength. At the time these studies were conducted neither ODC nor ADC had been cloned from a plant source. The use of a heterologous gene from yeast avoids problems of co-suppression, but this phenomenon was not recognised at the time these experiments were performed. A range of root lines was established and compared with a series of control lines developed using a similar vector containing the CAT (chloramphenicol acetyltransferase) gene. These root lines were screened for ODC enzyme activity and for three metabolically-related enzymes, ADC, putrescine *N*-methyltransferase (PMT) and *N*-methylputrescine oxidase (MPO). Several lines showed enhanced ODC activity over the control, which amounted to 2–3 fold at the time of maximal ODC activity in the controls but was 10–20 times higher than that in the controls late in the growth period when the ODC activity in the controls had fallen to low levels (Hamill *et al.*, 1990). In one line chosen for detailed study, this increase occurred without changes in ADC and MPO activity, although a small increase in PMT activity was found. In this line, the expression of the yeast gene in the plant system was confirmed by Northern blotting. A detailed analysis of the metabolic products of ODC activity showed that there was an elevated level of both the precursor, putrescine, and the end-product of the secondary pathway, nicotine, in this transformed line compared with the controls. However, there were no significant changes in the levels of either free or conjugated spermine and spermidine. Thus it is likely that increased availability of the precursor, putrescine, can in some cases stimulate its utilisation by the secondary pathway leading to nicotine, but that, in contrast, the polyamine branch is tightly controlled and largely unaffected by increased availability of precursor.

These experiments illustrate a salient fact relevant to all experiments to evaluate the metabolic influence of a transgene in hairy roots. This is that significant variation in the levels of enzymes and intermediates occurs between hairy root lines. Each root line is different, arising from a separate, unique transformation event, and this makes the selection of appropriate control lines difficult. Transformation via *Agrobacterium* vectors involves integration into sites within the chromosomes which is thought to be random. Thus, control lines can differ in the site of integration of the T-DNA and in the number of T-DNA sequences integrated. Such issues can make the analysis of the relatively small metabolic disturbance caused by the presence of the yeast ODC gene difficult to evaluate. In our case, we have compared transgenic lines bearing the ODC transgene with a series of control lines (up to 10) developed using the same vector in an identical way but bearing the CAT gene rather than the ODC gene.

Further studies on the expression of ODC genes from yeast into a low-nicotine-producing strain of *N. tabacum* have led to transgenic lines with nicotine contents up to 6-fold higher than in control lines (Lidgett *et al.*, 1994). In this investigation, an ADC gene cloned from oat by Bell and Malmberg (1990) was also used and it was shown that the expression of this gene in transformed lines of *N. tabacum* LAFC53 also led to enhanced levels of nicotine compared to control lines. More recently, an ODC gene has been cloned from a plant source for the first time (Michael *et al.*, 1996). This *Datura* ODC gene is now available to test whether overexpression of this gene in a plant system can influence either alkaloid or polyamine accumulation in transformed roots. The gene for PMT has also recently become available (Hibi *et al.*, 1994), and since this represents the first committed step into alkaloid formation as opposed to polyamine synthesis, the effects of overexpression of this particular gene would be interesting to discover.

Following feeding experiments in hairy roots of *Nicotiana* spp. demonstrating anabasine formation in response to exogenous cadaverine, a lysine decarboxylase (LDC) gene from *Hafnia alvei* under the direction of the cauliflower mosaic virus 35S promoter has been expressed in hairy root cultures of *N. glauca* (Fecker *et al.*, 1992). Two transformed root lines (of 54 putatively-transformed kanamycin-resistant lines) showed some elevated activity; the more pronounced of these produced about 10-fold more cadaverine (up to 0.1% of dry mass) than a control root line and about twice as much anabasine. Southern analysis showed correct integration of the lysine decarboxylase gene in most of the LDC-negative lines, but Northern analysis showed active transcription only in the line showing substantially-elevated LDC activity. The reason for the lack of active transcription in most of the transformed lines remains unclear. It was suggested that it could be a direct or indirect result of using a 35S promoter; earlier experiments had used the rbcS-promoter fused to the coding region of the rbcS-transit peptide and a very high overproduction of cadaverine (1% of dry mass) had then been obtained in some cases in tobacco leaves (Herminghaus *et al.*, 1991), although transport of cadaverine to the roots did not occur and elevated levels of anabasine were not produced.

Manipulating Degradation

The occurrence, or otherwise, of turnover, and the fluxes through the degradative pathways involved have important roles in determining the steady-state levels of secondary metabolites. In comparison with biosynthesis, degradation has so far received relatively little attention even at the biochemical level, let alone the molecular genetic one. In principle, however, genetic manipulation of degradative pathways is a valid alternative approach to increasing product accumulation.

As the basis for one series of experiments (M.J.C. Rhodes, A.J. Parr and J.M. Furze, unpublished), it was argued that the levels of ascorbic acid present in plant tissues might be potentially influenced by the activity of the enzyme ascorbate oxidase (AO) present in those tissues. Hairy roots of cucumber (*Cucumis sativus*) (Amselem and Tepfer, 1992; Rhodes *et al.*, 1994) were used as a model system. In preliminary experiments, an inverse correlation between AO activity and ascorbate levels as measured by HPLC was noticed during the growth of root cultures (Figure 2.4). This tended to support the initial hypothesis. The gene for cucumber AO, isolated by polymerase chain reaction technology, was then introduced into cucumber roots in both sense and antisense orientations. Many root lines containing the gene in the antisense orientation showed strongly depressed AO activity, down to 5% of control val-

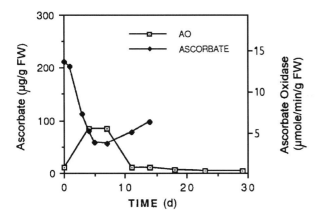

Figure 2.4. Levels of ascorbate and ascorbate oxidase during the growth of cucumber hairy root cultures.

ues. Despite the finding that, due to co-suppression (Napoli *et al.*, 1990), most of the lines containing the gene in the sense orientation also showed reduced rather than enhanced enzyme activity, it was still possible to generate some lines having elevated AO activities, the most extreme having a level 2000% that of unmanipulated roots. Analysis of ascorbate levels in all the sense and antisense lines unfortunately revealed no significant difference in concentration between lines with reduced, normal (i.e. around 10 μmole min^{-1} g^{-1} fresh weight), or enhanced enzyme activities (Table 2.1). This result implies that, under the conditions prevailing during our experiments, the levels of AO were not important in controlling the accumulation of ascorbate *in vivo*. This does not invalidate the general approach, but does emphasise that there are many factors which may need to be taken into account. Many enzyme activities may be present in gross excess (Stitt and Sonnewald, 1995) or, conversely, may not even contribute to flux so that their downregulation is of no physiological consequence. The compartmentation of enzymes and their substrates is another issue which may need to be addressed.

Table 2.1. The relationship between ascorbate oxidase activity and ascorbate levels in 4-d-old cultures of cucumber hairy roots. *n* is the number of cultures examined.

Ascorbate oxidase activity (μmol min^{-1} g^{-1} fresh weight)	Ascorbate level (μg g^{-1} fresh weight)
< 1	92 ± 41 (*n* = 3)
1 – 5	75 ± 60 (*n* = 4)
5 – 20	126 ± 97 (*n* = 4)
20 – 200	73 ± 33 (*n* = 4)

Physiological Transitions

One approach to understanding biosynthetic pathways and their regulation is to induce a change in flux by some empirical method, and then make a detailed examination of the biochemical changes which are associated with this altered flux. We have been investigating various physiological transitions which may induce changes in flux in certain metabolic pathways of hairy roots.

Elicitation in Hairy Roots

Plants respond to stress or challenge by pathogens by a variety of mechanisms, amongst which the induction of biosynthetic pathways for specific defence agents (the phytoalexins), coupled to enhancement of the synthesis of certain other compounds, is of major significance. Given the comparable structural and biochemical organisation of hairy roots and the roots of intact plants, it is likely that hairy roots in culture will respond in a similar manner to plants in the field. In the last few years there have been several studies of elicitation in hairy roots. Some of these have represented relatively empirical attempts to stimulate specific biochemical pathways for biotechnological exploitation, but fundamental work on the various phytoalexins and their biosynthetic pathways is also readily carried out using hairy roots.

The effects of elicitors on secondary metabolism in *Datura stramonium* have been studied by ourselves (Furze *et al.*, 1991). Various fungal extracts had little apparent effect on selected pathways studied, though a comprehensive search for active preparations was not made. On the other hand, treatment with metal ion abiotic elicitors, specifically Cu^{+2} or Cd^{+2} at 1 mM concentration, was found to induce rapid production and secretion of the typical solanaceous sesquiterpenoid phytoalexins (Furze *et al.*, 1991). Cd^{+2} favoured lubamin accumulation while Cu^{+2} favoured the accumulation of 3-hydroxylubamin; it is not yet clear whether this represents selective induction of the various biosynthetic enzymes, or whether it is a secondary effect of the metals on enzyme activity. Other workers have also reported normal induction of known phytoalexins in hairy root cultures, both by biotic and abiotic elicitors. For example, Signs and Flores (1990) reported induction of polyacetylene biosynthesis in *Bidens sulphureus* hairy roots. Robbins *et al.* (1991) demonstrated the production of isoflavans in *Lotus corniculatus* roots, and more recently they have compared the biochemical effects of different elicitor types (Robbins *et al.*, 1995). Compounds such as the thiophenes in hairy roots of *Tagetes* spp. have also been shown to be subject to elicitation (Mukundan and Hjortso, 1990).

In the studies of elicitation in *Datura* hairy roots, it was found that the constitutively-expressed tropane alkaloid pathway was little affected by biotic or abiotic elicitors, with total hyoscyamine levels remaining fairly constant (Furze *et al.*, 1991). This contrasts with some observations in suspension cultures (in which expression of the alkaloid pathway is at a much lower level, and is less predictable), where there are reports of positive effects of elicitation (Ballica *et al.*, 1993). In hairy roots, a range of metal ions did, however, produce an enhanced release of hyoscyamine into the growth medium, which at least in the case of Cu^{+2} treatments was found to be only transitory, the released alkaloid being resorbed over the course of the next few days (Furze *et al.*, 1991). This effect may be related in part to metal-ion-induced general membrane damage, but more specific effects on alkaloid transport might also have occurred (Furze *et al.*, 1991).

Hormonal Dedifferentiation

A major characteristic of transformed root lines is that they maintain their root phenotype over long periods in culture. In early experiments to grow hairy roots in fermenters it was observed that mechanical damage to the roots could lead to a wound response and the production of a callus-like growth at the root surface. In one such experiment in which transformed roots of *N. rustica* were grown in a 1-litre stirred-tank fermenter, the mechanical damage was such that at the end of the fermentation period, the culture had disorganised into a suspension of transformed cells (see Wilson, Chapter 16 of this volume). Analysis of these cells showed that they had only a very low content of nicotine compared with the original transformed root culture.

Subsequent studies showed that supplying auxin, particularly in the presence of a low level of cytokinin, could induce such a loss of root integrity in the absence of significant mechanical damage (Aird *et al.*, 1988, 1989; Robins *et al.*, 1992). Aird *et al.* (1989) showed that a culture of transformed roots of *N. rustica* in a medium containing 2 mg L^{-1} α-naphthaleneacetic acid (NAA) and 0.2 mg L^{-1} kinetin rapidly lost root integrity and by Day 28 of the growth period had become disorganised into a suspension of cells. This loss in integrity was accompanied by a fall in nicotine content from 3.2 mg g^{-1} dry weight to below 0.2 mg g^{-1} dry weight. After three further subcultures in hormone-containing medium, nicotine was undetectable. In contrast, roots maintained over these subcultures in hormone-free medium retained a high and constant nicotine content of between 3.1 and 4.0 mg g^{-1} dry weight. The loss of root integrity was shown to be reversed if roots were transferred back from hormone-containing medium to one free of hormones, and during this transition the capacity to synthesise nicotine at the normal levels was restored.

Cytogenetic analysis of the disorganised culture showed that loss of root integrity was associated with a loss of chromosome stability and, within such cultures, cells with a range of chromosomal numbers were found. If such cells were plated onto hormone-free medium, root lines with chromosome numbers varying from 38 to 89, compared to the euploid number of 48, developed from individual cells. These lines maintained their genotype stably over extended periods (Aird *et al.*, 1989). Similar studies on the relationship between root morphology and tropane alkaloid formation in transformed *Datura* roots showed a similar loss of root structure and ability to synthesise hyoscyamine when roots were transferred from hormone-free medium to one containing NAA and kinetin (Robins *et al.*, 1992). In both this study and that of Rhodes *et al.* (1989) on *Nicotiana*, changes in enzymes of alkaloid biosynthesis were studied in relation to the loss of alkaloid production associated with hormone-induced root disorganisation. It was shown that in *Datura* there was a dramatic reduction in PMT level within two days following transfer into hormone-containing medium. Lines which were maintained in the disorganised state for extended periods had less than 5% of the PMT and 20% of the MPO activity of normal transformed roots, while the activities of ODC and ADC were unchanged. In cultures which were returned to hormone-free medium, the levels of PMT and MPO were restored to their normal values. These results suggest that the expression of PMT, and to a lesser extent MPO, plays a crucial role in the coarse control of the tropane alkaloid biosynthetic pathway in *Datura*, and confirm similar findings in *Nicotiana* which again suggest an important regulatory role for the expression of PMT and MPO in nicotine alkaloid formation (Rhodes *et al.*, 1989). This is not, of course, to suggest that either PMT or MPO is necessarily rate-limiting to alkaloid formation when the alkaloid biosynthetic pathway is fully active.

Conclusions

Our experience has borne out our initial optimism that transformed roots in axenic culture would prove to be a good model system for studying aspects of secondary metabolism. In particular:

(1) The observed high rates of incorporation of label into tropane alkaloids in transformed roots of *Datura* and *Brugmansia* spp. are far higher than those obtained in comparable experiments using intact plants. This has enabled biosynthetic mechanisms to be investigated more easily.

It seems probable that this advantage will apply to many other pathways and plants.

(2)   The ease with which novel substrates and inhibitors can be supplied to (or withdrawn from) transformed roots has enabled experiments to be performed exploring the plasticity of plant secondary metabolism in a way which would be much more difficult with whole plants.

(3)   Experimental approaches are available both to upregulate and to downregulate secondary pathways in culture. These may be useful as tools to elucidate aspects of the regulation of expression of such pathways. For example, supplying auxin to the medium of transformed roots of both *Nicotiana* and *Datura* promotes downregulation of the pathways of alkaloid formation via an effect on root organisation. Similarly, supplying elicitors to the medium can induce upregulation of some, but not all, secondary pathways. These, however, are relatively blunt instruments, and it should be recognised that our knowledge of the regulation of most secondary metabolic pathways, especially at the level of fine control, is still in its infancy.

(4)   The ease with which transgenes can be integrated and expressed in hairy roots promises significant opportunities for a directed approach to the manipulation of secondary metabolism as expressed either in the roots themselves or in plants derived from them. As pointed out above, this approach remains hampered by our continuing lack of knowledge of the biochemical and genetic regulation of such pathways and the difficulty of identifying key targets for modification. However, the expression of heterologous genes may in practice offer more promising opportunities than the overexpression of endogenous genes, given the problems of co-suppression now widely recognised. Moreover, the tendency of plants to maintain metabolic homeostasis may dampen the end result of attempts to manipulate metabolic fluxes by overexpression of homologous transgenes, even if the overexpression itself is achieved. In such circumstances, attempts to establish novel branches of existing pathways or to use antisense methodology to downregulate genes seem more likely to prove successful.

References

Aird ELH, JD Hamill and MJC Rhodes (1988). Cytogenetic analysis of hairy root cultures from a number of plant species transformed by *Agrobacterium rhizogenes*. *Plant Cell Tiss Organ Cult* 15: 47–57.

Aird ELH, JD Hamill, RJ Robins and MJC Rhodes (1989). A study of the relationship between organisation, genetic stability and secondary metabolite production and the properties of variant lines of *Nicotiana rustica* hairy roots. In *Manipulating Secondary Metabolism in Culture*, edited by RJ Robins and MJC Rhodes, pp. 137–144. Cambridge University Press, Cambridge, UK.

Amselem J and M Tepfer (1992). Molecular basis for novel root phenotypes induced by *Agrobacterium rhizogenes* A4 on cucumber. *Plant Mol Biol* 19: 421–432.

Ballica R, DDY Ryu and CI Kado (1993). Tropane alkaloid production in *Datura stramonium* suspension cultures: elicitor and precursor effects. *Biotechnol Bioeng.* 41: 1075–1081.

Bell E and RL Malmberg (1990). Analysis of a cDNA encoding arginine decarboxylase from oat reveals similarity to the *Escherichia coli* arginine decarboxylase and evidence of protein processing. *Mol Gen Genet* 224: 431–436.

Boswell HD, AB Watson, NJ Walton and DJ Robins (1993). Formation of *N*-ethyl-*S*-nornicotine by transformed root cultures of *Nicotiana rustica*. *Phytochemistry* 34: 153–155.

Ellis BE, GW Kuroki and HA Stafford (eds) (1994). *Genetic Engineering of Plant Secondary Metabolism*. Recent Advances in Phytochemistry, vol 28. Plenum, New York.

Fecker LF, S Hillebrandt, C Rügenhagen, S Herminghaus, J Landsmann and J Berlin (1992). Metabolic effects of a bacterial lysine decarboxylase gene expressed in hairy root culture of *Nicotiana glauca*. *Biotechnol Lett* 14: 1035–1040.

Ford Y-Y, G Fox, RG Ratcliffe and RJ Robins (1994). *In vivo* [15]NMR studies of secondary metabolism in transformed root cultures of *Datura stramonium* and *Nicotiana tabacum*. *Phytochemistry* 36: 333–339.

Furze JM, MJC Rhodes, AJ Parr, RJ Robins, IM Whitehead and DR Threlfall (1991). Abiotic factors elicit sesquiterpenoid phytoalexin production but not alkaloid production in transformed root cultures of *Datura stramonium*. *Plant Cell Rep* 10: 111–114.

Hakamatsuka T, MF Hashim, Y Ebizuka and V Sankawa (1991). P-450-dependent oxidative rearrangement in isoflavone biosynthesis — reconstitution of P-450 and NADPH-P-450 reductase. *Tetrahedron* 47: 5969–5978.

Hamill JD, RJ Robins, AJ Parr, DM Evans, JD Furze and MJC Rhodes (1990). Overexpressing a yeast ornithine decarboxylase gene in transgenic roots of *Nicotiana rustica* can lead to enhanced nicotine accumulation. *Plant Mol Biol* 15: 27–38.

Herminghaus S, PH Schreier, JEG McCarthy, J Landsmann, J Botterman and J Berlin (1991). Expression of a bacterial lysine decarboxylase gene and transport of the protein into chloroplasts of transgenic tobacco. *Plant Mol Biol* 17: 475–486.

Hibi N, S Higashiguchi, T Hashimoto and Y Yamada (1994). Gene expression in tobacco low-nicotine mutants. *Plant Cell* 6: 723–735.

Leete E (1979). Aberrant biosynthesis of 5-fluoroanabasine from 5-fluoro [5,6-^{14}C, [13]C$_2$] nicotinic acid, established by means of carbon-13 nuclear magnetic resonance. *J Org Chem* 44: 165–168.

Leete E (1984). Chemistry of the tropane alkaloids. 34. 2-Migration of hydrogen during the biosynthesis of tropic acid from phenylalanine. *J Am Chem Soc* 106: 7271–7272.

Leete E (1987). Chemistry of the tropane alkaloids. 35. Stereochemistry of the 1,2-migration of the carboxyl group that occurs during the biosynthesis of tropic acid from phenylalanine. *Can J Chem* 65: 226–228.

Lidgett AJ, M Dobin, EB Fredericks, A Michael and JD Hamill (1994). Metabolic effects of over-expressing foreign ornithine decarboxylase (ODC) or arginine decarboxylase (ADC) genes in tobacco. *Abstr. 4th Internat Congr. Plant Molecular Biology*, Amsterdam, 1309.

McLauchlan WR, RA McKee and DM Evans (1993). The purification and immunocharacterisation of *N*-methylputrescine oxidase from transformed root cultures of *Nicotiana tabacum* L. cv SC58. *Planta* 191: 440–445.

Michael AJ and A Spena (1995). The plant oncogenes *rol*A, B and C from *Agrobacterium rhizogenes*. In *Methods in Molecular Biology*, edited by KMA Gartland and MR Davey, pp. 202–222, vol 44. *Agrobacterium* protocols. Humana, Totowa, NJ.

Michael AJ, JM Furze, MJC Rhodes and D Burtin (1996). Molecular cloning and functional identification of a plant ornithine decarboxylase cDNA. *Biochem J* 314: 241–248.

Mukundan U and M Hjortso (1990). Effect of fungal elicitor on thiophene production in hairy root cultures of *Tagetes patula*. *Appl Microbiol Biotechnol* 33: 145–147.

Napoli C, C Lemieux and R Jorgensen (1990). Introduction of a chimeric chalcone synthase gene into petunia results in reversible co-suppression of homologous genes in *trans*. *Plant Cell* 2: 279–289.

Parr AJ, J Payne, J Eagles, BT Chapman, RJ Robins and MJC Rhodes (1990). Variation in tropane alkaloid accumulation within the Solanaceae and strategies for its exploitation. *Phytochemistry* 29: 2545–2550.

Portsteffen A, B Dräger and A Nahrstedt (1994). The reduction of tropinone in *Datura stramonium* root cultures by 2 specific reductases. *Phytochemistry* 37: 391–400.

Rhodes MJC, M Hilton, AJ Parr, JD Hamill and RJ Robins (1986). Nicotine production by "hairy root" cultures of *Nicotiana rustica*: fermentation and product recovery. *Biotechnol Lett* 8: 415–420.

Rhodes MJC, AJ Parr, A Giulietti and ELH Aird (1994). Influence of exogenous hormones on the growth and secondary metabolite formation in transformed root cultures. *Plant Cell Tiss Organ Cult* 38: 143–151.

Rhodes MJC, RJ Robins, ELH Aird, J Payne, AJ Parr and NJ Walton (1989). Regulation of secondary metabolism in transformed root cultures. In *Primary and Secondary Metabolism of Plant Cell Cultures II*, edited by WGW Kurz, pp. 58–72. Springer-Verlag, Berlin.

Robbins MP, J Hartnoll and P Morris (1991). Phenylpropanoid defence responses in transgenic *Lotus corniculatus*. I. Glutathione elicitation of isoflavan phytoalexins in transformed root cultures. *Plant Cell Rep* 10: 59–62.

Robbins MP, B Thomas and P Morris (1995). Phenylpropanoid defence responses in transgenic *Lotus corniculatus*. II. Modelling plant defence responses in transgenic root cultures using thiol and carbohydrate elicitors. *J Exp Bot* 46: 513–524.

Robins RJ, ES Bent and MJC Rhodes (1992). Studies on the biosynthesis of tropane alkaloids by *Datura stramonium* L. transformed root cultures. 3. The relationship between morphological integrity and alkaloid biosynthesis. *Planta* 185: 385–390.

Robins RJ, P Bachmann, ACJ Peerless and S Rabot (1994a). Esterification reactions in the biosynthesis of tropane alkaloids in transformed root cultures. *Plant Cell Tiss. Organ Cult.* 38: 241–247.

Robins RJ, P Bachmann and JG Woolley (1994b). Biosynthesis of hyoscyamine involves an intramolecular rearrangement of littorine. *J Chem Soc Perkin Trans 1* 1994: 615–619.

Robins RJ, JG Woolley, M Ansarin, J Eagles and BE Goodfellow (1994c). Phenyllactic acid but not tropic acid is an intermediate in the biosynthesis of tropane alkaloids in *Datura* and *Brugmansia* root culture. *Planta* 194: 86–94.

Robins RJ, NCJE Chesters, D O'Hagan, AJ Parr, NJ Walton and JG Woolley (1995). The biosynthesis of hyoscyamine: the process by which littorine rearranges to hyoscyamine. *J Chem Soc Perkin Trans 1* 1995: 481–485.

Signs MW and HE Flores (1990). The biosynthetic potential of plant roots. *Bioessays* 12: 7–13.

Stitt, M and U Sonnewald (1995). Regulation of metabolism in transgenic plants. *Annu Rev Plant Physiol Plant Mol Biol* 46: 341–368.

Walton NJ and NJ Belshaw (1988). The effect of cadaverine on the formation of anabasine from lysine in hairy root cultures of *Nicotiana hesperis*. *Plant Cell Rep* 7: 115–118.

Walton NJ and WR McLauchlan (1990). Diamine oxidation and alkaloid production in transformed root cultures of *Nicotiana tabacum*. *Phytochemistry* 29: 1455–1457.

Walton NJ, RJ Robins and MJC Rhodes (1988). Perturbation of alkaloid production by cadaverine in hairy root cultures of *Nicotiana rustica*. *Plant Sci* 54: 125–131.

Walton NJ, ACJ Peerless, RJ Robins, MJC Rhodes, HD Boswell and DJ Robins (1994). Purification and properties of putrescine *N*-methyltransferase from transformed roots of *Datura stramonium* L. *Planta* 193: 9–15.

Watson AB, AM Brown, IJ Colquhoun, NJ Walton and DJ Robins (1990). Biosynthesis of anabasine in transformed root cultures of *Nicotiana* species. *J Chem Soc Perkin Trans 1* 1990: 2607–2610.

Wigle ID, LJJ Mestichelli and ID Spenser (1982). H-2 NMR-spectroscopy as a probe of the stereochemistry of biosynthetic reactions — the biosynthesis of nicotine. *J Chem Soc Chem Commun* 1982: 662–664.

Metabolic Engineering of Terpenoid Indole and Quinoline Alkaloid Biosynthesis in Hairy Root Cultures

3

Didier Hallard[1], Arjan Geerlings[1], Robert van der Heijden[1], M. Inês Lopes Cardoso[2], J. Harry C. Hoge[2] and Robert Verpoorte[1]

[1] *Division of Pharmacognosy, Leiden/Amsterdam Center for Drug Research, Leiden University, Gorlaeus Laboratories, PO Box 9502, 2300 RA Leiden, The Netherlands; email: Verpoort@lacdr.leidenuniv.nl*
[2] *Institute of Molecular Plant Sciences, Leiden University, Clusius Laboratory, PO Box 9505, 2300 RA Leiden, The Netherlands*

Introduction

Alkaloids constitute a large group of plant secondary metabolites. Many of these natural products are of pharmaceutical importance. The monomeric and dimeric terpenoid indole alkaloids produced by *Catharanthus roseus* serve as an example. The monomeric compounds ajmalicine and serpentine are used in the treatment of cardiac and circulatory diseases, while the dimeric compounds vincristine and vinblastine are used in the treatment of cancer. Another example are the quinoline alkaloids quinine and quinidine produced by *Cinchona* species. These compounds are used as drugs in the therapy of malaria and cardiac arrhythmias, respectively.

The biosynthesis of both terpenoid indole and quinoline alkaloids proceeds via the formation of strictosidine, a gluco-alkaloid assembled from tryptamine and secologanin. Somewhere beyond strictosidine formation, however, the biosynthetic pathway diverges into plant-species specific routes (Figure 3.1).

The biosynthesis of alkaloids in *Catharanthus* and *Cinchona* species is studied within the multidisciplinary project group Plant Cell Biotechnology of the inter-university collaboration Biotechnological Sciences Delft Leiden (Verpoorte *et al.*, 1991; Meijer *et al.*, 1993). These studies resulted among others in the purification of tryptophan decarboxylase (TDC, E.C. 4.1.1.28) and strictosidine synthase (SSS, E.C. 4.3.3.2.) from *C. roseus*, both key enzymes in alkaloid biosynthesis (Figure 3.1). Subsequently, cDNA clones coding for these enzymes were isolated from *C. roseus* (Goddijn, 1992; Pasquali *et al.*, 1992), and are currently being used in studies on the regulatory mechanisms controlling expression of the corresponding genes. These cDNA clones have also been expressed in heterologous *Nicotiana tabacum* plants and cell cultures (Goddijn *et al.*, 1993; Poulsen *et al.*, 1994; Hallard *et al.*, 1996). Recently, we introduced *C. roseus tdc* and *sss* cDNA genes into hairy root cultures of various plant species to find out if constitutive expression of these genes in (**1**) hairy roots of terpenoid indole and quinoline alkaloid accumulating plant species like *Catharanthus* and *Cinchona* might result in increased accumulation of alkaloids, and (**2**) hairy roots of plant species like *Weigela* and *Lonicera*, which contain secologanin but do not produce terpenoid indole or quinoline alkaloids, might result in the formation of strictosidine.

In this chapter we would like to present some initial results from these studies.

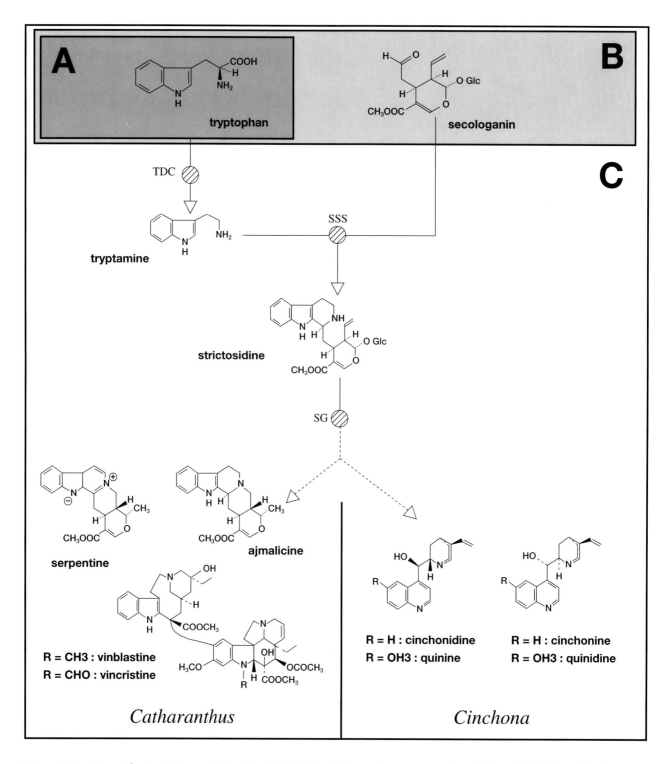

Figure 3.1. Biosynthesis of terpenoid indole alkaloids in *Catharanthus roseus* and quinoline alkaloids in *Cinchona* species. The boxes A, B and C embrace the compounds naturally produced by the plant species which are used for genetic modification of secondary metabolism. (A: *Nicotiana tabacum*; B: iridoid producing plants, *Weigela*, *Lonicera*; C: terpenoid indole and quinoline alkaloid producing plants). Dashed arrows indicate that the involved biosynthetic routes are not exactly known. TDC: tryptophan decarboxylase; SSS: strictosidine synthase; SG: strictosidine glucosidase.

Genetic Engineering of Secondary Metabolism in Hairy Root Cultures

Plants are able to accumulate high levels of alkaloids. For example, the bark of *Cinchona* trees may contain quinoline alkaloids up to 15% of its dry weight. Cell suspension cultures of *Cinchona ledgeriana* have been established, but proved to be poor producers of quinoline alkaloids (less than 1 µg g^{-1} dry weight). For such cultures it was found that accumulation of these compounds is correlated with the extent of cellular differentiation (Verpoorte *et al.*, 1987). In view of their cellular differentiation, hairy root cultures provide an interesting alternative as a production system. Hairy roots of *C. ledgeriana* have been reported to produce up to 50 µg g^{-1} fresh weight quinoline alkaloids (Hamill *et al.*, 1989). This is still much lower than in the bark of trees, but the plant as production system has the disadvantage that it takes 7 to 10 years before the bark can be harvested.

Hairy roots can be obtained by infecting wounded plant tissue with the soil bacterium *Agrobacterium rhizogenes*. During infection the bacterium transfers a small part of its large Ri (root-inducing) plasmid to plant cells, where the transferred DNA (T-DNA) becomes integrated in the chromosomal DNA and expressed. As a result, these transformed cells acquire the capacity to develop into hairy roots, the phenotype of which is characterised by extensive branching and plagiotropic growth on hormone-free culture media. When *A. rhizogenes* strains that carry a binary plant vector in addition to their Ri plasmid are used for infection of wound tissue, the T-DNA of the binary vector is also transferred to the plant cells. With such strains, hairy roots can be induced that are also transformed by the genes located in the T-DNA region of the binary vector. Genes of interest with regard to plant secondary metabolism can thus be introduced into hairy roots. Potentially, genetic modification of secondary metabolism is the most powerful method to obtain altered production yields (reviewed by Nessler *et al.*, 1994). In the last few years, hairy root cultures of *Nicotiana rustica* (Hamill *et al.*, 1990) and *Peganum harmala* (Berlin *et al.*, 1993) have been shown to produce increased amounts of the secondary metabolites nicotine and serotonin, respectively, when expressing transgenes from yeast (ornithine decarboxylase; *N. rustica*) or *C. roseus* (tryptophan decarboxylase; *P. harmala*).

Within our project group we have constructed binary plant vectors with T-DNA regions containing constitutive expression versions (CaMV35S promoter with double enhancer end *nos* terminator) of the already mentioned *tdc* and *sss* cDNA clones from *C. roseus*, together with an intron possessing β–glucuronidase (*gus* int) reporter gene and a hygromycin phosphotransferase (*hpt*) selection gene

Table 3.1. T-DNA regions of binary vector constructs.

Binary vector	Order and orientation of genes in T-DNA region	Plant markers
HG	hpt – gus ← – →	hyg/X-gluc
HTSG	hpt – tdc – sss – gus ← – ← – ← – →	hyg/X-gluc
HGST	hpt – gus – sss – tdc ← – → – → – →	hyg/X-gluc

Note: hpt: hygromycin phosphotransferase; gus: β-glucuronidase possessing an intron; tdc: tryptophan decarboxylase; sss: strictosidine synthase; hyg: hygromycin resistance; X-gluc: blue colouration upon incubation with 5-bromo-4-chloro-3-indolyl glucuronide.

(Lopes Cardoso and Hoge, unpublished). These binary vector constructs (Table 3.1) were used in conjunction with *A. rhizogenes* strain LBA 9402 to obtain *tdc* and *sss* gene transformed hairy roots of several plant species.

Introduction of Constitutive Expression Versions of *tdc* and *sss* cDNA Clones from *Catharanthus roseus* into Hairy Root Cultures of Various Plant Species

Nicotiana tabacum

Leaves of axenically grown *Nicotiana tabacum* cv Petit Havana SR1 plants were infected with *A. rhizogenes* strain LBA 9402 harbouring the binary vector constructs described in Table 3.1. Leaf midribs were wounded with a surgical knife that had been dipped into a fresh *A. rhizogenes* colony. This contamination procedure gave rise to wounds infected with different amounts of bacteria. After 3 d, the infected leaves were transferred to hormone-free Murashige and Skoog (MS) solid medium (Murashige and Skoog, 1962) containing 100 µg mL^{-1} Augmentin® to eliminate all the bacteria. Roots emerged from the infected wound tissue approximately 1 to 2 weeks after inoculation. Most developing roots grew rapidly and formed numerous branches in all directions. Upon excision and transfer to hormone-free solid MS medium, these roots continued to display the hairy root phenotype. Histochemical analysis for *gus* int reporter gene expression revealed that about one third of the induced hairy roots contained the binary vector T-DNA (uniform blue staining upon incubation with 5-bromo-4-chloro-3-indolyl glucuronide). Several T-DNA transformed hairy root lines were subcultured in hormone-free MS liquid medium and analysed for TDC and SSS activities

Table 3.2. Expression of *gus* int gene, growth, tryptamine levels and SSS activities recorded for different tobacco hairy root lines. Expression of *gus* int gene (blue colouration) and growth were recorded on solid medium, tryptamine levels and SSS activities in liquid medium.

Binary vector	Blue colouration	Growth	Tryptamine (nmol g^{-1} fresh weight)	SSS activity (pkat mg^{-1} protein)
HG	++	+++	–	–
	+++	+++	–	–
	+++	++	–	–
HTSG	+	+++	41	30
	+++	+++	142	35
	++	++	65	42
	++	+++	31	26
HGST	+++	++	34	39

Note: Blue colouration (+: uniform light blue; ++: uniform medium blue; +++: uniform dark blue). Growth (+: slow; ++: medium; +++: fast).

and tryptamine contents (Table 3.2). Wild-type tobacco is not known to display TDC activity or produce tryptamine. TDC activity and tryptamine were, as expected, undetectable in hairy roots transformed with HG T-DNA lacking the *tdc* and *sss* genes. TDC activity was also not detectable in the hairy roots of *N. tabacum* transformed with *C. roseus tdc* and *sss* genes. However, the presence of tryptamine in these roots demonstrated the expression of the *tdc* gene into an enzymatically active TDC protein. Although tryptamine is a foreign metabolite for tobacco it did not seem to affect growth of the root cultures. SSS activity could be detected in hairy roots transformed with HTSG and HGST T-DNAs (Table 3.2). However, no strictosidine could be formed because of the absence of secologanin in tobacco. Histochemical detection of GUS activity was shown to be a useful marker for selection of hairy roots transformed with binary vector T-DNA.

Iridoid Producing Plants

Seeds of different plant species known to accumulate secologanin (Jensen, 1991) were sterilised and allowed to germinate in the dark at 25°C on moistened cotton. Seedlings of *Lonicera xylosteum* and *Weigela 'Styriaca'* were obtained. Intact seedlings and their excised leaves were wounded and infected with *A. rhizogenes* strain LBA 9402 carrying HG or HTSG binary vector constructs.

Induction of roots from *Lonicera xylosteum* using the wounding and infection procedures described for tobacco was more successful with intact seedlings than excised leaves. This was probably due to a stress reaction resulting in a purple colouration observed after wounding and infection of the excised leaves. None of the 16 seedlings and 5 leaf-derived roots expressed the *gus* int reporter gene as

judged from histochemical staining, but some exhibited the hairy root phenotype. Two *L. xylosteum* hairy root lines developed from the seedlings were grown in liquid $^{1}/_{4}$ Gamborg's B5 medium (Gamborg *et al.*, 1968) containing $^{1}/_{4}$ macroelements (Figure 3.2). This medium appeared to allow slightly faster growth (doubling time: 16 d) than $^{1}/_{4}$ MS medium containing $^{1}/_{4}$ macroelements (doubling time: 19 d). Using a recently developed HPLC assay for secologanin (Hallard *et al.*, submitted) it was found that the two *L. xylosteum* hairy root cultures produced high levels of secologanin (up to 1.25 mg g^{-1} fresh weight). Cell suspension cultures of *Lonicera* species are unable to accumulate detectable amounts of secologanin. This observation encourages further efforts to obtain *L. xylosteum* hairy roots expressing the *tdc* and *sss* genes from *C. roseus*.

Roots of *Weigela 'Styriaca'* emerged from the inoculation sites on intact seedlings and excised seedling leaves. More roots developed from seedlings than from their excised leaves. Three root lines expressing the *gus* int gene and displaying the hairy root phenotype on solid MS medium were transferred to liquid $^{1}/_{4}$ MS medium. The medium was refreshed every two weeks. Unfortunately, the roots became brown and stopped growing after about 2 months. It cannot be excluded that formation of strictosidine or substances derived from strictosidine in roots not normally accumulating these compounds was toxic for growth of the *Weigela* roots.

Catharanthus **Species**

Several authors have described the establishment of hairy root cultures of *C. roseus* (Parr *et al.*, 1988; Toivonen *et al.*, 1989, 1990; Brillanceau *et al.*, 1989) and *Catharanthus trichophyllus* (Davioud *et al.*, 1989). Axenically grown plantlets were invariably used as

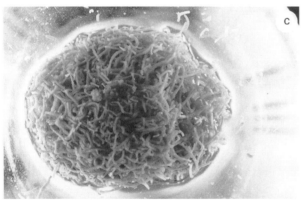

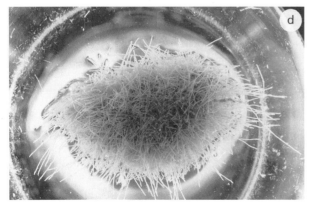

Figure 3.2. Morphology of hairy roots grown in liquid culture in the dark. (**a**) *Lonicera xylosteum*; (**b**) *Catharanthus roseus*; (**c**) *Catharanthus pusillus*; (**d**) *Cinchona ledgeriana*.

target material for induction of hairy roots. It has been reported that tissue cultured seedlings of *C. roseus* may represent the best plant material for obtaining hairy roots of this species (Hoekstra, 1993). We obtained abundant root induction using leaves excised from axenically grown *C. roseus* plantlets. As for tobacco, about one third of the roots stained blue in the histochemical analysis for *gus* int gene expression. On solid medium, roots displaying the hairy root phenotype had short bulbous tips. This root morphology for *C. roseus* was also described by Bhadra *et al.* (1993). When transferred to liquid medium, all hairy root lines of *C. roseus* eventually died. It was already known that transfer to liquid medium is the most unpredictable step in the establishment of hairy root cultures of *C. roseus* (Hoekstra, 1993). For this reason we tried to transform another *Catharanthus* species, *Catharanthus pusillus*.

Excised leaves of *C. pusillus* were wounded and infected using the procedure applied for tobacco. Roots emerged from the wound sites about 1 to 2 weeks after inoculation. These roots grew rapidly and showed *gus* int gene expression as indicated by a blue colouration in the histochemical GUS assay. The roots were introduced into liquid MS medium and cultured in the dark (Figure 3.2). Two *C. pusillus* hairy root cultures transformed with binary vector T-DNA were obtained: one with HG T-DNA and one with HTSG T-DNA. Both lines grew rapidly in liquid medium. The *C. pusillus* hairy root culture transformed with HTSG T-DNA did not have higher TDC and SSS activities than the one transformed with HG T-DNA. Measured TDC activities were 54.3 and 65.8 pkat mg^{-1} protein for the HG and HTSG lines, respectively. Their SSS activities were, respectively, 26.8 and 19.3 pkat mg^{-1} protein.

Cinchona Species

Cinchona species have been shown to be very recalcitrant to transformation by *Agrobacterium rhizogenes* (Hoekstra, 1993). Infection of axenic leaves from *C. ledgeriana* seedlings by a procedure similar to that described for tobacco resulted in a pronounced stress reaction visible as red pigmentation. To reduce this stress reaction, axenic seedlings were infected by wounding their leaves and stems with forceps contaminated with *A. rhizogenes*. About 6 weeks after infection, only one root emerged from a leaf of one of the seedlings infected with the strain carrying the binary vector HTSG. This root was transformed by the HTSG T-DNA since it expressed the *gus* int gene. After excision and subculturing on solid $^1/_4$ B5 medium where it exhibited the hairy root phenotype, the root was transferred to liquid $^1/_4$ B5 medium. Here the *C. ledgeriana* hairy root developed into a very compact structure with many short root tips on the outside and a rusty coloured tissue mass on the inside (Figure 3.2). The rusty colour of the hairy root line was probably due to the accumulation of anthraquinones. It is noteworthy that this particular phenotype of *C. ledgeriana* hairy roots was also seen by other authors (Hamill *et al.*, 1989).

Our *C. ledgeriana* hairy root was further characterised by dividing four subcultures of this line, each containing about 12 g fresh weight, into 16 equally sized portions, and transferring each portion to 25 mL fresh medium. Every 4 or 5 d, two cultures were harvested for measurement of growth, enzyme activities and alkaloid contents. A doubling time of about 15 d indicated that growth of the cultures was relatively slow. The very compact structure of the culture could play a role in this. Compared with *C. ledgeriana* cell suspension cultures (Skinner *et al.*, 1987) and seedlings (Aerts *et al.*, 1990), high TDC and SSS activities (about 10 times higher for both enzymes) were measured. Highest enzyme activities were detected 15 d after subculturing. TDC and SSS activities up to 4.1 and 450 pkat mg^{-1} protein, respectively, were recorded. The detected strictosidine glucosidase (SG) activity reached 300 pkat mg^{-1} protein 15 d after subculturing. Furthermore, high levels of tryptamine and strictosidine, up to 1200 and 1950 µg g^{-1} dry weight, respectively, were measured. These results indicate that production of quinoline alkaloids in the *C. ledgeriana* hairy root line might be limited by the availability of secologanin and the conversion of strictosidine. Biosynthesis of secologanin has previously been shown to be a rate-limiting step in hairy roots of *C. roseus* already established in our laboratories (Hoekstra, 1993). The SSS extracted from the *C. ledgeriana* hairy roots did not accept 5-methoxytryptamine as a substrate. This finding demonstrates that the SSS activity observed must result from expression of the introduced *sss* gene because, unlike SSS from *Cinchona* species, SSS from *C. roseus* is unable to catalyse the formation of 10-methoxystrictosidine (Stevens *et al.*, 1993). Despite its high enzyme activities, our *C. ledgeriana* hairy root line was found to accumulate amounts of quinoline alkaloids similar to those in the *C. ledgeriana* hairy root line described by Hamill *et al.* (1989). For quinine, levels up to 500 µg g^{-1} dry weight were found. Cinchonine and cinchonidine could not be detected separately; a total level of 400 µg g^{-1} dry weight was measured. Only the level of quinidine (up to 1000 µg g^{-1} dry weight) was found to be higher than in hairy roots of *C. ledgeriana* not transformed with the *tdc* and *sss* genes from *C. roseus* (Hamill *et al.*, 1989).

Conclusions and Perspectives

These initial results from our studies indicate that genetic engineering of secondary metabolism in hairy root cultures might be a strategy with prospects. However, several problems were encountered that could hamper application of this strategy. First, germination of disinfected seeds required to obtain young and microorganism-free plant material for induction of hairy roots may fail. Secondly, a poor or species-dependent rooting response to wounding and *A. rhizogenes* infection will restrict the possibilities for initiating hairy roots, in particular hairy roots transformed by binary vector T-DNA. Thirdly, successful establishment of a hairy root culture on solid medium does not guarantee that these roots will continue to grow when transferred to liquid medium. Application of assays for secologanin and strictosidine showed that substantial pools of these intermediates of terpenoid indole and quinoline alkaloid biosynthesis accumulate, respectively, in non-transformed hairy roots of *L. xylosteum* and *tdc* and *sss* transformed hairy roots of *C. ledgeriana*. Further attempts to obtain *L. xylosteum* hairy root lines that express TDC and SSS enzymes from *C. roseus* might be worthwhile, since this could result in accumulation of strictosidine. For hairy root lines that already accumulate strictosidine, like our transformed *C. ledgeriana* hairy root line, it will be interesting to try genetic modification of the biosynthetic steps beyond strictosidine. As depicted in Figure 3.1, biosynthetic pathways for terpenoid and quinoline indole alkaloids diverge into species-specific routes at some point beyond strictosidine. An important enzyme in this respect is strictosidine β-D-glucosidase (SG, E.C. 3.2.1.105). Recently this enzyme was purified from cell suspension cultures of *C. roseus* and *Tabernaemontana divaricata* (Luijendijk, 1995). Isolation of the corresponding genes may offer the opportunity to channel strictosidine towards production of a specific class of terpenoid indole or quinoline alkaloids.

References

Aerts RJ, T van der Leer, R van der Heijden and R Verpoorte (1990). Developmental regulation of alkaloid production in *Cinchona* seedlings. *J Plant Physiol* 136: 86–91.

Berlin J, C Rügenhagen, P Dietze, LF Fecker, OJM Goddijn and JHC Hoge (1993). Increased production of serotonin by suspension and root cultures of *Peganum harmala* transformed with a tryptophan decarboxylase cDNA clone from *Catharanthus roseus*. *Transgenic Res* 2: 336–344.

Bhadra R, S Vani and JV Shanks (1993). Production of indole alkaloids by selected hairy root lines of *Catharanthus roseus*. *Biotechnol Bioeng* 41: 581–592.

Brillanceau MH, C David and J Tempé (1989). Genetic transformation of *Catharanthus roseus* G. Don by *Agrobacterium rhizogenes*. *Plant Cell Rep* 8: 63–66.

Davioud E, C Kan, J Hamon, J Tempé and HP Husson (1989). Production of indole alkaloids by *in vitro* root cultures from *Catharanthus trichophyllus*. *Phytochemistry* 28: 2675–2680.

Gamborg OL, RA Miller and K Ijima (1968). Nutrient requirements of suspension cultures of soybean root cells. *Exp Cell Res* 50: 151–158.

Goddijn OJM (1992). *Regulation of Terpenoid Indole Alkaloid Biosynthesis in Catharanthus roseus: the Tryptophan Decarboxylase Gene*. Thesis, Leiden University, The Netherlands.

Goddijn OJM, PM van der Duyn Schouten, RA Schilperoort and JHC Hoge (1993). A chimaeric tryptophan decarboxylase gene as a novel selectable marker in plant cells. *Plant Mol Biol* 22: 907–912.

Hallard D, R van der Heijden, R Verpoorte, MI Lopes Cardoso, G Pasquali, J Memelink and JHC Hoge (1996). Suspension cultured transgenic cells of *Nicotiana tabacum* expressing tryptophan decarboxylase and strictosidine synthase cDNAs from *Catharanthus roseus* produce strictosidine upon feeding of secologanin. *Plant Cell Rep*, Accepted.

Hallard D, R van der Heijden, A Contin, EM Tomás Jiménez, W Snoeijer, R Verpoorte, SR Jensen, MI Lopes Cardoso, G Pasquali, J Memelink and JHC Hoge. An assay method for secologanin in plant tissues based on its enzymatic conversion into strictosidine. *Phytochem Anal*, Submitted.

Hamill JD, RJ Robins and MJC Rhodes (1989). Alkaloid production by transformed root cultures of *Cinchona ledgeriana*. *Planta Med* 55: 354–357.

Hamill JD, RJ Robins, AJ Parr, DM Evans, JM Furze and MJC Rhodes (1990). Over-expressing a yeast ornithine decarboxylase gene in transgenic roots of *Nicotiana rustica* can lead to enhanced nicotine accumulation. *Plant Mol Biol* 15: 27–38.

Hoekstra SS (1993). *Accumulation of Indole Alkaloids in Plant-Organ Cultures*. Thesis, Leiden University, The Netherlands.

Jensen SR (1991). Plant iridoids, their biosynthesis and distribution in angiosperms. In *Ecological Chemistry and Biochemistry of Plant Terpenoids*, edited by JB Harborne and FA Tomas-Barberan, pp. 133–158. Oxford: Oxford University Press, Oxford. *Proc Phytochem Soc Eur* 31.

Luijendijk TJC (1995). *Strictosidine Glucosidase in Indole Alkaloid Biosynthesis*. Thesis, Leiden University, The Netherlands.

Meijer AH, R Verpoorte and JHC Hoge (1993). Regulation of enzymes and genes involved in terpenoid indole alkaloid biosynthesis in *Catharanthus roseus*. *J Plant Res* Special Issue 3: 145–164.

Murashige T and F Skoog (1962). A revised medium for rapid growth and bioassays with tobacco tissue cultures. *Physiol Plant* 15: 473–497.

Nessler CL (1994). Metabolic engineering of plant secondary products. *Transgenic Res* 3: 109–115.

Parr AJ, ACJ Peerless, JD Hamill, NJ Walton, RJ Robins and MJC Rhodes (1988). Alkaloid production by transformed root cultures of *Catharanthus roseus*. *Plant Cell Rep* 7: 309–312.

Pasquali G, OJM Goddijn, A de Waal, R Verpoorte, RA Schilperoort, JHC Hoge and J Memelink (1992). Coordinated regulation of two indole alkaloid biosynthetic genes from *Catharanthus roseus* by auxin and elicitors. *Plant Mol Biol* 18: 1121–1131.

Poulsen C, OJM Goddijn, JHC Hoge and R Verpoorte (1994). Anthranilate synthase and chorismate mutase activities in transgenic tobacco plants overexpressing tryptophan decarboxylase from *Catharanthus roseus*. *Transgenic Res* 3: 43–49.

Skinner NJ, RJ Robins and MJC Rhodes (1987). Tryptophan decarboxylase, strictosidine synthase and alkaloid production by *Cinchona ledgeriana* cell suspension cultures. *Phytochemistry* 26: 721–725.

Stevens LH, C Giroud, EJM Pennings and R Verpoorte (1993). Purification and characterization of strictosidine synthase from a suspension culture of *Cinchona robusta*. *Phytochemistry* 33: 99–106.

Toivonen L, J Balsevich and WGW Kurz (1989). Indole alkaloid production by hairy root cultures of *Catharanthus roseus*. *Plant Cell Tiss Organ Cult* 18, 79–83.

Toivoinen L, M Ojala and V Kauppinen (1990). Indole alkaloid production by hairy root cultures of *C. roseus*. *Biotechnol Lett* 12: 519–524.

Verpoorte R, PAA Harkes and HJG ten Hoopen (1987). A biotechnical production process for *Cinchona* alkaloids: prospects and problems. *Acta Leidensia* 55: 29–44.

Verpoorte R, R van der Heijden, WM van Gulik and HJG ten Hoopen (1991). Plant biotechnology for the production of alkaloids: present status and prospects. In *The Alkaloids*, edited by A Brossi, pp. 1–187, vol 40. San Diego: Academic Press.

Characteristics of Selected Hairy Root Lines of *Catharanthus roseus*

4

Jacqueline V. Shanks and Rajiv Bhadra

Department of Chemical Engineering, MS 362, Institute of Biosciences and Bioengineering, Rice University, Houston, TX 77005-1892, USA; email: shanks@rice.edu

Introduction

The periwinkle *Catharanthus roseus* produces a number of terpenoid indole alkaloids, several of which have pharmaceutical activity. The most widely studied are the hypertension drugs ajmalicine and serpentine, and the extremely valuable anti-cancer agents vinblastine and vincristine. The use of plant cell biotechnology rather than plants for the production of these alkaloids has been a major area of research since the late 1960s. Extensive work on enhancement of productivity of cell suspension cultures has been performed that encompasses selection of high-yielding cell lines, optimisation of growth and production media, and addition of elicitors/agents. Ajmalicine, serpentine and catharanthine, a monomeric precursor of the dimer vinblastine, can be produced in cell suspensions at higher yields than in the intact plant (Verpoorte *et al.*, 1991). However, the high-producing cell lines tend to be unstable (Deus-Neumann and Zenk, 1984). The production of vinblastine and vincristine and their other monomeric precursor, vindoline, in cell suspension culture has been an elusive goal. Research in the last decade has shown that the synthesis of these alkaloids is more complex than previously

known: light and developmental regulation appear to play key roles (Meijer *et al.*, 1993). Cell suspensions exposed to light and photoautotrophic cell suspensions still do not produce vindoline or the dimers.

The potential of hairy roots of *C. roseus* for the production of these indole alkaloids has been a subject of investigation since an initial report in a review by Flores in 1987. Hairy roots tend to be genetically and biochemically stable, an advantage over cell suspensions for the development of a commercial process. Since hairy roots are differentiated tissues, production of the alkaloid dimers may be more plausible in this system. In 1988, Parr *et al.* reported an intriguing result: the detection by immunoassay of a small level of vinblastine in *C. roseus* hairy root cultures, although a year later, Toivonen *et al.* (1989) could not detect this compound in their hairy root clones. Whether this difference was due to cultivar variation was not clear, since at that time only a few cultivars of *C. roseus* had been investigated (Brillanceau *et al.*, 1989; Davioud *et al.*, 1989; Parr *et al.*, 1988; Toivonen *et al.*, 1989). In 1989, we initiated work to establish hairy root clones from several cultivars of *C. roseus* and to evaluate the productivity of this plant tissue culture system, and published our initial findings on the resulting five heterotrophic clones that grew well in liquid culture (Bhadra *et al.*, 1991, 1993). Ajmalicine, serpentine, and catharanthine were present in our clones, as reported for other *C. roseus* hairy root cultures (Ciau-Uitz *et al.*, 1994; Jung *et al.*, 1992; Parr *et al.*, 1988; Toivonen *et al.*, 1989). However, the presence of vindoline or the target bisindole dimers has not been confirmed in our clones or in other clones produced since Parr *et al.* (1988). All the hairy root clones of *C. roseus* reported to date have been cultured heterotrophically in the dark or in dim light.

For hairy root cultures to compete commercially with plants as a source of medicinal indole alkaloids, productivity will have to be enhanced as in the case of cell suspensions. The potential for improvement of indole alkaloid production in *C. roseus* hairy roots using strategies developed for cell suspensions has just begun to be assessed (Jung *et al.*, 1992; Saenz-Carbonell *et al.*, 1993; Toivonen *et al.*, 1991, 1992a, 1992b; Vasquez-Flota *et al.*, 1994). Certain paradigms have emerged from the cell suspension literature for the influence of nutrient concentrations and light on indole alkaloid production. For example, generally accepted phenomena are that increased sucrose levels enhance ajmalicine and serpentine production, and that light reduces ajmalicine formation while stimulating serpentine and catharanthine production (Verpoorte *et al.*, 1991).

This chapter examines the influence of sucrose levels and photoheterotrophy on indole alkaloid production in our five hairy root clones. Heterotrophic *C. roseus* hairy roots were successfully adapted for growth under different sucrose concentrations and photoheterotrophic conditions. Photoheterotrophy does not appear to stimulate vindoline biosynthesis, as determined by HPLC with photodiode array detection. Tabersonine, a member of the vindoline biosynthetic pathway, as well as ajmalicine, serpentine and catharanthine were monitored in these clones, thus providing information for several branches of the indole alkaloid pathways. Similarities and differences in growth and alkaloid production among our five clones are highlighted and discussed with respect to the trends observed in cell suspensions.

Biosynthetic Pathways of Indole Alkaloids

Identification, characterisation and regulation of the genes and enzymes of the indole alkaloid pathways is still an active area of research, which has been reviewed in detail by Meijer *et al.* (1993). Figure 4.1 is a schematic of the biosynthetic pathways of the terpenoid indole alkaloids. The enzymes for unidentified hypothesised steps are denoted by dashed pathway arrows. The shikimate and mevalonate pathways from primary metabolism feed the terpenoid indole alkaloid pathways. These alkaloids contain an indole moiety, provided by tryptamine, and the terpenoid portion provided by secologanin. Tryptamine is produced from tryptophan by the enzyme tryptophan decarboxylase (TDC). The shikimate and mevalonate pathways converge with the condensation of tryptamine and secologanin by the enzyme strictosidine synthase (SSS) to form strictosidine, which is the central precursor of all terpenoid indole alkaloids. Strictosidine is converted to 4,21-dehydrogeissoschizine, which may be converted to geissoschizine or diverted to two branches: the "cathenamine" pathway leading to ajmalicine and serpentine; and the "bisindole" pathway forming catharanthine, tabersonine, vindoline and the bisindoles, vinblastine and vincristine. In the cathenamine pathway, ajmalicine is converted in a single step to serpentine by a peroxidase. The exact steps and enzymes for the bisindole pathway between 4,21-dehydrogeissoschizine and tabersonine and catharanthine have not been identified. Stemmadenine is the likely branch-point in the bisindole pathway: the biosynthesis of vindoline from tabersonine is one fork, and the formation of catharanthine is the other. Most of the enzymes in the former branch have been determined. The two branches converge with the coupling of catharanthine and vindoline by a peroxidase to form vinblastine.

Development of Transformed Lines

Heterotrophic Clones

Our strategy to obtain hairy root clones used growth as the primary criterion for selection. Fast growth appeared to be the primary factor of importance for overall productivity of *C. roseus* clones developed by others (Brillanceau *et al.*, 1989; Davioud *et al.*, 1989; Parr *et al.*, 1988; Toivonen *et al.*, 1989). We established hairy root cultures of *C. roseus* by the infection of seedlings from a total of four cultivars: Little Bright Eye (LBE), Rose Carpet (RC), Little Linda (LL), and Snow Carpet (SC), with *Agrobacterium rhizogenes* ATCC 15834 (Bhadra *et al.*, 1991, 1993). Approximately 150 transformants were evaluated for growth in three stages: after transformation, after two or three passages on solid medium, and following four generations in liquid culture (Gamborg's B5/2 medium with 3% w/v sucrose) in the dark. Fast-growing clones from liquid-adapted cultures were chosen for subsequent analysis of alkaloid productivity. The net result was a total of five fast-growing clones: LBE-4-2, LBE-6-1, LL-3-6, SC-3-1 and SC-3-6.

These fast-growing clones had similar morphologies characterised by thin, straight and regular branches with thin tips. Exponential phase growth

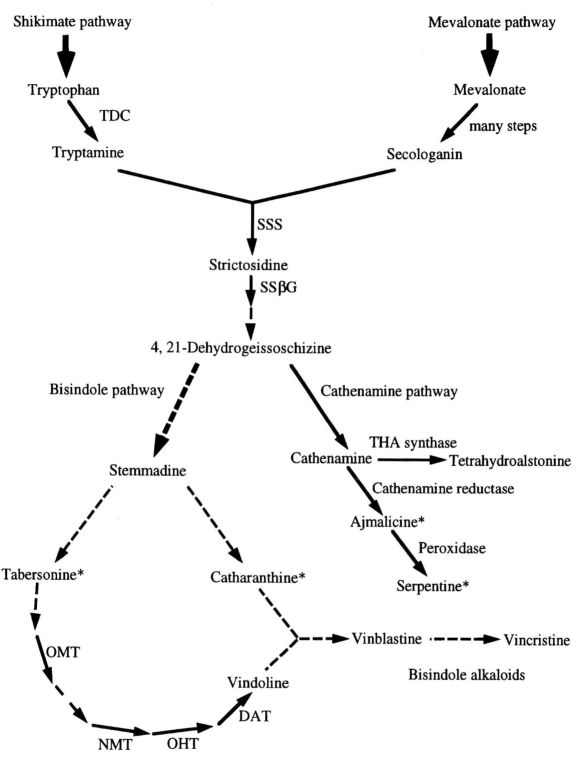

Figure 4.1. Schematic of monoterpenoid indole alkaloid pathways in *Catharanthus roseus*. The mevalonate and shikimate pathways supply the necessary building blocks. Dashed lines represent hypothesised steps in the biosynthetic pathway. Indole alkaloids marked with asterisks are measured in this study. Abbreviations are: TDC, tryptophan decarboxylase; SSS, strictosidine synthase; SSβG, strictosidine β-glucosidase; THA, tetrahydroalstonine; OMT, *O*-methyl transferase; NMT, *N*-methyl transferase; OHT, 2-oxoglutarate-dependent dioxygenase; DAT, deactylvindoline *O*-acetyltransferase.

was observed in liquid culture, with fresh weight doubling times for clones LBE-4-2, LBE-6-1 and SC-3-6 of 3.1, 3.7 and 4.1 d, respectively. These doubling times (3–4 d) fall within the range of those reported for cell suspensions (1.5–5 d) of *C. roseus* (Drapeau *et al.*, 1986; Morris, 1986b; Payne *et al.*, 1988; Scragg *et al.*, 1989).

Adaptation to Different Levels of Sucrose

For comparison of growth and alkaloid production between clones grown on B5/2 nutrients supplemented with various levels of sucrose, a uniform basis was adopted for (1) inoculum conditions; (2) culture conditions; and (3) adaptation procedures. All cultures were started with an inoculum optimised for growth rate: 5 root tips, each 35–40 mm long in 50 mL antibiotic-free medium in a 250 mL Erlenmeyer flask (Bhadra and Shanks, 1995), and grown in the dark. Culture adaptation is a particularly important distinction from the studies (Jung *et al.*, 1992; Parr *et al.*, 1988; Toivonen *et al.*, 1991; Vasquez-Flota *et al.*, 1994) that highlight production-induction schemes in which production is induced by immediate transfer to a different medium. Hairy roots were adapted to each new sucrose level for at least 2–3 generations. Clone LBE-6-1 was adapted to sucrose levels of 1, 2, 3, 4, 5, 6, 7 and 8%, while the remaining four were adapted to sucrose levels of 2, 3, 4 and 5%. Biomass and alkaloid analyses were performed similar to Bhadra *et al.* (1993) on roots harvested at 40–45 d. Additional measurements of our extracts, with mass spectrometry and HPLC photodiode array detection of TLC separated compounds, concur with our original identification for the compounds ajmalicine, serpentine, catharanthine and tabersonine (Vani and Shanks, in preparation).

Photoheterotrophic Adaptation

The five hairy root clones of *C. roseus* (LBE-4-2, LBE-6-1, LL-3-6, SC-3-1 and SC-3-6) were adapted to photoheterotrophic conditions through successive generations of increased photoperiod and reduced exogenous sugar. Photoheterotrophic adaptation was started in maintenance medium for heterotrophic cultures (Gamborg's B5/2 nutrients and 3% sucrose), atmospheric gas phase, and a photoperiod of 4 hours per day with inocula from heterotrophic cultures (grown in the dark). The light source was fluorescent white light (cool white, 2500–4500 lux). Optimised inoculum conditions as described in the previous section were also used in the photoheterotrophic adaptations. The photoperiod was increased in successive generations to 8, 12 and, finally, 18 h d^{-1}. However, 12 h d^{-1} appeared optimal for intensity and extent of greening of root tissue, and hairy root growth was suppressed with longer photoperiods. After 5 generations of sustained photo-

heterotrophic growth in maintenance medium, the initial sucrose concentration was reduced to 2% and the roots maintained thereafter as photoheterotrophic "maintenance" cultures. Although sucrose levels of 0.5–1.0 % are usually considered optimal for high chlorophyll content and photomixotrophic growth of cell suspensions (Horn and Widholm, 1984), 2% sucrose was satisfactory in our case for growth and tissue greening in photoheterotrophy.

The first change observed in photoheterotrophic growth with 3% sucrose was a thicker hairy root morphology than dark-grown or heterotrophic cultures. Green pigmentation was observed in clone LBE-6-1 after 2 generations of photoheterotrophy, and after 3 generations in clone LBE-4-2, restricted mainly to sections of older roots located in the centre of the root bunch. Clones SC-3-1, SC-3-6 and LL-3-6 grew without any visible greening. Increased greening of roots with 2% sucrose was perceptible only in clone LBE-6-1 compared with heterotrophic cultures. The rest of the hairy root clones, LBE-4-2, SC-3-1, SC-3-6 and LL-3-6, appeared a hue of greenish-brown distinct from their dark-grown heterotrophic counterparts. Propagation of the "green" root sections only, i.e. subculture from the mature regions of photoheterotrophic cultures, did not result in faster or more widespread greening of any hairy root clone. Only clone LBE-6-1 could be sustained for a generation with 1% sucrose under illumination. Subsequently, the 2%-sucrose photoheterotrophic cultures designated as photoheterotrophic "maintenance" cultures, were analysed at 40–45 d for growth and alkaloid levels as were the sucrose-adapted cultures.

Effect of Sucrose Levels on Alkaloid Production

Role of Sucrose

The carbohydrate source supplies carbon and energy in non-photosynthetic cultures. Sucrose is the usual carbon source utilised, although glucose and fructose are also used with lesser frequency. The rule of high sucrose levels for the enhanced production of indole alkaloids is generally accepted for *C. roseus* cell suspensions. Different parameters can be used to assess production capability: specific yield (mg g^{-1} dry weight), volumetric yield (mg L^{-1} reactor volume), average volumetric productivity (mg L^{-1} d^{-1}), and instantaneous volumetric productivity. Increased biomass yields, which can be measured as dry weight, fresh weight or cell number, contribute in part to the enhanced productivity, though biomass yields can decrease at sucrose levels of 8–10% (Verpoorte *et al.*, 1991). Nonetheless, Verpoorte *et al.* (1991) concluded that despite the different

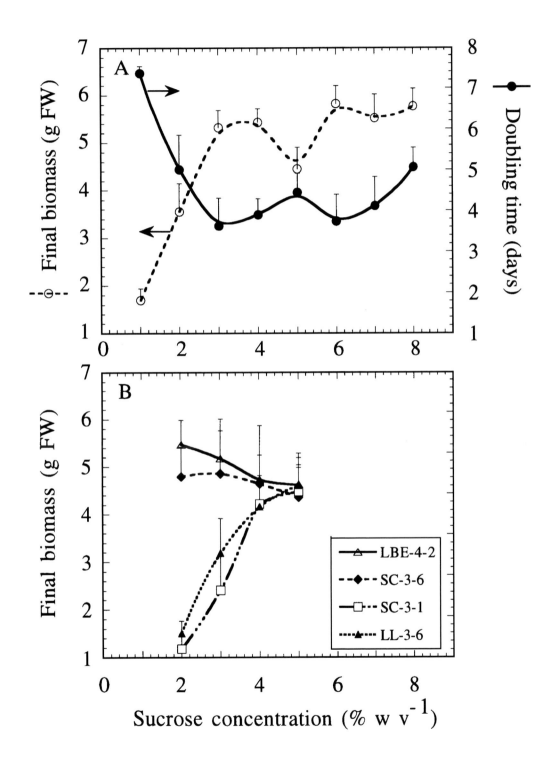

Figure 4.2. The effect of initial sucrose level on growth rate (doubling time) and final biomass of *C. roseus* hairy root clones: (A) LBE-6-1; and (B) LBE-4-2 (Δ), SC-3-1 (□), SC-3-6 (◆), and LL-3-6 (▲). Hairy root clones were cultured heterotrophically in media containing Gamborg's B5/2 salts and sucrose as indicated. Error bars are standard deviations and data are the average of at least 3 replicates.

measurement bases, production of ajmalicine and serpentine can be enhanced by a factor 2 or 3 by increasing the concentration of sucrose relative to standard growth media. In cell suspension studies, whether productivity is assessed for cells that were adapted to a higher sucrose level or for cells switched to a higher sucrose medium for an "induction" effect, the general conclusion still holds. In this chapter, growth and alkaloid production are examined for hairy root clones adapted to growth in sucrose concentrations ranging from 1–8% or 2–5%.

Growth Rate and Biomass Yield

The effect of initial sucrose concentration over the range of 1 to 8% on the maximum specific growth rate (expressed as doubling times) and final biomass level of heterotrophic hairy root cultures is shown for clone LBE-6-1 in Figure 4.2(A). Exponential growth was observed at all sucrose levels tested, and doubling times were measured for data from Day 2 through 18 in the culture period. Final biomass levels were obtained between Days 40 and 45, during stationary phase. The initial sucrose level affected growth rate and final biomass levels ($1-\alpha \geq 0.99$; standard analysis of variance, Scheffe F-test: ANOVA). The most rapid growth with doubling times of 3.6–4.4 d was obtained with initial sucrose concentrations of 3–7%. The biomass growth rate decreased at both ends of the sucrose spectrum, i.e. at 1 and 2% sucrose, and at 8% sucrose. The lowest growth rate or slowest doubling time: 7.4 ± 0.3 d, was obtained with 1% sucrose. The variation of final biomass level with initial sucrose concentration generally mirrors the doubling time profile. The final biomass increased with increasing initial sucrose concentration from 1–3%, then maintained level from 3–8%.

For the remainder of the clones: LBE-4-2, SC-3-1, SC-3-6 and LL-3-6, final biomass levels are shown in Figure 4.2(B). In the sucrose range we tested, the final biomass levels of SC-3-1 and LL-3-6 followed an upward trend with increasing initial sucrose concentration ($1-\alpha > 99\%$). The final biomass levels of clones LBE-4-2 and SC-3-6 were insensitive to initial sucrose concentration over this range.

These studies show clearly that sucrose level can affect growth rate and final biomass concentration of *C. roseus* hairy roots, as has been observed for cell suspensions, although clonal variability and variation with cultivar exist. The effect of sucrose level on doubling time of *C. roseus* hairy roots has not been previously reported, and shows a wide range of optimal sucrose levels, between 3 and 7%. In the case of *C. roseus* cell suspensions, the specific growth rate is often less sensitive to initial sucrose concentration (in Gamborg's basal salts) than is final biomass concentration (Fowler, 1982; Merillon et al., 1984;

Morris, 1986a). Growth rates of cell suspensions remain fairly constant between 2 and 6% initial sucrose levels, while final biomass increases in the range 2–8% sucrose. In other studies of *C. roseus* hairy roots, final biomass has been measured in terms of fresh weight or dry weight, for sucrose levels of 1–7% in Parr et al. (1988), and narrower ranges in others (Toivonen et al., 1991; Vasquez-Flota et al., 1994), with conflicting results. In all of these studies, roots were not adapted to the changed sucrose level, and one root clone was examined in each work. Only Parr et al. (1988) observed a decrease in final biomass (2-fold) as sucrose level increased from 1 to 7%. Toivonen et al. (1991) reported a maximum of 50% increase in dry weight with increasing sucrose levels from 3% to 6%, and Vasquez-Flota et al. (1994) reported no effect of sucrose level on biomass. All growth media were similar to that used in this study. Final biomass increased between 2- and 4-fold from the lowest sucrose levels tested for clones LBE-6-1, SC-3-1 and LL-3-6, but essentially no effect was seen for clones LBE-4-2 and SC-3-6. Clonal variation and cultivar variation, it appears, can explain some differences in the general trend of increased biomass with increased sucrose levels.

The possibility of utilising another carbon source besides sucrose for the growth of *C. roseus* hairy roots has been tested on a few clones for one growth cycle in fructose or glucose (Jung et al. 1992) and glucose, galactose and fructose (Vasquez-Flota et al., 1994), with similar or reduced growth observed. Adaptation of hairy roots of *C. roseus* to fructose or glucose was attempted with all our clones, but was judged largely unsuccessful since there was no evidence of long-term generational stability. Most clones yielded low growth for only one or two generations. The characteristic description of these unsuccessful cultures includes sparse root bunches with mostly short root tips, a few rapidly dividing long root tips, and considerable callus. Only clone LBE-4-2 responded after 15 generations to adaptation to 2% fructose, but a low final biomass (1.1 ± 0.2 g fresh weight at 40–45 d) was obtained.

Alkaloid Production

The production capabilities of our clones were assessed by specific and volumetric alkaloid yields, since the experiments were end-point analyses. A semi-quantitative guide was used to analyse and present simply the vast amount of data for "optimal" sucrose levels. When scanning for optimal sucrose levels, alkaloid values within 25% of the maximum were considered equivalent. Only the values of yields for the optimal sucrose levels are listed in Table 4.1. In the case where standard analysis of variance (ANOVA) yielded significance at the 0.05 level or higher ($\alpha < 0.05$), pairs of sucrose treat-

Table 4.1. Sucrose levels (% w/v) for maximum values of specific yield[a] and volumetric yield[a] of indole alkaloids. Corresponding maximum value of yield for optimal sucrose level[b] is given below in parenthesis.

| Hairy root clone[c] | Cathenamine Pathway | | | | Bisindole Pathway | | | |
| | Ajmalicine | | Serpentine | | Catharanthine | | Tabersonine | |
	Specific yield (mg g^{-1} dry weight)	Volumetric yield (mg L^{-1})	Specific yield (mg g^{-1} dry weight)	Volumetric yield (mg L^{-1})	Specific yield (mg g^{-1} dry weight)	Volumetric yield (mg L^{-1})	Specific yield (mg g^{-1} dry weight)	Volumetric yield (mg L^{-1})
LBE-6-1	6% (0.9)	6% (9.4)	2–3%; 6% (1.9)	6–7% (20.4)	5–6% (0.5)	5–6% (4.8)	2%* (0.6)	2%* (3.0)
LBE-4-2	2–4% (0.7)	2% (6.2)	4–5%* (1.1)	5%* (13.0)	2% (0.3)	2%; 5% (3.2)	3%* (0.9)	3% (5.4)
SC-3-1	3–5% (1.2)	3–5% (7.5)	2–5% (1.1)	3–5% (6.6)	4%* (1.0)	4%* (6.8)	3%* (0.6)	3%* (3.3)
SC-3-6	2–5% (1.6)	5% (21.0)	3–5%* (1.4)	4–5% (15.8)	3–5% (1.0)	3%; 5% (10.4)	2% (0.2)	2% (1.9)
LL-3-6	2% (1.4)	5% (11.5)	2%; 4% (1.6)	4–5% (12.8)	2–5% (0.5)	4–5% (3.8)	2–4% (1.8)	4% (1.6)

[a] Alkaloid yields are the average of at least duplicate flasks.
[b] Optimal sucrose levels noted for yields within 25% of the maximum value of yield
[c] LBE-6-1 examined for sucrose levels 1–8%; all other clones 2–5%.
* Significant at the 97% confidence level or higher ($1 - \alpha \geq 0.97$).

ments were tested for statistical significance using the unpaired *t*-test; these results are labelled with an asterisk.

Cathenamine pathway

The specific and volumetric yields for ajmalicine and serpentine are shown in Figure 4.3(A) for clone LBE-6-1 over the range of 1–8% initial sucrose concentration. Specific yields of ajmalicine and serpentine were fairly stable for the entire range of sucrose concentrations except for possibly 2% and 6%. When biomass yield is factored into the assessment as given by volumetric yield, ajmalicine and serpentine have maximum values at 6% and 7% sucrose, respectively.

The optimal sucrose levels and corresponding yields of ajmalicine and serpentine for all clones are compiled in Table 4.1. Specific yields did not vary significantly with sucrose levels between 2 and 5%. Volumetric yields, however, tended to peak at the higher sucrose levels tested, and for clones LBE-6-1, SC-3-1 and LL-3-6, likely reflected the significant trends with biomass yield accounted earlier. For all the clones tested, at most a 3-fold increase in volumetric yields was observed in comparison to standard growth media with 3% sucrose (data not shown).

The maximum specific yields and volumetric yields for ajmalicine and serpentine from our cultures are in the range 0.7–2 mg g^{-1} dry weight and 6–21 mg L^{-1}, respectively. These levels compare favourably with maximum levels observed for *C. roseus* hairy roots, but are lower by an order of magnitude than those observed for optimised, high producing cell suspension cultures (Verpoorte *et al.*, 1991); the productivity in these cell suspension cultures was unstable, however. LBE-6-1 is the only hairy root clone for which the specific level of serpentine is consistently higher than that of ajmalicine at all sucrose concentrations (data not shown). With the other clones, either the reverse is true or the specific yields of the two indole alkaloids are comparable.

Bisindole pathway

The specific and volumetric yields of catharanthine and tabersonine are shown in Figure 4.3(B) for LBE-6-1. Catharanthine yields are highest at 5% sucrose. Tabersonine shows a striking trend as yields decrease with initial sucrose levels. The maximum yield values for all clones are tabulated in Table 4.1. Tabersonine tends to have a single sucrose level for the optimum. A consistent trend for the optimal yields of catharanthine did not emerge; however, initial sucrose concentration had a strong effect on tabersonine accumulation. In all clones except

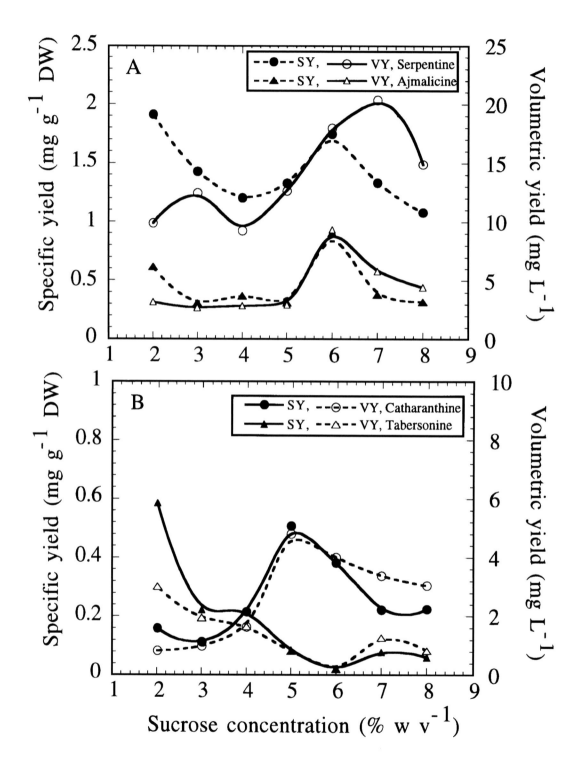

Figure 4.3. The effect of initial sucrose level on specific and volumetric alkaloid yields for *C. roseus* hairy root clone LBE-6-1. (A) serpentine and ajmalicine and (B) catharanthine and tabersonine. SY = specific yield; VY = volumetric yield. Hairy root clones were cultured heterotrophically in media containing Gamborg's B5/2 salts and sucrose as indicated. Data are the average of at least 2 replicates.

LL-3-6, the maximum specific and volumetric yields of tabersonine occur in 2%- or 3%-sucrose cultures. Our studies on the dependence of catharanthine and tabersonine yields of *C. roseus* hairy roots on sucrose levels is unique, particularly since tabersonine is usually not measured or reported. In the case of *C. roseus* cell suspensions, Smith *et al.* (1987) report the relative insensitivity of yields of catharanthine to initial sucrose concentration (3–6%). Catharanthine specific and volumetric yields in our clones are similar to other *C. roseus* hairy roots for growth in 3% sucrose (Ciau-Uitz *et al.*, 1994, Toivonen *et al.*, 1989).

Effect of Photoheterotrophy on Alkaloid Production

Role of Light and Differentiation

Light has been used as a means to stimulate alkaloid production in cell suspension cultures. The effect of light on the regulation of alkaloid biosynthesis has focused downstream of 4,21-dehydrogeissoschizine on the cathenamine and bisindole pathways since the activities of the enzymes TDC and SSS in the early part of the terpenoid indole alkaloid pipeline do not appear to be influenced by exposure to light (Meijer *et al.*, 1993).

The specific and volumetric yields of serpentine (Drapeau *et al.*, 1987; Knobloch *et al.*, 1982) in cell suspensions of *C. roseus* were found to increase several-fold with a concomitant decrease in ajmalicine when exposed to light. Ajmalicine can be converted to serpentine by basic peroxidases in the vacuole, and the activity of these peroxidases correlated positively to serpentine accumulation in light-grown cells (Meijer *et al.*, 1993). However, the effect of light on serpentine content may not be explained adequately by the conversion of ajmalicine to serpentine, since the magnitude of the serpentine increase was significantly larger than the decrease in ajmalicine, while in another case of enhanced serpentine yields, the specific yields of ajmalicine remained the same (Drapeau *et al.*, 1987).

Catharanthine readily accumulates in cell suspension cultures and specific yields reportedly increase upon exposure to light (Drapeau *et al.*, 1987; Park *et al.*, 1990). Tabersonine has been detected in heterotrophic cell suspensions (Kurz *et al.*, 1980), but the absence of vindoline and the bisindole alkaloids both in dark-grown and light-exposed cultures has been routinely reported (De Luca *et al.*, 1985; Eilert *et al.*, 1987a, 1987b; Fahn *et al.*, 1985). Photoautotrophic cell suspension cultures of *C. roseus* that possess chloroplasts still did not synthesise vindoline (Tyler *et al.*, 1986).

The evidence for the light-inducibility and developmental regulation of the biosynthetic pathway leading from tabersonine to vindoline in *C. roseus* includes studies on precursors and intermediates of the reaction network (Balsevich *et al.*, 1986; De Luca *et al.*, 1986), as well as on enzyme activities (De Carolis *et al.*, 1990; De Luca and Cutler, 1987; De Luca *et al.*, 1988) in seedlings. Vindoline reaches higher levels in light-grown compared to etiolated seedlings of *C. roseus* (Balsevich *et al.*, 1986; De Luca *et al.*, 1986). At the same time, tabersonine is depleted and only trace levels of other intermediates of the pathway are detected (Balsevich *et al.*, 1986). The activities of the late enzymes of vindoline biosynthesis, NMT, OHT and DAT, are highly regulated. The activity of the third-to-last enzyme in vindoline biosynthesis, NMT, is associated with the thylakoid membrane of chloroplasts in leaves (De Luca and Cutler, 1987), but its activity increases by only 30% in light-grown seedlings (De Luca *et al.*, 1988). NMT is also localised in the endoplasmic reticulum, and in etioplasts of etiolated seedlings (De Luca *et al.*, 1988), hence these NMT enzymes may be regulated differently. Light strongly increases the activities of the cytoplasmic enzymes, OHT and DAT (De Carolis *et al.*, 1990; De Luca *et al.*, 1988). In heterotrophic cell suspension cultures, activities of NMT and DAT were not detected (De Luca and Cutler, 1987).

The studies in this section address the photoheterotrophic culture of hairy roots of *C. roseus*. The inducibility of DAT and OHT by illumination (De Carolis *et al.*, 1990; De Luca *et al.*, 1988), as well as the chloroplast- or etioplast-associated activity of NMT (De Luca and Cutler, 1987), serve as incentives for the possible light induction of vindoline synthesis in hairy roots of *C. roseus*. However, we did not detect vindoline in our clones. These clones are likely not photosynthetically active, unlike the photo-

Table 4.2. The final biomass of hairy root clones of *C. roseus* under photoheterotrophic and heterotrophic conditions in Gamborg's B5/2 nutrients and 2% sucrose.

Clone	Final biomass (g fresh weight)[a,b]	
	Photoheterotrophy	Heterotrophy
LBE-6-1	6.1 ± 1.2*	3.6 ± 0.9
LBE-4-2	6.3 ± 1.4	5.5 ± 0.5
SC-3-1	5.1 ± 0.5*	1.2 ± 0.3
SC-3-6	7.6 ± 0.6*	4.8 ± 0.6
LL-3-6	5.6 ± 0.3*	1.5 ± 0.3

[a] Cultures were harvested at 40–45 d.
[b] Data are the average of at least 3 flasks.
* Denotes significant effect ($1-\alpha \geq 0.99$) of photoheterotrophy relative to heterotrophy.

autotrophic and photomixotrophic hairy roots of *D. stramonium* studied by Flores *et al.* (1993), so the possibility of vindoline production in photoautotrophic clones of *C. roseus* cannot be ruled out. The results of photoheterotrophy on growth and alkaloid production are presented below.

Biomass and Alkaloid Yields

The effect of photoheterotrophy on the biomass yields of hairy roots of *C. roseus* is summarised in Table 4.2. Photoheterotrophy enhanced the biomass accumulation of most hairy root clones ($1-\alpha > 0.99$). LBE-4-2 was the only hairy root clone that did not show any statistically significant effect on final fresh weight. The final fresh weight was approximately 300% greater than in heterotrophic cultures in the case of SC-3-1 and LL-3-6, and 60–70% greater in the case of LBE-6-1 and SC-3-6.

The values for the alkaloid yields in both photoheterotrophic cultures and heterotrophic controls for the cathenamine and bisindole pathways are summarised in Table 4.3. Table 4.4 summarises the direction (positive, negative, or none) of the effects of photoheterotrophy using the 25% cutoff as described for the sucrose studies. Results with statistical significance (using the unpaired *t*-test) are denoted by an asterisk.

Serpentine volumetric yields increased with photoheterotrophy for all clones, whereas the specific yields increased for only LBE-4-2 and SC-3-6, hence it was mainly a biomass effect. Ajmalicine specific and volumetric yields either remained the same or increased during photoheterotrophy for all clones except LBE-4-2, for which the yields sharply decreased. For clone LBE-4-2, the increase of serpentine was likely mainly due to conversion from ajmalicine. (The increase in specific yield of serpentine is roughly equivalent to the decrease in ajmalicine specific yield.) This is the only clone where the effect of light on ajmalicine to serpentine conversion is evident.

Few generalisations emerge on the effect of photoheterotrophy on the specific yields of alkaloids in the bisindole pathway, perhaps due to clonal and cultivar variation. The volumetric yield of catharanthine was enhanced in photoheterotrophic cultures of LBE-6-1 and SC-3-6 and decreased in LBE-4-2 and SC-3-1. Photoheterotrophy decreased tabersonine accumulation in LBE-6-1, LBE-4-2, and SC-3-6, but the opposite was true in the other two clones. Unfortunately, we do not have data on the metabolites between tabersonine and vindoline, so it is not obvious whether the decrease in tabersonine levels is an indication of build-up of these metabolites.

Table 4.3. Specific yields[a] and volumetric yields[a] of indole alkaloids for heterotrophic and photoheterotrophic hairy root clones.

| Hairy root clone | Cathenamine Pathway | | | | Bisindole Pathway | | | |
| | Ajmalicine | | Serpentine | | Catharanthine | | Tabersonine | |
	Specific yield (mg g^{-1} dry weight)	Volumetric yield (mg L^{-1})	Specific yield (mg g^{-1} dry weight)	Volumetric yield (mg L^{-1})	Specific yield (mg g^{-1} dry weight)	Volumetric yield (mg L^{-1})	Specific yield (mg g^{-1} dry weight)	Volumetric yield (mg L^{-1})
LBE-6-1								
Heterotrophic	0.6	3.2	1.9	9.9	0.16	0.8	0.6	3.0
Photoheterotrophic	0.9	11.6	1.7	21.8	0.3	4.1	0.15	2.0
LBE-4-2								
Heterotrophic	0.7	6.2	0.6	5.8	0.3	2.9	0.4	3.5
Photoheterotrophic	0.2	1.2	1.4	11.1	0.1	1.0	0.06	0.5
SC-3-1								
Heterotrophic	0.5	5.3	1.2	3.0	0.3	3.2	0.03	0.4
Photoheterotrophic	0.9	7.5	1.1	9.2	0.2	1.3	0.2	1.7
SC-3-6								
Heterotrophic	1.4	12.3	0.9	8.1	0.5	4.0	0.2	1.8
Photoheterotrophic	1.7	14.4	1.4	12.0	0.7	5.6	0.07	0.6
LL-3-6								
Heterotrophic	1.4	6.3	1.6	7.2	0.4	2.0	0.2	0.7
Photoheterotrophic	1.3	12.0	1.1	10.3	0.2	1.9	0.2	1.4

[a] Alkaloid yields are the mean of at least duplicate flasks.

Table 4.4. Effect of photoheterotrophy[a] on indole alkaloid yields.

Clone	Cathenamine Pathway				Bisindole Pathway			
	Ajmalicine		Serpentine		Catharanthine		Tabersonine	
	Specific yield	Volumetric yield	Specific yield	Volumetric yield	Specific yield	Volumetric yield	Specific yield	Volumetric yield
LBE-6-1	+*	+*	~	+*	+*	+*	−*	−
LBE-4-2	−*	−*	+*	+*	−*	−*	−*	−*
SC-3-1	+*	+	~	+	~/−	−	+*	+*
SC-3-6	~/+	~/+	+*	+*	~/+	+	−	−
LL-3-6	~	+*	−	+	−	~	~	+

[a] Effect of photoheterotrophy noted if more than 25% different than heterotrophic condition.
+ Denotes positive effect.
− Denotes negative effect.
~ No effect of photoheterotrophy.
~/+ or ~/− Approximately a 25% difference, positive or negative, respectively.
* Significant at the 90% confidence level or higher $(1-\alpha \geq 0.90)$.

The results in Tables 4.3 and 4.4 amply illustrate the complexity of the metabolic pathways and the difficulty in predicting the effect of a chosen perturbation such as light on alkaloid productivity. Rigorous analysis for alkaloid pathways is difficult. Analysis is limited since quantitative measurements are labour intensive and not all the reaction steps in Figure 4.1 are known. For example, alkaloid catabolism has been little studied and likely occurs (Meijer *et al.*, 1993). An understanding of the terpenoid indole alkaloid pathways may eventually lead to the successful genetic manipulation of plant tissue cultures for enhanced productivity (Meijer *et al.*, 1993). Transient kinetic studies in which the levels of several indole alkaloids are measured simultaneously should complement studies on the regulation of enzymes and genes in these pathways. The genetic and biochemical stability of hairy roots make them an attractive system for such coordinated metabolic studies.

Conclusion

The studies accounted in this chapter fulfill an important process objective of examining the influence of carbon source and light on the biosynthetic output of *C. roseus* hairy roots. Both sucrose levels and photoheterotrophy influence growth. Volumetric yields of the major indole alkaloid pathways, the cathenamine pathway and the bisindole pathway, can be affected by the level of sucrose, mainly through final biomass yields. Tabersonine is the only alkaloid with a striking trend of yield with sucrose level, decreasing with increasing sucrose concentration.

Exposure to light positively affects growth, as reflected in biomass yields. Variation in the root lines, whether due to clonal or cultivar differences, was key to the effect of light on fluxes through the alkaloid pathways. Though serpentine production in most hairy root clones was enhanced by photoheterotrophy, only one clone out of five clearly showed the light-associated conversion of ajmalicine to serpentine. These studies amply reiterate the rich complexity of secondary metabolic pathways.

Acknowledgments

This work was funded in part by the National Science Foundation (NSF) Young Investigator Award to J.V. Shanks (BCS-9257938), NSF (BES-9411928), and the Robert A. Welch Foundation. Standards of catharanthine and vindoline were gifts from Eli Lilly and Company; tabersonine was a gift from Prof. H. Hamada, Okayama Univ., Japan. The authors thank John Morgan for his careful review of the manuscript.

References

Balsevich J, V De Luca and WGW Kurz (1986). Altered alkaloid pattern in dark grown seedlings of *Catharanthus roseus*. The isolation and characterization of 4-desacetoxyvindoline: a novel indole alkaloid and proposed precursor of vindoline. *Heterocyc* 24: 2415–2421.

Bhadra R and JV Shanks (1995). Statistical design of the effect of inoculum conditions on growth of hairy root cultures of *Catharanthus roseus*. *Biotechnol Techniq* 9: 681–686.

Bhadra R, C Ho and JV Shanks (1991). Growth characteristics of hairy root lines of *Catharanthus roseus*. *In Vitro Cell Dev Biol* 27: 109A.

Bhadra R, SN Vani and JV Shanks (1993). Production of indole alkaloids by selected hairy root lines of *Catharanthus roseus*. *Biotechnol Bioeng* 41: 581–592.

Brillanceau M-H, C David and J Tempé (1989). Genetic transformation of *Catharanthus roseus* G. Don by *Agrobacterium rhizogenes*. *Plant Cell Rep* 8: 63–66.

Ciau-Uitz R, ML Miranda-Ham, J Coello-Coello, B Chi, LM Pacheco and VM Loyola-Vargas (1994). Indole alkaloid production by transformed and non-transformed root cultures of *Catharanthus roseus*. *In Vitro Cell Dev Biol* 30P: 84–88.

Davioud E, C Kan, J Hamon, J Tempé and H-P Husson (1989). Production of indole alkaloids by in vitro root cultures from *Catharanthus trichophyllus*. *Phytochemistry* 28: 2675–2680.

De Carolis E, F Chan, J Balsevich and V De Luca (1990). Isolation and characterization of a 2-oxyglutarate dependent dioxygenase involved in the second-to-last step in vindoline biosynthesis. *Plant Physiol* 94: 1323–1329.

De Luca, V and AJ Cutler (1987). Subcellular localization of enzymes involved in indole alkaloid biosynthesis in *Catharanthus roseus*. *Plant Physiol* 85: 1099–1102.

De Luca V, J Balsevich and WGW Kurz (1985). Acetyl coenzyme A:deacetylvindoline *O*-acetyltransferase, a novel enzyme from *Catharanthus*. *J Plant Physiol* 121: 417–428.

De Luca V, J Balsevich, RT Tyler, U Eilert, BD Panchuk and WGW Kurz (1986). Biosynthesis of indole alkaloids: developmental regulation of the biosynthetic pathway from tabersonine to vindoline in *Catharanthus roseus*. *J Plant Physiol* 125: 147–156.

De Luca V, JA Fernandez, D Campbell and WGW Kurz (1988). Developmental regulation of enzymes of indole alkaloid biosynthesis in *Catharanthus roseus*. *Plant Physiol* 86: 447–450.

Deus-Neumann, B and MH Zenk (1984). Instability of indole alkaloid production in *Catharanthus roseus* cell suspension cultures. *Planta Med* 50: 427–431.

Drapeau D, HW Blanch and CR Wilke (1986). Growth kinetics of *Dioscorea deltoidea* and *Catharanthus roseus* in batch culture. *Biotechnol Bioeng* 28: 1555–1563.

Drapeau D, HW Blanch and CR Wilke (1987). Ajmalicine, serpentine and catharanthine accumulation in *Catharanthus roseus* bioreactor cultures. *Planta Med* 53: 373–376.

Eilert U, V De Luca, F Constabel and WGW Kurz (1987a). Elicitor-mediated induction of tryptophan decarboxylase and strictosidine synthase activities in cell suspension cultures of *Catharanthus roseus*. *Arch Biochem Biophys* 254: 491–497.

Eilert U, V DeLuca, WGW Kurz and F Constabel (1987b). Alkaloid formation by habituated and tumorous cell suspension cultures of *Catharanthus roseus*. *Plant Cell Rep* 6: 271–274.

Fahn W, H Gundlach, B Deus-Neumann and J Stockigt (1985). Late enzymes of vindoline biosynthesis. Acetyl-CoA:17-*O*-deacetylvindoline 17-*O*-acetyl-transferase. *Plant Cell Rep* 4: 333–336.

Flores HE, MW Hoy and JJ Pickard (1987). Secondary metabolites from root cultures. *Trends in Biotechnol* 5: 64–69.

Flores HE, Y Dai, JL Cuello, IE Maldonado-Mendoza and VM Loyola-Vargas (1993). Green roots: photosynthesis and photoautotrophy in an underground plant organ. *Plant Physiol* 101: 363–371.

Fowler MW (1982). Substrate utilization by plant-cell cultures. *J Chem Tech Biotechnol* 32: 338–346.

Horn ME and JM Widholm (1984). Aspects of photosynthetic plant tissue cultures. In *Applications of Genetic Engineering to Crop Improvement*, edited by GB Collins and JB Petolino, pp. 113–161. Dordrecht: Martinus Nijhoff/Dr. W. Junk.

Jung KH, SS Kwak, SW Kim, H Lee, CY Choi and JR Liu (1992). Improvement of the catharanthine productivity in hairy root cultures of *Catharanthus roseus* by using monosaccharides as a carbon source. *Biotechnol Lett* 14: 695–700.

Knobloch K-H, G Bast and J Berlin (1982). Medium- and light-induced formation of serpentine and anthocyanins in cell suspension cultures of *Catharanthus roseus*. *Phytochemistry* 21: 591–594.

Kurz WGW, KB Chatson, F Constabel, JP Kutney, LSL Choi, P Kolodziejczyk, SK Sleigh, KL Stuart and BR Worth (1980). Alkaloid production in *Catharanthus roseus* cell cultures. IV. Characterization of the 953 cell line. *Helvetica Chimica Acta* 63: 1891–1895.

Meijer AH, R Verpoorte and JHC Hoge (1993). Regulation of enzymes and genes involved in terpenoid indole alkaloid biosynthesis in *Catharanthus roseus*. *J Plant Res* 3: 145–164.

Merillon JM, M Rideau and JC Cheneiux (1984). Influence of sucrose on levels of ajmalicine, serpentine and tryptamine in *Catharanthus roseus* cells *in vitro*. *Planta Med* 50: 497–502.

Morris P (1986a). Regulation of product synthesis in cell cultures of *Catharanthus roseus*. III. Alkaloid metabolism in cultured leaf tissue and primary callus. *Planta Med* 52: 127–132.

Morris P (1986b). Long term stability of alkaloid productivity in cell suspension cultures of *Catharanthus roseus*. In *Secondary Metabolism in Plant Cell Cultures*, edited by P MorrisP AH Scragg, A Stafford and MW Fowler, pp. 257–262. Cambridge: Cambridge University Press.

Park HH, SK Choi, JK Kang and HY Lee (1990). Enhancement of producing catharanthine by suspension growth of *Catharanthus roseus*. *Biotechnol Lett* 12: 603–608.

Parr AJ, ACJ Peerless, JD Hamill, NJ Walton, RJ Robins and MJC Rhodes (1988). Alkaloid production by transformed root cultures of *Catharanthus roseus*. *Plant Cell Rep* 7: 309–312.

Payne GF, NN Payne, ML Shuler and M Asada (1988). *In situ* adsorption for enhanced alkaloid production by *Catharanthus roseus*. *Biotechnol Lett* 10: 187–192.

Saenz-Carbonell LA, IE Maldonado-Mendoza, O Moreno-Valenzula, R Ciau-Uitz, M Lopez-Meyer, C Oropeza and VM Loyola-Vargas (1993). Effect of the medium pH on the release of secondary metabolites from roots of *Datura stramonium*, *Catharanthus roseus*, and *Tagetes patula* cultured *in vitro*. *Appl Biochem Biotechnol* 38: 257–267.

Scragg AH, RC Cresswell, S Ashton, A York, PA Bond and MW Fowler (1989). Growth and alkaloid production in bioreactors by a selected *Catharanthus roseus* cell line. *Enzyme Microb Technol* 11: 329–333.

Smith JI, AA Quesnel, NJ Smart, M Misawa and WGW Kurz (1987). The development of a single-stage growth and indole alkaloid production medium for *Catharanthus roseus* (L.). G. Don suspension cultures. *Enzyme Microb Technol* 9: 466–469.

Toivonen L, J Balsevich and WGW Kurz (1989). Indole alkaloid production by hairy root cultures of *Catharanthus roseus*. *Plant Cell Tiss Organ Cult* 18: 79–83.

Toivonen L, M Ojala and V Kauppinen (1991). Studies on the optimization of growth and alkaloid production by hairy root cultures of *Catharanthus roseus*. *Biotechnol Bioeng* 37: 673–680.

Toivonen L, S Laakso and H Rosenqvist (1992a). The effect of temperature on hairy root cultures of *Catharanthus roseus*: growth, indole alkaloid accumulation and membrane lipid composition. *Plant Cell Rep* 11: 395–399.

Toivonen L, S Laakso and H Rosenqvist (1992b). The effect of temperature on growth, indole alkaloid accumulation and lipid composition of *Catharanthus roseus* cell suspension cultures. *Plant Cell Rep* 11: 390–394.

Tyler RT, WGW Kurz and BD Panchuk (1986). Photoautotrophic cell suspension cultures of periwinkle (*Catharanthus roseus* (L.) G Don): transition from heterotrophic to photoautotrophic growth. *Plant Cell Rep* 3: 195–198.

Vasquez-Flota F, O Moreno-Valenzuela, ML Miranda-Ham, J Coello-Coello and VM Loyola-Vargas (1994). Catharanthine and ajmalicine synthesis in *Catharanthus roseus* hairy root cultures. *Plant Cell Tiss Organ Cult* 38: 273–279.

Verpoorte R, R van der Heijden, WM van Gulik and HJG ten Hoopen (1991). Plant biotechnology for the production of alkaloids: present status and prospects. In *The Alkaloids*, edited by A. Brossi, vol 40, pp. 1–187. San Diego: Academic Press.

Alteration of Metabolite Formation and Morphological Properties of Hairy Roots by Environmental Stimuli

5

Setsuji Tone, Masahito Taya and Masahiro Kino-oka

Department of Chemical Engineering, Faculty of Engineering Science, Osaka University, 1-3 Machikaneyama-cho, Toyonaka, Osaka 560, Japan; email: taya@cheng.es.osaka-u.ac.jp

Introduction

Hairy roots are caused by the infection of plant cells with a soil bacterium, *Agrobacterium rhizogenes*, and are associated with the integration of transfer-DNA from the bacterial root-inducing plasmid into chromosomal DNA in plant cells (White and Sinkar, 1987). Hairy roots can grow as fast as unorganised plant cells such as calli, and maintain a stable phenotype as a differentiated root organ as reflected in their high productivity of root-derived metabolites. Hairy roots are promising materials not only for the production of useful metabolites but also for the breeding of transgenic plants.

For effective production of metabolites by hairy roots, enhanced growth rates and product yields are primary requirements. In submerged bioreactor culture of hairy roots, it is desirable that metabolites of interest be available extracellularly. When hairy roots are utilised as embryo-like inclusions for artificial seeds, it is also desired that the root morphology be altered to suspended cell aggregates because an aggregate type of morphology facilitates cell entrapment in gel materials of artificial seeds.

In this chapter, changes in hairy root metabolite formation and morphology caused by physical and chemical stimuli including nutrient limitation, illumination and hormone dosage are discussed from a viewpoint of expanding the biotechnological application of hairy root cultures.

Production and Release of Metabolites in Hairy Root Cultures

Response to Phosphorus Limitation

It is well known that medium constituents exert notable effects on growth and metabolite formation by cultured plant cells (Fujita *et al.*, 1981; Kino-oka *et al.*, 1991). In culture of red beet (*Beta vulgaris*) hairy roots, phosphate is a key nutrient for increasing the accumulation of betanin pigment in the roots (Taya *et al.*, 1994b). Figure 5.1 shows root growth, pigment content in the cells and total amount of pigment produced in the culture with Murashige and Skoog (MS) medium (Murashige and Skoog, 1962) containing between 0 and 2.5 mol m^{-3} phosphate. Normal root growth with enhanced pigment formation was achieved even in medium without phosphate. In higher plant cells, it is recognised that the availability of phosphorus stored in cells supports growth under phosphate-deficient culture conditions (Bieleski, 1973).

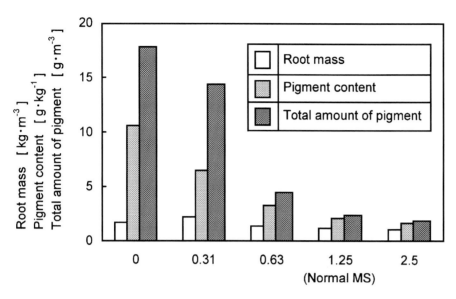

In phosphate-free medium, pigment levels in red beet hairy roots tend to increase towards the end of the culture when the phosphorus content in the cells is reduced. As shown in Figure 5.2, the pigment content increases remarkably with decreasing phosphorus levels. This inverse relationship between metabolite and phosphorus contents has been reported also for cultures of other plant cells (Table 5.1). As a reason for the stimulated metabolite formation at lower phosphorus levels in plant cells, Fischer *et al.* (1993) suggested that the activities of

enzymes involved in the metabolism of aromatic amino acids (precursors for many secondary metabolites) were regulated by intracellular phosphorus.

Response to Oxygen Starvation

In plant cell cultures, the metabolites of interest are most frequently accumulated in cells without secretion, thereby limiting continuous metabolite production. Several procedures for releasing intracellular products from plant cells have been proposed, including addition of inorganic chemicals or organic solvents, and exposure of cells to electricity (Endress, 1994).

Hairy roots of red beet and madder (*Rubia tinctorum*) have been found to release pigments into the medium when the roots are kept at a low level of dissolved oxygen (DO) (Taya *et al.*, 1992; Kino-oka *et al.*, 1994). The effect of DO level on pigment leakage from red beet hairy roots was examined by Kino-oka *et al.* (1992). As shown in Figure 5.3, at DO levels of 1–5 ppm, concentrations of pigment in the medium were negligible after several hours of DO control, whereas significant pigment leakage was obtained when the DO level was kept at zero.

Based on this finding, hairy roots were cultivated in a medium with fructose as a carbon source using a column-type reactor operated without direct aeration but in which oxygen-enriched medium was circulated instead. As a gradient of DO concentration is generated in this reactor due to oxygen uptake by the hairy roots, pigment release from roots was expected to occur in locations where the DO concentration was reduced. To recover the released pigment, the culture broth was circulated through an adsorbent column packed with a synthetic resin.

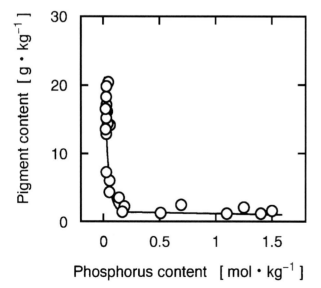

Figure 5.2. Inverse relationship between contents of pigment and phosphorus in red beet hairy roots cultivated in phosphate-free medium.

Table 5.1. Examples of stimulated metabolite formation in plant cells by cultivating under phosphate-deficient conditions.

Plant	Metabolite	Reference
Hairy roots		
Ajuga reptans var. *atropurpurea*	20-Hydroxyecdysone	Uozumi *et al.* (1995)
Beta vulgaris (red beet)	Betanin	Taya *et al.* (1994b)
Catharanthus roseus	Indole alkaloids	Toivonen *et al.* (1991)
Hyoscyamus muticus	Solavetivone	Pannuri *et al.* (1993)
Suspension cells		
Catharanthus roseus	Alkaloids and phenolics	Knobloch *et al.* (1982)
Coffea arabica	Caffeine	Bramble *et al.* (1991)
Nicotiana tabacum	Cinnanoyl putrescine	Knobloch *et al.* (1981)
Vitis sp.	Anthocyanins	Yamakawa *et al.* (1983)

Details of the culture equipment and operation are available in the literature (Kino-oka *et al.*, 1995).

Figure 5.4 shows the time course of red beet hairy root culture associated with pigment release using the column-type reactor. Oxygen-enriched medium was introduced into the reactor at a flow rate of 16.2 L h^{-1} during the early period of culture to ensure active growth of the roots. When the root mass reached about 9 kg m^{-3} at 260 h, the flow rate of medium was changed to 7.8 L h^{-1}. In response to

this change, the hairy roots began to release pigment into the medium and the DO concentration at the outlet of the reactor became nearly zero. Further release of pigment occurred during the subsequent culture period of 260–672 h.

Response to Light Irradiation

Hairy roots are characterised as an underground plant organ and are thus expected to produce metabolites found in the roots of field-grown plants. It

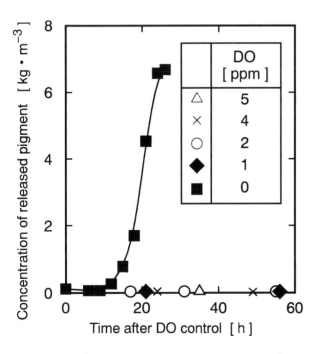

Figure 5.3. Change of pigment concentration in the medium during culture of red beet hairy roots with DO control at various levels.

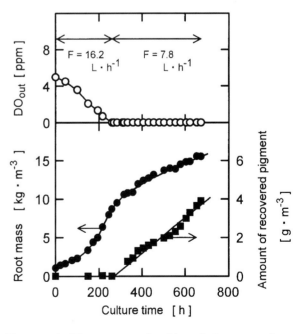

Figure 5.4. Time course of red beet hairy root culture in a column-type reactor associated with pigment release. F = medium flow rate in the reactor, DO$_{out}$ = DO concentration at the outlet of the reactor.

Table 5.2. Green hairy roots derived from various plants.

Hairy roots	Metabolite detected	Reference
Acmella oppositifolia	Polyacetylene	Flores *et al.* (1993)
Bidens sulphureus	Polyacetylene	Flores *et al.* (1987)
Datura stramonium	Alkaloids	Flores and Curtis (1992)
Digitalis lanata	Cardenolides	Yoshimatsu *et al.* (1990)
Hyoscyamus albus	Alkaloids	Aoki *et al.* (1992)
Ipomoea aquatica (pak-bung)	SOD and POD	Taya *et al.* (1994a)
Lippia dulcis	Terpenes	Sauerwein *et al.* (1991)

SOD: superoxide dismutase; POD: peroxidase.

was reported recently, however, that some hairy roots acquire certain functions of above-ground plant organs in response to light associated with greening in colour. Table 5.2 lists hairy root cultures derived from various plants which have exhibited greening after exposure to light.

In our research, it was found that hairy roots of pak-bung (*Ipomoea aquatica*) turned green when cultured under continuous light, and maintained their branched root morphology with the formation of chlorophyll in cells (Taya *et al.*, 1994a). Figure 5.5 shows the ultrastructure of the green hairy root cells. A plastid with a chloroplast-like structure was observed, being associated with thylakoid membranes and grana stacks.

Figure 5.6 shows the time course of green pak-bung hairy root culture with and without light irradiation. The green hairy roots cultivated under light exhibited increased root growth and activities of superoxide dismutase (SOD) and peroxidase (POD). These enzymes are known to have important roles in scavenging toxic oxidants such as superoxide and peroxide in cells. Induction of these enzymes is associated with photochemical reactions in chloroplasts because chloroplasts are involved in the formation of toxic oxidants. The enhanced activities of SOD and POD in green pak-bung hairy roots under light may be closely linked to the formation of chlorophyll and the development of a thylakoid membrane system.

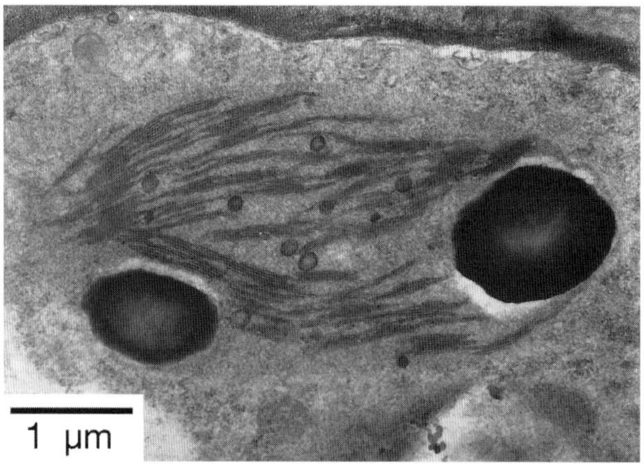

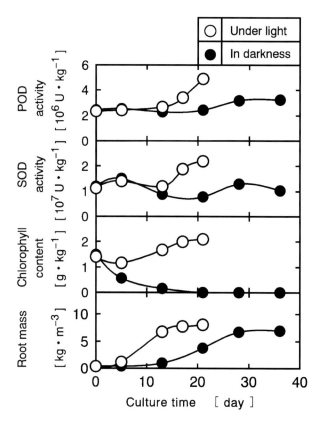

Figure 5.5. Electron micrograph showing the cellular ultrastructure of green pak-bung hairy roots cultivated under light.

Figure 5.6. Time course of green pak-bung hairy root culture with and without exposure to light.

Hormone-Induced Morphology Change in Hairy Roots

Response to Auxin and Cytokinin

Auxin and cytokinin are known to be involved in regulating morphogenesis and organogenesis in many plant tissue cultures (Repunte *et al.*, 1993). To achieve a reversible change in morphology between hairy roots and cell aggregates, we examined the effects of these hormones on the growth and morphological properties of horseradish (*Armoracia rusticana*) hairy roots.

The combined addition of auxin (1-naphthaleneacetic acid, NAA: 1.0 or 2.0 g m^{-3}) and cytokinin (6-benzylaminopurine, BAP: 0.5 g m^{-3}) to the medium was effective for cell aggregate formation from horseradish hairy roots. Figure 5.7A shows roots cultivated in medium with NAA and BAP; Figure 5.7B shows roots cultivated without added hormones. The morphology of the hairy roots changed significantly, forming compact, non-friable cell aggregates in the presence of both hormones.

Cell aggregates separated from the hairy roots were cultivated to examine the regeneration of adventitious roots. Figure 5.8A shows cells cultivated in medium containing 2.0 g m^{-3} NAA; Figure 5.8B shows cells cultured without added hormones. It can be seen that more adventitious roots appeared from cell aggregates in the presence of NAA. This

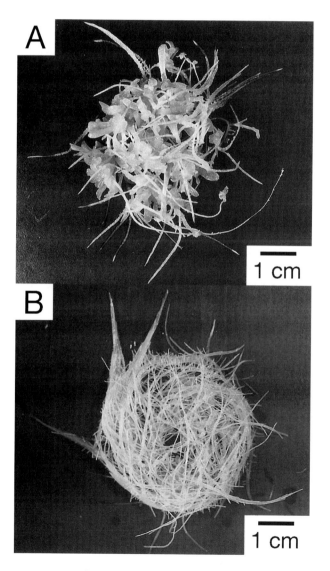

Figure 5.7. Photographs showing morphology change of horseradish hairy roots. A. Culture in medium containing 1.0 g m^{-3} NAA and 0.5 g m^{-3} BAP. B. Culture without added hormones.

Figure 5.8. Photographs showing the development of adventitious roots from horseradish cell aggregates. A. Culture with 2.0 g m^{-3} NAA. B. Culture without added hormones.

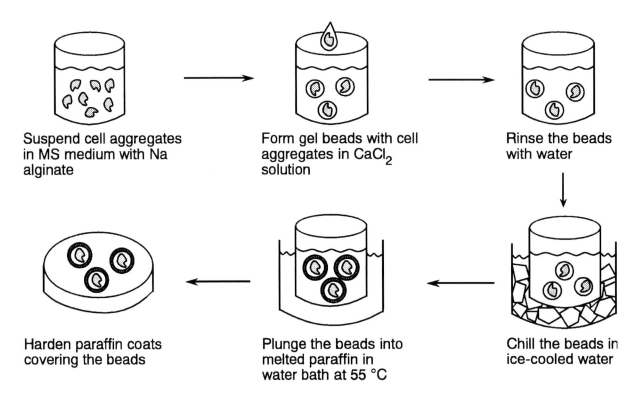

Figure 5.9. Procedure to encapsulate horseradish cell aggregates in alginate gel covered with a paraffin coat.

hormone also enhanced hairy root growth while stimulating the proliferation of lateral roots.

Based on these findings, kinetic models were presented to elucidate the behaviour of reversible morphology change in hairy roots, and a fair approximation of the respective growths of cell aggregates and adventitious roots was verified by considering the synergistic effects of BAP and NAA (Repunte et al., 1994 and 1995b).

Preparation of Artificial Seeds Using Cell Aggregates

Cell aggregates derived from horseradish hairy roots could be used as embryo-like inclusions for production of artificial seeds by entrapping them in alginate gel capsules covered with a thin paraffin coat (Repunte et al., 1995a). The procedure to encapsulate the cell aggregates is illustrated in Figure 5.9. Samples of aggregates were suspended in MS medium containing sodium alginate. Aggregates surrounded by alginate were dropped one at a time into $CaCl_2$ solution. The resultant capsules were rinsed with water, dipped into ice-cooled water, and plunged into melted paraffin in a water bath at 55°C. The capsules were removed and the surrounding paraffin coats solidified at room temperature.

For the practical utilisation of artificial seeds, it is desired that encapsulated embryos do not germi-

nate during storage or transportation. To test this, the cell aggregates embedded in alginate gel with and without paraffin coats were stored at 25°C for 60 d. Root emergence was not observed from any of the capsules stored with the paraffin coats. In contrast, roots were found to emerge from capsules without paraffin coats. Coating the alginate surface with paraffin appears to be a successful means for suppressing root regeneration from cell aggregates during storage.

As shown in Figure 5.10, the frequency of root regeneration from cell aggregates stored in paraffin-coated capsules for up to 60 d was similar to that from aggregates not subjected to a storage period. This indicates that serious loss of root regeneration potential does not occur during storage of the aggregates. Subsequently, formation of plantlets was investigated by cultivating the aggregates stored in capsules with paraffin coats at 25°C for 60 d. Two or three shoots appeared on the regenerated roots after they were transferred to light conditions; these shoots then developed into plantlets (Figure 5.11).

As pointed out by Fujii et al. (1987), a hydrophobic thin coat over a gel surface plays an important role in preventing the rupture and rapid desiccation of artificial seeds. It is also expected that the coat acts as a shield against microbial and chemical contamination. In our experiments, it was found that

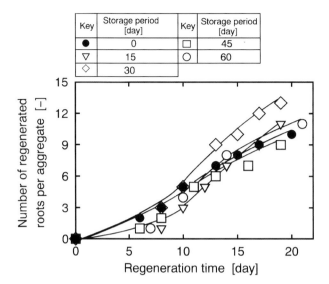

Figure 5.10. Time course of root regeneration from horseradish cell aggregates after storage in capsules with paraffin coats.

capsules with paraffin coats which became infected by microorganisms could be sterilised in sodium hypochlorite solution without causing damage to the aggregates.

Prospects

From an engineering point of view, plant cell culture processes can be classified into two categories of "mass production" and "mass propagation". The former process is concerned with the large-scale culture of plant cells for the efficient production of industrially valuable phytochemicals. We are now

2 cm

Figure 5.11. Photograph showing horseradish plantlets developing from cell aggregates after storage in capsules with paraffin coats for 60 d.

investigating the construction of a novel bioreactor suitable for hairy roots, with optimisation of culture conditions based on the findings presented in this chapter of enhanced synthesis and release of pigment. The latter process focuses on the clonal propagation of elite plants and preparation of artificial seeds for the preservation and delivery of plants. We have demonstrated that horseradish hairy roots can undergo a reversible morphology change caused by hormone treatment, and have described procedures for preparation of artificial seeds using the derived cell aggregates. In principle, this morphology change can be induced in hairy roots of other plants and thus further studies in this respect are in progress.

References

Aoki T, Y Toda and K Shimomura (1992). Characteristics of green hairy roots of *Hyoscyamus albus* transformed with *Agrobacterium rhizogenes* strain A4. *Plant Tiss Cult Lett* 9: 214–219.

Bieleski RL (1973). Phosphate pools, phosphate transport, and phosphate availability. *Ann Rev Plant Physiol* 24: 225–252.

Bramble JL, DJ Graves and P Brodelius (1991). Calcium and phosphate effects on growth and alkaloid production in *Coffea arabica*: experimental results and mathematical model. *Biotechnol Bioeng* 37: 859–868.

Endress R (1994). *Plant Cell Biotechnology*. Springer-Verlag, Berlin.

Fischer RS, CA Bonner, ME Theodorou, WC Plaxton, G Hrazdine and RA Jensen (1993). Response of aromatic pathway enzyme of plant suspension cells to phosphate limitation. *Bioorg Med Chem Lett* 3: 1415–1420.

Flores HE and WR Curtis (1992). Approaches to understanding and manipulating the biosynthetic potential of plant roots. *Ann NY Acad Sci* 665: 188–209.

Flores HE, MW Hoy and JJ Pickard (1987). Secondary metabolites from root cultures. *Trends in Bitechnol* 5: 64–69.

Flores HE, Y Doi, JL Cuello, IE Maldonado-Mendoza and VM Loyola-Vargas (1993). Green roots: Photosynthesis and photoautotrophy in an underground plant organ. *Plant Physiol* 101: 363–371.

Fujii JAA, DT Slade, K Redenbaugh and KA Walker (1978). Artificial seeds for plant propagation. *Bio/Technol* 5: 335–339.

Fujita Y, Y Hara, T Ogino and C Suga (1981). Production of shikonin derivatives by cell suspension cultures of *Lithospermum erythrorhizon*. I. Effect of nitrogen sources on the production of shikonin derivatives. *Plant Cell Rep* 1: 59–60.

Kino-oka M, M Taya and S Tone (1991). Production of superoxide dismutase from plant hairy roots by considering the effect of nitrogen source in their cultures. *Kagaku Kogaku Ronbunshu* 17: 1012–1018.

Kino-oka M, Y Hongo, M Taya and S Tone (1992). Culture of red beet hairy root in bioreactor and recovery of pigment released from the cells by repeated treatment of oxygen starvation. *J Chem Eng Japan* 25: 490–495.

Kino-oka M, K Mine, M Taya, S Tone and T Ichi (1994). Production and release of anthraquinone pigments by hairy roots of madder (*Rubia tinctorum* L.). under improved culture conditions. *J Ferment Bioeng* 77: 103–106.

Kino-oka M, M Taya and S Tone (1995). Culture of red beet hairy roots in a column-type reactor associated with pigment release. *Plant Tiss Cult Lett* 12: 201–204.

Knobloch K-H, G Beutnagel and J Berlin (1981). Influence of accumulated phosphate on culture growth and formation of cinnamoyl putrescines in medium-induced cell suspension cultures of *Nicotiana tabacum*. *Planta* 153: 582–585.

Knobloch K-H, G Bast and J Berlin (1982). Medium- and light-induced formation of serpentine and anthocyanins in cell suspension cultures of *Catharanthus roseus*. *Phytochemistry* 21: 591–594.

Murashige T and F Skoog (1962). A revised medium for rapid growth and bioassays with tobacco tissue cultures. *Physiol Plant* 15: 473–497.

Pannuri S, GR Reddy, D McNeill and WR Curtis (1993). Interpreting the role of phosphorus and growth rate in enhanced fungal induction of sesquiterpenes from *Hyoscyamus muticus* root cultures. *Appl Microbiol Biotechnol* 38: 550–555.

Repunte VP, M Kino-oka, M Taya and S Tone (1993). Reversible morphology change of horseradish hairy roots cultivated in phytohormone-containing media. *J Ferment Bioeng* 75: 271–275.

Repunte VP, S Shimamura, M Taya and S Tone (1994). Characteristics and growth kinetics of adventitious roots emerging out of cell aggregates derived from horseradish hairy roots. *J Chem Eng Japan* 27: 523–528.

Repunte VP, M Taya and S Tone (1995a). Preparation of artificial seeds using cell aggregates from horseradish hairy roots encapsulated in alginate gel with paraffin coat. *J Ferment Bioeng* 79: 83–86.

Repunte VP, S Shimamura, M Taya and S Tone (1995b). Kinetic analysis of the process of hormone-induced cell aggregate formation from horseradish hairy roots. *J Chem Eng Japan* 28: 847–850.

Sauerwein M, T Yamazaki and K Shimomura (1991). Hernandulcin in hairy root cultures of *Lippia dulcis*. *Plant Cell Rep* 9: 579–581.

Taya M, K Mine, M Kino-oka, S Tone and T Ichi (1992). Production and release of pigments by culture of transformed hairy root of red beet. *J Ferment Bioeng* 73: 31–36.

Taya M, H Sato, M Kino-oka and S Tone (1994a). Characterization of pak-bung green hairy roots cultivated under light irradiation. *J Ferment Bioeng* 78: 42–48.

Taya M, K Yakura, M Kino-oka and S Tone (1994b). Influence of medium constituents on enhancement of pigment production by batch culture of red beet hairy roots. *J Ferment Bioeng* 77: 215–217.

Toivonen L, M Ojala and V Kauppinen (1991). Studies on the optimization of growth and indole alkaloid production by hairy root cultures of *Catharanthus roseus*. *Biotechnol Bioeng* 37: 673–680.

Uozumi N, S Makino and T Kobayashi (1995). 20-Hydroxyecdysone production in *Ajuga* hairy root controlling intracellular phosphate content based on kinetic model. *J Ferment Bioeng* 80: 362–368.

White FF and VP Sinkar (1987). Molecular analysis of root induction by *Agrobacterium rhizogenes*. In *Plant DNA Infectious Agents*, edited by Th Horn and J Schell, pp. 149–177. Berlin: Springer-Verlag.

Yamakawa T, S Kato, K Ishida, T Kodama and Y Minoda (1983). Production of anthocyanins by *Vitis* cells in suspension culture. *Agric Biol Chem* 47: 2185–2191.

Yoshimatsu K, M Satake and K Shimomura (1990). Determination of cardenoides in hairy root cultures of *Digitalis lanata* by enzyme-linked immunosorbent assay. *J Nat Prod* 53: 1498–1502.

Production of Ginsenosides in Ginseng Hairy Root Cultures

6

Takafumi Yoshikawa

School of Pharmaceutical Sciences, Kitasato University, Minato-ku, Tokyo 108, Japan;
email: yoshikawat@platinum.pharm.kitasato-u.ac.jp

Introduction

Ginseng (*Panax ginseng* C. A. Meyer) is a herbaceous plant belonging to the Araliaceae family. Ginseng root is a favourite remedy of Chinese and Japanese medicine that has long been used as a tonic. Chemical and pharmacological studies of ginseng have confirmed that there are 10 or more ginseng saponins, including the ginsenosides Ra, Rb_1, Rb_2, Rb_3, Rc and Rd which contain protopanaxadiol as the sapogenin, and Re, Rf, Rg_1, Rg_2, Rh_1 and Rh_2 which contain protopanaxatriol. Of those ginsenosides, Rb_1 and Rg_1 are the most active. Other components are also considered to be active ingredients in the prevention and/or treatment of several diseases. Thus, ginseng and/or its extracts continue to be consumed as a health food or drink worldwide. However, ginseng root is very expensive due to a difficult and time consuming (5–7 years) growth process. Therefore, as early as 1967, studies have examined the production of saponins, the crude and/or active ingredients, through cell and tissue culture methods. It has already been shown that ginseng callus produces saponins which are chemically and pharmacologically similar to those of the original plant (Furuya *et al.*, 1983a, 1983b). Additionally, various culture methods for effective saponin production have been compared. A newly isolated strain, which was induced in 1978 and coded as Pg-3, was examined to determine growth and saponin production in large-scale culture such as a jar fermenter or tank culture (Furuya *et al.*, 1983c, 1984). Recently, the Pg-3 strain was cultured in a 20-ton tank (Nitto Denko Co.), and has been used in commercial products (Ushiyama *et al.*, 1986).

An attempt was made to increase the yield of ginsenosides by inducing a hairy root transformed with an Ri plasmid of *Agrobacterium rhizogenes* to establish a ginseng hairy root culture. In liquid medium, the hairy roots developed numerous lateral branches and grew rapidly, while producing large quantities of saponins resembling those of native roots (Yoshikawa and Furuya, 1987).

Abbreviations

D: 2,4-dichlorophenoxyacetic acid; IBA: indole-3-butyric acid; IAA: indole-3-acetic acid; NAA: 1-naphthaleneacetic acid; K: kinetin; P: N-phenyl-N'-(4-pyridyl)urea.

Production of Ginsenosides by Various Callus Cultures

Pg-1 Callus

In 1967, a Pg-1 callus was induced on Murashige and Skoog (MS) agar medium containing 0.1 mg L^{-1} D from the petiole of 2-year-old ginseng. The callus was subcultured at 4-week intervals using the same medium at 25°C in the dark. Pg-1 calli maintained the ability to produce saponins and differentiate into organs. The Pg-1 callus was transferred to MS medium containing 1 mg L^{-1} K without D (Pg-1 K1 callus) under illumination (2500–4000 lux, 16 h d^{-1}, by a warm fluorescent light in a phytotron cabinet). The Pg-1 K1 callus gradually generated shoots and small roots. Roots were selected and subcultured in the dark on solid MS medium containing 1 mg L^{-1} IBA instead of K1, where they continued to form roots only in the dark (Pg-1 IBA1 roots).

The IBA1 roots had the greatest saponin production. However, it was thought that large lumps which formed during the suspension culture process might pose a problem for mass cultivation, particularly for transferring the cells to a suspension or jar fermenter for large-scale culture. Consequently, a new callus, the Pg-3 callus, was induced and established as an effective strain for large-scale culture under the optimal conditions established in the Pg-1 callus studies.

Pg-3 Callus

In 1978, slices of 5-year-old Korean ginseng root were placed on MS agar medium supplemented with 1 mg L^{-1} D and 0.1 mg L^{-1} K (called the DK medium). The developing Pg-3 callus was transferred to the same medium, maintained at 25°C in the dark, and subcultured at intervals of 4 weeks (Pg-3 DK callus). After the third subculture, the callus was transferred to MS agar medium containing 2 mg L^{-1} IBA and 0.1 mg L^{-1} K (B2K medium) to exclude 2,4-D which may be toxic to humans due to its use as an agricultural chemical. The callus was kept at 25°C in the dark and subcultured at intervals of 4 weeks for about a year. Eventually, substantial growth was achieved in small 3–5 mm aggregates (Pg-3 B2K callus).

The newly isolated Pg-3 B2K cell aggregates, with a lower degree of differentiation than Pg-1 roots on the same B2K medium, were examined in large-scale liquid culture to determine optimal culture conditions for growth and saponin production.

Determination of Saponins

The crude saponins obtained from each callus or tissue were spotted together with standard samples of ginsenosides Rb_1 and Rg_1 on a Merck silica gel TLC plate 60 F254 and developed with an upper layer of n–BuOH–AcOEt–H_2O (5:1:4). The saponin spots were detectable when sprayed with 10% H_2SO_4 followed by heating at 105°C for 10 min. Saponin spots were compared with the standards by densitometry using a Shimadzu CS-910 dual-wavelength TLC scanner. Rb group content was calculated as the total of ginsenosides Ra, Rb_1, Rb_2, Rc and Rd having protopanaxadiol as the sapogenin; Rg group content was calculated as the total of ginsenosides Re, Rf, Rg_1, Rg_2 and Rh_1 having protopanaxatriol as the sapogenin. TLC densitometric analysis of total saponins in each culture revealed the presence of pure ginsenosides. The average of two measurements on each of two or three different cultures was calculated. The IBA series cultures had the highest saponin content.

Ginsenoside content was also determined by HPLC under the following conditions: Senshu Pak NP 10×300 mm column; mobile phases of 22% CH_3CN for ginsenosides with protopanaxatriol and 33% CH_3CN for ginsenosides with protopanaxadiol; flow rates of 5 mL min^{-1} for the former and 4 mL min^{-1} for the latter; and detection by photodiode array (UV, 202 nm). Development of the HPLC method has allowed more accurate determination of saponin content. However, it is extremely time intensive since ginsenosides must be treated separately based on their sapogenin content (either protopanaxadiol or protopanaxatriol) and the assay time for each sample is about 1 h. Consequently, repetition on a large number of samples would take too long.

Saponin Production Using Various Culture Strains and Conditions

Values shown in Table 6.1 are the average of all data obtained under the previously described conditions. The effects of illumination (2500–4000 lux) on saponin production and growth were evaluated only through Rb / Rg group ratios which are thought to provide a qualitative evaluation. Differences in the ratio were primarily related to decreases in the amount of Rb group saponins under illumination.

For liquid cultures, IBA1 roots were used and cultured on either a gyratory or a reciprocal shaker. Liquid cultures on a gyratory shaker experienced greater growth than those on a reciprocal shaker. However, the saponin content was lower in cultures on a gyratory shaker than on a reciprocal shaker. The gyratory cultures produced soft, brownish cell aggregates which generated many roots and showed a 1.8 times greater growth rate and an increased production index compared with reciprocal cultures. The cell aggregates cultured on a reciprocal shaker were more rigid and showed lower differentiation

Table 6.1. Growth and saponin content in different ginseng culture strains under various conditions.

Culture condition	Growth ratio*	Dry weight (g per 100 g fresh weight)	Saponin content** (mg per 100 g fresh weight)			Rb/Rg	Total saponin content (% dry weight)	Production index***
			Rb	Rg	Total			
Static dark								
DK callus	4.3	2.48	2.4	8.4	10.8	0.29	0.44	46
IBA1 root	3.4	3.09	27.6	25.1	52.7	1.10	1.71	179
K1 shoot	3.1	2.59	31.1	12.1	43.2	2.57	1.67	133
Static light								
K1 shoot	3.6	2.70	25.8	14.8	40.6	1.74	1.50	140
Liquid								
IBA1 root								
reciprocal	1.9	3.76	40.5	29.7	70.2	1.36	1.87	134
gyratory	3.4	3.28	40.4	26.5	66.9	1.52	2.04	226
IBA2 K0.1	4.7	4.95	32.8	44.2	77.0	0.74	1.56	359
IBA5 P0.1	8.2	3.44	25.3	17.0	42.3	1.49	1.23	346
Plant								
aerial (stem, leaf)		9.53	21.7	62.1	83.8	0.35	0.88	
root		23.91	59.0	37.3	96.3	1.58	0.40	

* The growth ratio was determined by the increase in fresh weight after 4 weeks of culture.
** The Rb group content was calculated from the total amount of ginsenosides with protopanaxadiol as the sapogenin; the Rg group content was calculated from the total amount of ginsenosides with protopanaxatriol as the sapogenin. The values are the averages of each tissue in 2 flasks cultured in media supplemented with every plant growth regulator.
*** The production index is the product of the growth ratio and total saponin content.

of root. The production index, i.e. the product of the growth ratio and total saponin content, is thought to indicate the productivity of a culture. Gyratory cultures produced 1.26 times the amount of saponins of static cultures (IBA1 roots). Gyratory cultures offered the best conditions for IBA1 roots, producing qualitatively almost the same saponin as the roots of the original plant. Total saponin content measured as a percentage of dry mass was five times higher in gyratory suspension cultures than in plant roots.

The independent or combined effects of auxins and cytokinins on saponin content and growth in static cultures using IBA1 roots were examined. D, NAA, IAA, and IBA were used as auxins, and K and P were used as cytokinins. An increase in the growth ratio and in the amount of saponins was observed when a combination of IBA and K was used. Various suspension culture conditions were used to establish the best medium for saponin con-

tent and growth of the IBA1 roots. The combination of IBA and K in liquid culture showed a lower growth ratio and produced much higher saponin than the combination of IBA and P. Specifically, the combination of 2 mg L^{-1} IBA and 0.1 mg L^{-1} K resulted in the best production index, while the combination of 5 mg L^{-1} IBA and 0.1 mg L^{-1} P produced the best growth ratio of 8.19. The production index was 359 using 2 mg L^{-1} IBA and 0.1 mg L^{-1} K, and 346 using 5 mg L^{-1} IBA and 0.1 mg L^{-1} P.

Saponin content in the 5-year-old Korean ginseng root is shown in the last row of Table 6.1 to allow comparison with the cultures. Under the optimal condition of a combination of 2 mg L^{-1} IBA and 0.1 mg L^{-1} K, the saponin content detected in liquid culture, particularly in the protopanaxatriol group saponins, was similar to that in the root. However, the content of Rb$_1$ in protopanaxadiol group saponins was higher in culture than in the original root, while Rc and Rd were lower.

Figure 6.1. (a) *Panax ginseng* hairy roots induced from Pg-4 callus after infection with *Agrobacterium rhizogenes*. (b) An enlargement of (a). (c)–(e) Hairy roots cultured in hormone-free liquid MS medium. (f) Non-transformed Pg-1 IBA1 roots cultured on MS agar medium supplemented with 2 mg L^{-1} IBA and 0.1 mg L^{-1} K (B2K medium). (g)–(h) Non-transformed Pg-1 IBA1 roots cultured in B2K liquid medium (same medium as (f) without agar). (Reproduced from T Yoshikawa and T Furuya, 1987, Saponin production by cultures of *Panax ginseng* transformed with *Agrobacterium rhizogenes*. *Plant Cell Reports* 6: 449–453, Figure 6.1; with permission from Springer-Verlag, Berlin.)

Induction of a Ginseng Hairy Root
(Yoshikawa and Furuya, 1987)

A ginseng hairy root culture was established from Pg-4 callus, which was induced on DK medium from a 2-year-old root cultivated in Japan in 1980, by infection with *Agrobacterium rhizogenes*. The induction of the hairy roots was carried out using the following procedure. After the Pg-4 calli (small rigid cell aggregates) were cultured in liquid B2K medium for 3 weeks, the aggregates were collected by filtration using a 20 mesh nylon cloth. The aggregates were incubated in liquid MS medium containing 2% cellulase, 0.5% macerozyme and 0.3 M mannitol at 30°C for 2 h. Protoplasts of the Pg-4 cells formed only on the surface parts of the cell aggregates. The aggregates were inoculated into an *A. rhizogenes* bacterial suspension that was cultured in YEB medium (Vervliet *et al.*, 1975) for 3 d. Fifteen hours after infection, the aggregates were collected using a nylon cloth and thoroughly washed with sterilised water. To eliminate the bacteria, the aggregates were incubated in liquid MS medium supplemented with vancomycin, carbenicillin and tetracycline at 25°C for 4 d. After washing with sterilised water, the aggregates were placed on MS agar medium without phytohormones and cultured at 25°C in the dark. Approximately 4 weeks after infection, roots began to protrude from the surface of the calli (Figures 6.1a and 6.1b). These protrusions further developed into fine roots 1 to 2 cm long, and were removed from the callus. The roots were then cul-

tured in hormone-free liquid medium on a gyratory shaker. The roots grew rapidly, developing numerous lateral branches in the liquid medium (Figures 6.1c–6.1e). Analysis for asepsis was performed by homogenising the cultured hairy roots in sterilised distilled water, followed by incubation on YEB agar medium for a week at 25°C. Hairy roots that did not form bacterial colonies on the agar plates were designated as axenic hairy root cultures.

Characteristics of Hairy Root Cultures

The morphological properties of transformed hairy roots were compared with those of ordinary cultured roots induced under hormonal control from Pg-1 DK calli (IBA1 roots). Pg-1 roots were cultured in liquid MS medium supplemented with 2 mg L^{-1} IBA and 0.1 mg L^{-1} K, called B2K medium. They grew rapidly without branching, and sometimes formed callus-like aggregates and/or shoots and leaves. Conversely, hairy root cultures grew with extensive lateral branching and never differentiated into tissues other than a fine root. Subsequently, in long-term culture over 4 weeks, the hairy root cultures developed colourless aerial roots in liquid medium, whereas ordinary cultured roots formed a light yellow, ball-like aggregate without aerial roots. Hairy roots could grow in hormone-free medium (growth ratio: 3.07), whereas ordinary cultured roots could not (growth ratio: 1.26; Table 6.2).

Table 6.2. Growth and saponin contents in callus, ordinary cultured roots, hairy roots, and native root of *Panax ginseng*. (Adapted from Yoshikawa and Furuya, 1987.)

Tissue and medium	Growth ratio*	Dry weight (g per 100 g fresh weight)	Saponin content** (mg per 100 g fresh weight)			Rb/Rg	Total saponin content (% dry weight)	Production index***
			Rb	Rg	Total			
Callus								
IBA2 K0.1	2.9	5.97	28.3	10.6	38.9	2.67	0.65	111
Ordinary roots								
hormone-free	1.3	5.31	15.3	4.9	20.2	3.15	0.38	25
IBA2 K0.1	4.0	5.57	25.7	25.0	50.7	1.03	0.91	200
Hairy roots								
hormone-free	3.1	10.09	24.8	11.1	35.9	2.24	0.36	110
K0.1	2.3	10.29	21.7	14.6	36.3	1.48	0.35	81
IBA2	5.1	9.62	39.6	26.6	66.2	1.49	0.69	338
IBA0.5 K0.1	4.3	10.45	56.3	40.0	96.3	1.41	0.92	414
IBA2 K0.1	6.2	10.58	55.7	44.5	100.2	1.25	0.95	621
Native root		23.91	59.0	37.3	96.3	1.58	0.40	

Each tissue was cultured in liquid MS medium containing the hormone(s) shown on a gyratory shaker at 25°C. Each value is the average from the five different cultures. See Table 6.1 for footnotes *–***.

Saponin Production in Hairy Root Cultures

Production of ginseng saponins in hairy root cultures was examined. Saponin contents in both the hairy root cultures and ordinary cultured roots are shown in Table 6.2. The ratio of Rb group content to Rg group content was used as an indicator for qualitative evaluation of the hairy roots. The saponin content of hairy roots cultured in hormone-free medium was similar to that of the original callus when compared on a fresh weight basis and by the composition of the saponin constituents. However, in medium supplemented with IBA, the hairy root cultures produced a higher saponin content (66.2 mg per 100 g fresh weight) than the original callus (38.9 mg) and ordinary cultured roots (50.7 mg). Furthermore, growth of the hairy roots was stimulated by a factor of up to 2.0 in terms of the growth ratio when cultured in medium supplemented with IBA. Finally, the production index of hairy roots cultured in B2K medium was 621.2, the highest of all culture conditions studied. The saponin content in the hairy root culture (0.95% dry weight) was 2.4 times higher than in the native root (0.40%), and the qualitative evaluation was almost the same as the native root. When extracts of ginseng hairy root are taken by mouth as a health food or medicine, the bacterial plasmid is also taken at the same time. Therefore, a toxicological test was carried out. As a result, toxicological safety in the hairy root cultures was confirmed through two mutagenicity tests, the Ames test and a chromosome aberration test.

Application of a Bioreactor

When the hairy roots were cultured in liquid medium supplemented with IBA and kinetin (B2K medium), they produced a larger amount of saponin than the original Pg-4 cell aggregates or regenerated ordinary roots. Both the ordinary roots and the aggregates are capable of glycosylating various compounds such as phenylcarboxylic acid to form glycosides (Furuya et al., 1989). Therefore, it was expected that the hairy roots would also have the same high potential to glycosylate (Ushiyama et al., 1989). To examine this, glycosylation of various organic compounds demonstrated to have high glycosylation activity, such as phenylcarboxylic acid, digitoxigenin (Kawaguchi et al., 1990) and 18β-glycyrrhetinic acid (Asada et al., 1993), was investigated using ginseng hairy root cultures.

Hairy root cultures converted digitoxigenin to five glucosides, and 18β-glycyrrhetinic acid to four glucosides and two malonyl derivatives. The ability to convert (RS)-2-phenylpropionic acid (PPA) was compared among three different cultures: cell aggregates, regenerated ordinary roots, and hairy roots. The cell aggregates showed a lower glycosylation

ability than either of the root cultures, producing only a glucosyl-ester. Both the ordinary root and hairy root cultures showed a greater glycosylation potential, nearly a 100% conversion ratio, with about half of the glucosyl-ester excreted into the medium.

Hairy roots were especially adaptable to bioreactor culture because they developed an extensive network of lateral branches and never formed callus-like tissues, which were frequently generated from the ordinary roots and caused blockages in the outlet filter of the reactor. Consequently, to form glycosides from PPA in a bioreactor, hairy roots were used. A revised reversed flask-type bioreactor was used. Using this method, it was possible to conduct a continuous reaction over about 2 months by exchanging the reaction mixture every 3 d. The hairy roots used in the bioreactor continued to convert for about 2 months with a maximum conversion ratio of 45% and an average ratio of 15% (Yoshikawa et al., 1993).

Conclusions

Through the use of selected ginseng cultured tissues, specifically hairy root cultures, and by means of revised culture methods for production, several useful compounds were produced, including ginsenosides. Using this system, further investigations into the production of useful compounds should be conducted to improve the quality and efficiency of the system.

References

Asada Y, H Saito, T Yoshikawa, K Sakamoto and T Furuya (1993). Biotransformation of 18β-glycyrrhetinic acid by ginseng hairy root culture. Phytochemistry 34: 1049–1052.

Furuya T, T Yoshikawa, T Ishii and K Kajii (1983a). Effects of auxins on growth and saponin production in callus cultures of Panax ginseng. Planta Med 47: 183–187.

Furuya T, T Yoshikawa, T Ishii and K Kajii (1983b). Regulation of saponin production in callus cultures of Panax ginseng. Planta Med 47: 200–204.

Furuya T, T Yoshikawa, Y Orihara and H Oda (1983c). Saponin production in cell suspension cultures of Panax ginseng. Planta Med 48: 83–87.

Furuya T, T Yoshikawa, Y Orihara and H Oda (1984). Studies of the culture condition for Panax ginseng cells in jar fermentors. J Nat Prod 47: 70–75.

Furuya T, M Ushiyama, Y Asada and T Yoshikawa (1989). Biotransformation of 2-phenylpropionic acid in root culture of Panax ginseng. Phytochemistry 28: 483–487.

Kawaguchi K, M Hirotani, T Yoshikawa and T Furuya (1990). Biotransformation of digitoxigenin by ginseng hairy root cultures. Phytochemistry 29: 837–843.

Ushiyama K, H Oda and Y Miyamoto (1986). Large scale tissue culture of Panax ginseng root. Proc VIth Internat Congr Plant Tissue and Cell Culture, Minnesota, p. 252.

Ushiyama M, Y Asada, T Yoshikawa and T Furuya (1989). Biotransformation of aromatic carboxylic acids by root culture of *Panax ginseng*. *Phytochemistry* 28: 1859–1869.

Vervliet G, M Holsters, H Teuchy, M van Montagu and J Schell (1975). Characterization of different plaque-forming and defective temperate phages in *Agrobacterium* strains. *J Gen Virol* 26: 33–48.

Yoshikawa T and T Furuya (1987). Saponin production by cultures of *Panax ginseng* transformed with *Agrobacterium rhizogenes*. *Plant Cell Rep* 6: 449–453.

Yoshikawa T, Y Asada and T Furuya (1993). Continuous production of glycosides by a bioreactor using ginseng hairy root culture. *Appl Microbiol Biotechnol* 39: 460–464.

Co-Culture of Hairy Roots and Shooty Teratomas

7

M. Ahkam Subroto[1], M.G.P. Mahagamasekera[1], Kian H. Kwok[1], John D. Hamill[2] and Pauline M. Doran[1]

[1] *Department of Biotechnology, University of New South Wales, Sydney NSW 2052, Australia; email: p.doran@unsw.edu.au*
[2] *Department of Genetics and Developmental Biology, Monash University, Clayton, Melbourne VIC 3168, Australia*

Introduction

Hairy roots have been demonstrated over the past 10 years to have great potential for production of plant secondary metabolites. In most cases, the range and concentrations of compounds formed by hairy roots are very similar to those produced by roots of the parent plant. This characteristic gives hairy roots an important intrinsic advantage compared with dedifferentiated tissues by overcoming problems with low yields of secondary metabolites in callus and suspension cultures. However, it also means that products formed in other plant organs cannot usually be obtained from hairy roots: to date, few hairy root cultures have been found capable of producing significant levels of compounds synthesised in the aerial parts of plants. Some success in producing leaf-derived compounds *in vitro* has been achieved using differentiated shoots; for example, scented and flavoured oils (Charlwood and Moustou, 1988; Charlwood *et al.*, 1988), and specific metabolites produced by *Catharanthus roseus*, *Chrysanthemum cinerariaefolium*, *Rauwolfia serpentina*, *Cinchona* spp., *Digitalis* spp. and *Dioscorea composita* (reviewed in Heble, 1985) have been obtained from shoot cultures. Although less information is available about the performance of shoot cultures compared with

roots, it seems likely that shoots are also largely restricted to producing substances wholly synthesised in the leaves of the parent plant (Hamilton *et al.*, 1986; Kitamura, 1988).

For some secondary compounds, production in plants requires participation of the roots *and* the leaves. As illustrated in Figure 7.1, organ-specific enzymes in the roots may produce a metabolic precursor which is translocated to the aerial parts of the plant for conversion to another product by the leaves. If expression and activity of the enzymes involved in this process retain their organ specificity *in vitro*, neither root cultures nor shoot cultures will be successful for *de novo* synthesis of the final product. Root cultures would be capable of producing the precursor but would be unable to convert it; shoot cultures could not supply the precursor even though the biosynthetic machinery to transform it would be available.

One solution to this problem is genetic manipulation of either the root culture or the shoot culture to ensure production of the essential enzymes normally found in the other organ (Hashimoto *et al.*, 1993). However, given the complexity of secondary metabolic pathways and the possible involvement of specific organelles such as chloroplasts in the biosynthesis, this strategy is not realistic at the present

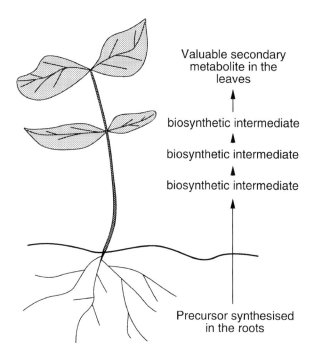

Valuable secondary
metabolite in the
leaves

↑

biosynthetic intermediate

↑

biosynthetic intermediate

↑

biosynthetic intermediate

↑

Precursor synthesised
in the roots

Figure 7.1. Synthesis of some secondary compounds requires the participation of the leaves as well as the roots of the plant. Precursor compounds may be synthesised in the roots and later modified in the leaves (as shown); alternatively, the precursor may be synthesised in the shoots and transported to the roots for biotransformation.

time for many applications. Another approach is co-culture of roots and shoots, in which precursor produced in one organ is released into the shared nutrient medium and taken up by the other organ for bioconversion. Root–shoot co-culture in Petri dishes has previously been used by Saito *et al.* (1989) for biotransformation studies of pyridine alkaloids in *Nicotiana tabacum*; however the potential of this system for large-scale production of plant compounds has not been examined. Because untransformed roots and shoots *in vitro* normally require significantly different levels of exogenous hormones to maintain their respective differentiated states, co-culture in the same medium is not feasible unless the organs are autotrophic in plant growth regulators such as auxin and cytokinin. Fortunately, a characteristic of genetically transformed plant organs induced by agrobacterial infection is hormone autotrophy. Thus, root–shoot co-culture becomes possible using hairy roots and their genetically transformed shoot counterparts, 'shooty teratomas'.

Shooty Teratomas

Genetically transformed shooty teratomas are produced by infection of plant material with *Agrobac-*

terium tumefaciens. Four approaches have been successful in inducing shooty teratomas (Saito *et al.*, 1989; Spencer *et al.*, 1993a): (1) using wild-type nopaline strains of *A. tumefaciens* such as T37 and C58; (2) using strains with transposon insertion mutations in either of the auxin genes in the T-DNA of octopine strains; (3) using disarmed strains carrying a binary vector with the wild-type cytokinin gene (*ipt*) coding for isopentenyltransferase isolated from *A. tumefaciens* T-DNA; and (4) using disarmed strains carrying a binary vector with a construct designed to overexpress the *ipt* coding sequence under the control of the CaMV35S promoter. The molecular basis of shooty teratoma formation has not yet been positively determined; however shoot development which sometimes occurs after infection of tissues with nopaline strains is probably caused by an imbalance in the cytokinin:auxin ratio due to increased transcription of gene 5 in tumours induced by such strains. Gene 5 encodes for indole-3-lactate synthase and the resulting indole-3-lactate (ILA) may compete with auxin for auxin binding proteins, thus leading to an apparent increase in the cytokinin:auxin ratio *in vivo* (Körber *et al.*, 1991; reviewed in Hamill, 1993). The range of plant species for which teratomas have been produced and cultured *in vitro* is much smaller than the corresponding range for hairy roots (Table 7.1; compare with Table 1.1 in Hamill and Lidgett, Chapter 1). To date, members of the Solanaceae family have been the most amenable to shooty teratoma development.

Shooty teratomas can be grown in the absence of exogenous hormones either on solid or in liquid medium. For reactor-scale applications, the shoots must remain fully differentiated and callus-free, and must be capable of growing in liquid culture without browning or hyperhydricity (vitrification). Like hairy roots, shooty teratomas display a wide range of morphologies as described by Subroto *et al.* (1996a). For example, transformation of the same species may produce shoots with compact clusters of stems and leaves (Figure 7.2a), relatively long stems and narrow leaves (Figure 7.2b), or short thick stems with broad leaves (Figure 7.2c); all the shoots shown in Figure 7.2 were cultured under the same conditions. Shoot morphology affects properties such as shear sensitivity and the ease of nutrient and oxygen transfer within the teratoma, and may also therefore influence reactor design.

Biotransformation of Tropane Alkaloids

One natural product system possessing the features represented in Figure 7.1 is the conversion *in planta* of hyoscyamine to scopolamine. Hyoscyamine (the levorotatory component of racemic atropine) and scopolamine are tropane alkaloids which affect the parasympathetic nervous system and exhibit a wide

Table 7.1. Plant species reported to develop shooty teratomas after infection with various strains of *Agrobacterium tumefaciens*.

Plant species	*A. tumefaciens* strain	*A. tumefaciens* chromosomal background	Key genetic constitution	Opine	Reference
Nicotiana tabacum	T37	T37	pTiT37 T-DNA	Nopaline	Gresshoff *et al.* (1979)
Nicotiana tabacum	LBA4060	Ach5	*aux* 1 gene inactivated	Octopine	Ooms *et al.* (1981)
Nicotiana tabacum	LBA1501	C58	*aux* 2 gene inactivated	Octopine	Ooms *et al.* (1981)
Nicotiana tabacum	A348 *tms* mutant	C58	*aux* 1 gene inactivated	Octopine	Garfinkel *et al.* (1981)
Nicotiana tabacum	AC34-8*	Ach5	*ipt* cds fused to CaMV35S promoter	–	Smigocki and Owens (1988)
Nicotiana tabacum	pGV3845	C58	*aux* 1 gene inactivated	Nopaline	Saito *et al.* (1989)
Nicotiana langsdorffii	T37	T37	pTiT37 T-DNA	Nopaline	Gresshoff *et al.* (1979)
Nicotiana langsdorffii	C58	C58	pTiC58 T-DNA	Nopaline	Gresshoff *et al.* (1979)
Nicotiana plumbaginifolia	AC34-8*	Ach5	*ipt* cds fused to CaMV35S promoter	–	Smigocki and Owens (1988)
Atropa belladonna	pGV2215	C58	*aux* 2 gene inactivated	Octopine	Saito *et al.* (1991)
Atropa belladonna	T37	T37	pTiT37 T-DNA	Nopaline	Subroto *et al.* (1996a)
Atropa belladonna	C58	C58	pTiC58 T-DNA	Nopaline	Subroto *et al.* (1996a)
Atropa belladonna	LBA4404+pFIH10*ipt**	Ach5	*ipt* cds fused to CaMV35S promoter	–	Subroto *et al.* (1996a)
Mentha citrata	T37	T37	pTiT37 T-DNA	Nopaline	Spencer *et al.* (1990)
Mentha piperita	T37	T37	pTiT37 T-DNA	Nopaline	Spencer *et al.* (1993a)
Mentha piperita	LBA4404+pFIH10*ipt**	Ach5	*ipt* cds fused to CaMV35S promoter	–	Spencer *et al.* (1993b)
Mentha piperita	C58/3+pFIH10*ipt**	C58	*ipt* cds fused to CaMV35S promoter	–	Spencer *et al.* (1993b)
Solanum eleagnifolium	T37	T37	pTiT37 T-DNA	Nopaline	Alvarez *et al.* (1994)
Solanum dulcamara	C58	C58	pTiC58 T-DNA	Nopaline	Ehmke *et al.* (1995)
Solanum aviculare	T37	T37	pTiT37 T-DNA	Nopaline	Subroto *et al.* (1996a)
Pimpinella anisum	T37	T37	pTiT37 T-DNA	Nopaline	Salem and Charlwood (1995)
Duboisia leichhardtii × *D. myoporoides*	T37	T37	pTiT37 T-DNA	Nopaline	Subroto *et al.* (1996a)
Artemisia annua	T37	T37	pTiT37 T-DNA	Nopaline	Paniego and Giulietti (1996)

* Disarmed strain containing a binary vector with CaMV35S–*ipt* construct.

range of pharmaceutical activity useful in many medical treatments. As scopolamine produces fewer side-effects and has a greater efficacy than hyoscyamine in certain medical applications, it is considered the more valuable drug. Hyoscyamine has been shown in several studies to be synthesised in the roots of producing plants (West and Mika, 1957; Waller and Nowacki, 1978); however, it may be subsequently transported from the root to the aerial parts of the plant for storage and/or transformation

to other compounds. Hyoscyamine is converted to scopolamine according to the following pathway (Yamada and Hashimoto, 1988):

Hyoscyamine $\xrightarrow[\text{(H6H)}]{\substack{\text{Hyoscyamine} \\ \text{6}\beta\text{-hydroxylase}}}$ 6β-Hydroxyhyoscyamine

$\xrightarrow{\text{H6H}}$ Scopolamine

Figure 7.2. *Atropa belladonna* shooty teratomas displaying a range of morphologies under the same culture conditions. (a) Compact clusters of stems and leaves after transformation with *A. tumefaciens* strain T37. (b) Shooty teratomas with long stems and narrow leaves after transformation with *A. tumefaciens* strain C58. (c) Shooty teratomas with short thick stems and broad leaves after transformation with *A. tumefaciens* strain 4404+pFIH10*ipt*.

The enzyme, hyoscyamine 6β-hydroxylase (H6H), catalyses both reactions of the two-step process. In plants producing scopolamine, the main site of scopolamine synthesis depends on whether H6H is expressed in the roots or only in the leaves. In some species, significant levels of scopolamine are found only in the aerial parts as the roots produce little or no H6H. In these cases, hyoscyamine is translocated from the roots through the xylem for bioconversion to scopolamine in the leaves (Yun *et al.*, 1992).

Co-Culture Experiments

Experiments were carried out to investigate the feasibility of root–shoot co-culture for biotransformation of secondary metabolites, using the hyoscyamine–scopolamine conversion as a model system (Subroto *et al.*, 1996b). The co-cultures were performed either in shake flasks or in air-sparged fermenters with total capacity up to 10 L. Hairy roots of *Atropa belladonna* and shooty teratomas of either *A. belladonna*

Figure 7.3. (a) Co-culture of *A. belladonna* hairy roots and shooty teratomas in a 1.25-L sparged fermenter. (b) Co-culture of *A. belladonna* hairy roots and *Duboisia* hybrid shooty teratomas in dual 1-L shake flasks. (c) Co-culture of *A. belladonna* hairy roots and shooty teratomas in dual 5-L sparged fermenters.

or a *Duboisia* hybrid (*D. leichhardtii* × *D. myoporoides*) were used. The hairy roots were transformed with *Agrobacterium rhizogenes* A4; the shooty teratomas were transformed with *A. tumefaciens* T37. All cultures were grown at 25°C using hormone-free Murashige and Skoog (MS) medium containing 3% sucrose. The shoots were maintained in screw-cap jars on solid medium containing 0.2% Phytagel (Sigma, USA) with continuous overhead illumination (18 W fluorescent lamps; Osram, Australia) at an irradiance of ca. 1.7 W m^{-2}. The hairy roots were maintained in liquid medium in shake flasks. Except when co-cultured in the same vessel as the shooty teratomas, the roots were grown in the dark.

A range of reaction vessel configurations was used for the co-culture experiments, including single 1-L shake flasks, dual 1-L shake flasks, single 1.25-L air-sparged fermenters, and dual 5-L air-sparged fermenters. The dual vessel and single fermenter systems are illustrated in Figure 7.3. The dual flask and fermenter systems were inoculated with roots and shoots in separate vessels; the single flasks and fermenters were inoculated with roots and shoots together. The dual flasks were connected at the bottom by 1.25-cm i.d. silicone tubing which allowed interchange of the medium (but not the biomass) between the flasks during shaking on an orbital platform. The dual fermenters were connected using silicone tubing and a peristaltic pump which recirculated medium between the vessels. Separate control cultures of roots and shoots were also grown in single 1-L flasks or 1.25-L bioreactors.

Alkaloid Production

For co-culture of hairy roots and shooty teratomas to be effective for scopolamine production, four processes must take place. First, hyoscyamine must be synthesised in the roots; second, hyoscyamine must be released from the root tissue into the medium; third, hyoscyamine must be taken up from the medium by the shoots; and finally, hyoscyamine in the shoots must be enzymatically converted to scopolamine. In the experiments described above, after 28–29 d co-culture in the various reactor vessels, the roots, shoots and medium were analysed for tropane alkaloids using HPLC (Subroto *et al.,*

Table 7.2. Alkaloid levels after 28–29 d co-culture of hairy roots and shooty teratomas in shake flasks and air-sparged bioreactors. H: hyoscyamine/atropine; S: scopolamine; ND: none detected; ± indicates maximum error from duplicate flasks or standard deviation from triplicate flasks. (Adapted from Subroto *et al.*, 1996b.)

Culture	Initial liquid volume	Root:shoot inoculum ratio (g g^{-1} dry weight)	Species	Organ	Intracellular alkaloid (mg g^{-1} dry weight)	Total alkaloid (mg)
Control	300 mL	1:0	*A. belladonna*	Roots	H = 0.65 ± 0.12 S = ND	H = 0.53 S = ND
Control	300 mL	0:1	*A. belladonna*	Shoots	H = ND S = ND	H = ND S = ND
Control	1 L	0:1	*Duboisia* hybrid	Shoots	H = ND S = ND*	H = ND S = ND
Single flask co-culture	300 mL	0.67:1	*A. belladonna*	Roots	H = 0.75 ± 0.25 S = 0.12 ± 0.05	H = 1.5 S = 0.95
			A. belladonna	Shoots	H = 0.17 ± 0.03 S = 0.18 ± 0.03	
Dual flask co-culture	2 × 300 mL	0.69:1	*A. belladonna*	Roots	H = 0.21 ± 0.03 S = ND	H = 2.6 S = 5.0
			A. belladonna	Shoots	H = 0.28 ± 0.05 S = 0.84 ± 0.08	
Dual flask co-culture (inter-generic)	2 × 300 mL	1:1	*A. belladonna*	Roots	H = 0.91 ± 0.19 S = 0.13 ± 0.02	H = 2.8 S = 7.2
			Duboisia hybrid	Shoots	H = 0.03 ± 0.05 S = 2.7 ± 1.7	
Single bioreactor co-culture	1 L	0.17:1	*A. belladonna*	Roots	H = 0.78 S = 0.06	H = 2.4 S = 2.8
			A. belladonna	Shoots	H = 0.08 S = 0.22	
Dual bioreactor co-culture	2 × 3 L	0.35:1	*A. belladonna*	Roots	H = 0.17 S = 0.04	H = 62 S = 11
			A. belladonna	Shoots	H = 0.22 S = 0.15	

* Trace levels of scopolamine observed using HPLC could not be confirmed by GC-MS.

1996b). The results are shown in Table 7.2. The control *A. belladonna* hairy root cultures produced hyoscyamine, but no scopolamine was detected. This result is consistent with the low activity of H6H in roots of this species. Also as expected, the control shoot cultures of *A. belladonna* and the *Duboisia* hybrid contained neither hyoscyamine nor scopolamine, reflecting the inability of the shoots to synthesise the hyoscyamine precursor. In contrast, the various co-cultures of roots and shoots contained relatively high levels of scopolamine. *A. belladonna* shoots accumulated up to 0.84 mg g^{-1} dry weight scopolamine in dual flask co-culture, or 3–11 times the average levels reported for leaves of the whole plant (Hartmann *et al.*, 1986; Simola *et al.*, 1988). The *Duboisia* hybrid shoots contained the highest concentrations of scopolamine (average of 2.7 mg g^{-1} dry weight), reflecting the scopolamine-rich nature of *Duboisia* species. The beneficial effects of co-culture compared with separate root and shoot cultures is clear from the data shown in Table 7.2. As scopolamine levels in *A. belladonna* and *Duboisia* spp. callus and cell suspensions have been limited previously to trace quantities (Hamilton *et al.*, 1986; Kitamura, 1988), these results demonstrate that root–shoot co-culture has considerable potential for expanding the range of metabolites which can be produced at high levels *in vitro*. The success of co-culture using transformed organs from different plant genera shows that, provided the roots and shoots require similar nutrients and culture conditions and do not secrete inhibitory by-products into the medium, biotransformations which do not normally occur *in vivo* could be carried out *in vitro* with the potential for commercial exploitation.

Unlike the *A. belladonna* shooty teratomas which adapted well to growth in liquid medium, some *Duboisia* hybrid shoots tended to vitrify and revert to callus or cell suspensions after 20–25 d of culture. Just as hairy roots are known to lose their ability to synthesise secondary metabolites when their morphology is disrupted (Robins *et al.*, 1991), this morphological instability in the *Duboisia* shooty teratomas is likely to have reduced scopolamine production. Variation in the extent of callusing in different flasks also affected the reproducibility of measurements from replicate cultures. Further work testing different medium and/or culture conditions is needed to minimise problems with callusing of this species in submerged culture.

Effect of the Root:Shoot Inoculum Ratio

Because the roots and shoots in co-culture systems compete with each other for nutrients, and as the relative amounts of biomass present determine the balance between precursor source and sink effects, the ratio of root and shoot fresh weights used for inoculation could be expected to influence co-culture performance. This hypothesis was borne out experimentally, as the root:shoot inoculum ratio was found to exert a significant influence on scopolamine synthesis, biomass accumulation and alkaloid profiles. The effect of inoculum ratio was examined using *A. belladonna* hairy roots and *Duboisia* hybrid shooty teratomas. The highest root:shoot inoculum ratio tested (1:1) produced scopolamine levels in the shoots which were ca. 17 times higher than those in cultures inoculated at the lowest ratio (0.2:1). As shown in Table 7.3, after 14 d co-culture a greater concentration of hyoscyamine was detected in the medium at the 1:1 inoculum ratio compared with 0.2:1, although concentrations in the root biomass were not significantly different. The reasons for these

effects are not clear at present, but the results indicate the responsiveness of root–shoot co-cultures to operating parameters not normally considered with single-organ tissue culture systems.

Conclusion

A. belladonna hairy roots and *A. belladonna* and *Duboisia* hybrid shooty teratomas have been used to demonstrate the effectiveness of plant organ co-culture for synthesis and biotransformation of secondary metabolites. Production of scopolamine was significantly improved in co-culture compared with separate root and shoot systems. The ratio of roots: shoots used to inoculate the co-cultures had a pronounced influence on alkaloid synthesis. Other factors, such as the relative timing of inoculation and/or harvest of the roots and shoots, could also produce interesting effects. Application of co-cultures to bioconversion systems involving more than one enzyme would be valuable in demonstrating the utility of this approach for enhancing the production of metabolites not easily manipulated using genetic engineering techniques. Although untested at the present time, it is also likely that root–shoot co-culture would be useful for biotransformation of precursor compounds synthesised in the shoots and requiring modification in the roots.

Acknowledgments

We thank Dr William Griffin of the University of Queensland, Australia, for providing untransformed *Duboisia* hybrid plant material. P.M.D. acknowledges support from an Australian Research Council (ARC) Queen Elizabeth II Research Fellowship.

References

Alvarez MA, J Rodríguez Talou, NB Paniego and AM Giulietti (1994). Solasodine production in transformed organ cultures (roots and shoots) of *Solanum eleagnifolium* Cav. *Biotechnol Lett* 16: 393–396.

Table 7.3. Effect of the root:shoot inoculum ratio on hyoscyamine levels after 14 d co-culture of *A. belladonna* hairy roots and *Duboisia* hybrid shooty teratomas in dual 300-mL shake flasks. ± indicates standard deviation from triplicate flasks.

Root:shoot inoculum ratio (g:g fresh weight)	Hyoscyamine in the medium (mg L^{-1})	Hyoscyamine in the roots (mg g^{-1} dry weight)
1:1	0.80 ± 0.02	1.63 ± 0.08
0.7:1	0.23 ± 0.01	1.53 ± 0.11
0.5:1	0.20 ± 0.00	1.37 ± 0.76
0.2:1	0.12 ± 0.04	1.78 ± 0.42

Charlwood BV and C Moustou (1988). Essential oil accumulation in shoot-proliferation cultures of *Pelargonium* spp. In *Manipulating Secondary Metabolism in Culture*, edited by RJ Robins and MJC Rhodes, pp. 187–194. Cambridge: Cambridge University Press.

Charlwood KA, S Brown and BV Charlwood (1988). The accumulation of flavour compounds by cultures of *Zingiber officinale*. In *Manipulating Secondary Metabolism in Culture*, edited by RJ Robins and MJC Rhodes, pp. 195–200. Cambridge: Cambridge University Press.

Ehmke A, D Ohmstede and U Eilert (1995). Steroidal glycoalkaloids in cell and shoot teratoma cultures of *Solanum dulcamara*. *Plant Cell Tiss Organ Cult* 43: 191–197.

Garfinkel DJ, RB Simpson, LW Ream, FF White, MP Gordon and EW Nester (1981). Genetic analysis of crown gall: fine structure map of the T-DNA by site-directed mutagenesis. *Cell* 27: 143–153.

Gresshoff PM, ML Skotnicki and BG Rolfe (1979). Crown gall teratoma formation is plasmid and plant controlled. *J Bacteriol* 137: 1020–1021.

Hamill JD (1993). Alterations in auxin and cytokinin metabolism of higher plants due to expression of specific genes from pathogenic bacteria: a review. *Aust J Plant Physiol* 20: 405–423.

Hamilton RM, J Lang, H Pedersen and C-K Chin (1986). Secondary metabolites from organized *Atropa belladonna* cultures. *Biotechnol Bioeng Symp* 17: 685–698.

Hartmann T, L Witte, F Oprach and G Toppel (1986). Reinvestigation of the alkaloid composition of *Atropa belladonna* plants, root cultures, and cell suspension cultures. *Planta Med* 52: 390–395.

Hashimoto T, D-J Yun and Y Yamada (1993). Production of tropane alkaloids in genetically engineered root cultures. *Phytochemistry* 32: 713–718.

Heble MR (1985). Multiple shoot cultures: a viable alternative *in vitro* system for the production of known and new biologically active plant constituents. In *Primary and Secondary Metabolism of Plant Cell Cultures*, edited by K-H Neumann, W Barz and E Reinhard, pp. 281–289. Berlin: Springer-Verlag.

Kitamura Y (1988). *Duboisia* spp.: *in vitro* regeneration, and the production of tropane and pyridine alkaloids. In *Biotechnology in Agriculture and Forestry 4: Medicinal and Aromatic Plants I*, edited by YPS Bajaj, pp. 419–436. Berlin: Springer-Verlag.

Körber H, N Strizhov, D Staiger, J Feldwisch, O Olsson, G Sandberg, K Palme, J Schell and C Koncz (1991). T-DNA gene 5 of *Agrobacterium* modulates auxin response by autoregulated synthesis of a growth hormone antagonist in plants. *EMBO J* 10: 3983–3991.

Ooms G, PJJ Hooykaas, G Moolenaar and RA Schilperoort (1981). Crown gall plant tumors of abnormal morphology, induced by *Agrobacterium tumefaciens* carrying mutated octopine Ti plasmids; analysis of T-DNA functions. *Gene* 14: 33–50.

Paniego NB and AM Giulietti (1996). Artemisinin production by *Artemisia annua* L.-transformed organ cultures. *Enzyme Microb Technol* 18: 526–530.

Robins RJ, EG Bent and MJC Rhodes (1991). Studies on the biosynthesis of tropane alkaloids by *Datura stramonium* L. transformed root cultures. 3. The relationship between morphological integrity and alkaloid biosynthesis. *Planta* 185: 385–390.

Saito K, I Murakoshi, D Inzé and M van Montagu (1989). Biotransformation of nicotine alkaloids by tobacco shooty teratomas induced by a Ti plasmid mutant. *Plant Cell Rep* 7: 607–610.

Saito K, M Yamazaki, A Kawaguchi and I Murakoshi (1991). Metabolism of solanaceous alkaloids in transgenic plant teratomas integrated with genetically engineered genes. *Tetrahedron* 47: 5955–5968.

Salem KMSA and BV Charlwood (1995). Accumulation of essential oils by *Agrobacterium tumefaciens*-transformed shoot cultures of *Pimpinella anisum*. *Plant Cell Tiss Organ Cult* 40: 209–215.

Simola LK, S Nieminen, A Huhtikangas, M Ylinen, T Naaranlahti and M Lounasmaa (1988). Tropane alkaloids from *Atropa belladonna*, Part II. Interaction of origin, age, and environment in alkaloid production of callus cultures. *J Nat Prod* 51: 234–242.

Smigocki AC and LD Owens (1988). Cytokinin gene fused with a strong promoter enhances shoot organogenesis and zeatin levels in transformed plant cells. *Proc Nat Acad Sci USA* 85: 5131–5135.

Spencer A, JD Hamill and MJC Rhodes (1990). Production of terpenes by differentiated shoot cultures of *Mentha citrata* transformed with *Agrobacterium tumefaciens* T37. *Plant Cell Rep* 8: 601–604.

Spencer A, JD Hamill and MJC Rhodes (1993a). *In vitro* biosynthesis of monoterpenes by *Agrobacterium* transformed shoot cultures of two *Mentha* species. *Phytochemistry* 32: 911–919.

Spencer A, JD Hamill and MJC Rhodes (1993b). Transformation in *Mentha* species (mint). In *Biotechnology in Agriculture and Forestry 22, Plant Protoplasts and Genetic Engineering III*, edited by YPS Bajaj, pp. 278–293. Berlin: Springer-Verlag.

Subroto MA, JD Hamill and PM Doran (1996a). Development of shooty teratomas from several solanaceous plants: growth kinetics, stoichiometry and alkaloid production. *J Biotechnol* 45: 45–57.

Subroto MA, KH Kwok, JD Hamill and PM Doran (1996b). Coculture of genetically transformed roots and shoots for synthesis, translocation, and biotransformation of secondary metabolites. *Biotechnol Bioeng* 49: 481–494.

Waller GR and EK Nowacki (1978). *Alkaloid Biology and Metabolism in Plants*. Plenum, New York.

West FR and ES Mika (1957). Synthesis of atropine by isolated roots and root-callus cultures of belladonna. *Bot Gaz* 119: 50–54.

Yamada Y and T Hashimoto (1988). Biosynthesis of tropane alkaloids. In *Applications of Plant Cell and Tissue Culture*, edited by G Bock and J Marsh, pp. 199–212. Chichester: Wiley.

Yun D-J, T Hashimoto and Y Yamada (1992). Metabolic engineering of medicinal plants: transgenic *Atropa belladonna* with an improved alkaloid composition. *Proc Nat Acad Sci USA* 89: 11799–11803.

Hairy Roots as an Expression System for Production of Antibodies

8

Raviwan Wongsamuth and Pauline M. Doran

Department of Biotechnology, University of New South Wales, Sydney, NSW 2052, Australia;
email: p.doran@unsw.edu.au

Introduction

This chapter summarises recent work with hairy roots initiated from transgenic tobacco plants expressing a full-length mouse IgG$_1$ monoclonal antibody. Although synthesis and assembly of antibody proteins have been demonstrated previously in several plant systems, characteristics of growth and antibody production in cultures suitable for large-scale production remain largely untested. In this study, properties such as the long-term stability of antibody expression in hairy roots, variation between hairy root clones, the time course of antibody accumulation in batch culture, and the effect of medium composition on growth, antibody accumulation and antibody secretion were investigated.

Following recent advances in plant genetic engineering, plants and plant cells have been applied for expression of a range of foreign proteins. Some of this work has been aimed at controlling or directing plant metabolism; bacterial, yeast and foreign plant genes have been introduced into plant cells for direct manipulation of enzyme activity in secondary pathways (Hamill *et al.*, 1990; Berlin *et al.*, 1993; Hashimoto *et al.*, 1993). Other applications focus on the potential of plants and plant cell culture for

large-scale production of foreign proteins of direct commercial value (Hogue *et al.*, 1990; Gao *et al.*, 1991; Gao and Lee, 1992; Domansky *et al.*, 1995). Of these proteins, antibodies and antibody fragments have received most attention to date.

Research over the last 20–30 years has revealed the structure of antibodies, the cellular apparatus needed for their synthesis, assembly and secretion, and the mechanisms by which they bind to foreign material in the body as a biochemical defence against pathogens. Knowledge of these processes has increased our understanding of the immune system and allowed exploitation of antibodies in a wide variety of medical and industrial applications. Antibodies bind with high affinity to specific molecules and are applied extensively in medical diagnosis, in commercial assays, as therapeutic agents, and as binding moieties in affinity separations. Monoclonal antibodies are currently produced in large-scale fermentation systems using hybridoma cells formed by fusion of animal lymphocytes with myeloma cells. However, in recent years, several heterologous systems including bacteria, yeast, insect cells and non-lymphoid mammalian cells have been tested for antibody production (Better *et al.*, 1988; Horwitz *et al.*, 1988; Hasemann and Capra, 1990; Weidle *et al.*, 1987).

Assembly of complete antibodies containing heavy and light chains and constant and variable regions has not been achieved in prokaryotes; once polypeptides have formed aggregates as inclusion bodies in bacteria such as *E. coli*, proper folding of protein molecules is difficult to achieve. Yields of antibody proteins in many heterologous systems have been very low, of the order of 0.2–2.0 µg mL^{-1} (Better *et al.*, 1988; Horwitz *et al.*, 1988; Weidle *et al.*, 1987) compared with 50–120 µg mL^{-1} in hybridoma cultures (Bibila and Flickinger, 1991; Miller *et al.*, 1988; Renard *et al.*, 1988).

Research into antibody production in plant systems has several goals and areas of potential application. These include modulation of metabolic pathways in plants using antibodies to bind to key metabolic intermediates or effector molecules (Owen *et al.*, 1992), and development of crops with in-built resistance to viruses and pathogens (Tavladoraki *et al.*, 1993). Plant systems are also being considered as an alternative to animal cell culture as a commercial source of monoclonal antibodies. So far, plants, callus and plant cell suspensions have been used for antibody expression (De Neve *et al.*, 1993; Düring *et al.*, 1990; Firek *et al.*, 1993; Hiatt *et al.*, 1989; Ma *et al.*, 1994; Owen *et al.*, 1992; Wahl *et al.*, 1995), but not hairy roots. In this chapter we report the results from experiments with hairy roots initiated from transgenic tobacco seedlings expressing Guy's 13 antibody. Guy's 13 is a murine IgG$_1$ monoclonal antibody which binds to a surface protein of *Streptococcus mutans*, a causative agent of dental caries in humans. Expression of Guy's 13 in plants has been considered for production of a topical therapeutic agent to prevent colonisation of *S. mutans* on teeth (Ma *et al.*, 1987, 1994). The aim of this work was to demonstrate the effectiveness of hairy roots for antibody synthesis, assembly and secretion, and to examine the effects of culture conditions on antibody production.

Characteristics of Antibody Accumulation in Tobacco Hairy Roots

Seeds of transgenic *Nicotiana tabacum* plants expressing Guy's 13 antibody were provided by Dr A. Hiatt and Scripps Research Institute, La Jolla, California. The transformation procedures and gene constructs used to induce antibody expression in the tobacco plants are described by Ma *et al.* (1994). Eight antibody-producing hairy roots were initiated from the tobacco seedlings using *Agrobacterium rhizogenes* strains 8196 and 15834. Hairy roots were also developed from wild-type *N. tabacum* using *A. rhizogenes* strain 15834. The root cultures were cultured at 25°C in either Murashige and Skoog (MS) or Gamborg's

B5 medium containing 3% sucrose in 250-mL shake flasks on orbital shakers operated at 100 rpm in the dark. Antibody levels in the biomass and medium were quantified by ELISA using goat anti-mouse antibodies specific for γ and κ chains; only antibody proteins assembled into γ–κ chain complexes were measured. Expression of complete IgG$_1$ antibody was verified using Western blotting (Wongsamuth and Doran, 1997).

Growth and antibody accumulation by wild-type and 8 antibody-producing hairy root clones were compared after 21 d culture in MS or B5 medium. As shown in Figure 8.1a, synthesis of antibody by Clones 1–8 did not retard growth compared with wild-type hairy roots. Roots clones cultured in B5 medium produced more biomass than those in MS medium, although there was little difference for the wild-type cultures. Growth variation between the clones in B5 medium was significant as indicated by analysis of variance at the 0.05 confidence level. Total antibody accumulation (intracellular + extracellular) for the transgenic clones is shown in Figure 8.1b. Average total antibody levels in B5 medium (0.67–0.89 mg, or 13–18 µg mL^{-1}) were higher than in MS medium (0.41–0.75 mg, or 8–15 µg mL^{-1}) for all the clones tested; analysis of variance showed that there was no significant difference in total antibody levels between the clones in B5 medium at the 0.05 confidence level. Intracellular antibody accounted for between 1.0 and 1.8% of total protein in the cells. As indicated in Figure 8.1c, antibody secretion into the medium accounted for up to 14% of the antibody accumulated; antibody release was greater in B5 than in MS medium.

The total antibody titres observed in the hairy root cultures are significantly higher than the yields of antibody proteins reported for many other heterologous systems (Better *et al.*, 1988; Horwitz *et al.*, 1988; Weidle *et al.*, 1987), and are also an improvement on values reported previously for several plant systems (De Neve *et al.*, 1993; Firek *et al.*, 1993; Owen *et al.*, 1992; Wahl *et al.*, 1995). The hairy roots in this work accumulated 15–30% of the antibody titres typically obtained in batch hybridoma cultures (Miller *et al.*, 1988; Bibila and Flickinger, 1991). In contrast with hybridoma cells, however, most of the antibody remained associated with the root tissue rather than being secreted extracellularly.

Secretion of antibody into the medium is beneficial for downstream processing as it eliminates the need for harvesting of the biomass and avoids the additional cost of separating antibody from other intracellular proteins. Clone 8 was therefore of particular interest because of its tendency to secrete a higher proportion of antibody into the medium. Time courses for growth, sugar uptake, total antibody

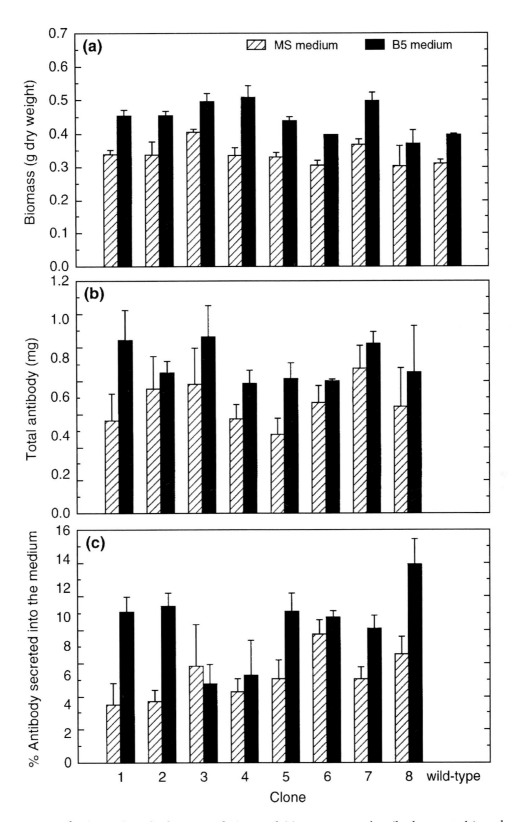

Figure 8.1. (a) Growth, (b) total antibody accumulation, and (c) percentage of antibody secreted into the medium for 8 antibody-producing clones and wild-type hairy roots after 21 d growth in MS and B5 media in shake flasks. The initial liquid volume was 50 mL.

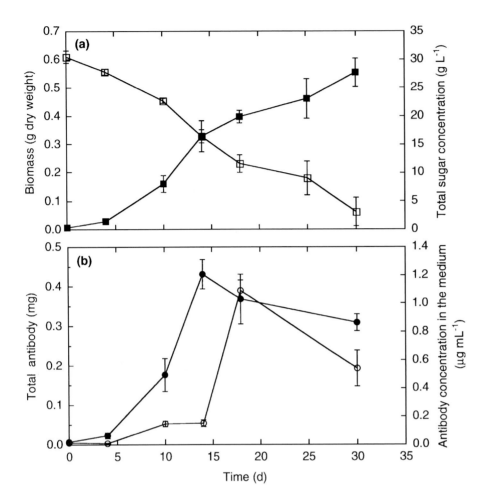

Figure 8.2. (a) Biomass (■) and sugar concentration (□), and (b) total antibody accumulation (●) and concentration of secreted antibody (○) during 30 d culture of Clone 8 hairy roots in B5 medium in shake flasks. The initial liquid volume was 50 mL.

accumulation and antibody secretion by Clone 8 are shown in Figure 8.2. As shown in Figure 8.2a, after 30 d culture 0.55 ± 0.05 g dry weight of biomass was produced, corresponding to 11 g L^{-1} based on the initial medium volume of 50 mL. Growth was exponential only for the first ca. 10 d; growth continued after this time but with decreasing specific growth rate. The initial specific growth rate during the exponential phase was 0.30 ± 0.01 d^{-1} (doubling time: 2.3 d). For comparison, the initial specific growth rate of wild-type hairy roots was virtually identical at 0.31 ± 0.01 d^{-1}. As shown in Figure 8.2a, reductions in the total sugar (sucrose, glucose and fructose) concentration in the medium mirrored the pattern of growth; the final biomass yield (Y_{XS}) was 0.43 g g^{-1} dry weight. Sucrose was completely hydrolysed to fructose and glucose by Day 18, and glucose was taken up preferentially. As indicated in Figure 8.2b, assembled antibody accumulated in the culture only during the first 14 d; after this time, antibody degradation reduced both intracellular and extracellular levels. A sharp rise in secreted antibody concentration from 2–4% of the total antibody

present to 12% occurred between Days 14 and 18; the maximum extracellular antibody concentration after 18 d culture was 1.1 µg mL^{-1}.

The decline in total antibody levels in Figure 8.2b distinguishes the pattern of antibody accumulation in hairy roots from the continually rising levels obtained in hybridoma cultures. Antibody degradation was much more significant in the hairy roots. The reason for the sharp reduction in antibody levels after 14 d culture is unclear; however it is possible that protease activity may increase significantly in the roots at the end of exponential growth as the cells adapt to changes in the culture environment.

Long-term stability of culture characteristics including growth, antibody accumulation and antibody secretion are essential for commercial production processes and are also highly desirable for experimental purposes. Variation in these properties with time was monitored for Clone 8 over a period of 7 months starting one year after initiation of the hairy root cultures (Wongsamuth and Doran, 1997). The amount of biomass produced in 30-d shake-flask cultures was relatively stable with a standard deviation about the mean of ca. 11%. Total antibody accumulation and extracellular antibody concentration varied to a greater extent, ranging 25 and 34%, respectively, around their mean values. These results indicate that some variation in antibody production and secretion by the hairy roots can be expected from culture to culture. However, as none of the clones lost their ability to express assembled antibody over the entire 19-month experimental period, hairy root cultures can be considered to have a reasonable degree of long-term stability.

Culture Treatments Aimed at Increasing Antibody Accumulation and Secretion

Several treatments were tested using selected hairy root clones to determine the effect of medium additives and initial medium pH on antibody accumulation and secretion. The experiments were carried out in shake flasks.

Gelatin

Flasks containing B5 medium with 0–9 g L^{-1} gelatin were inoculated with Clone 8 hairy roots and cultured for 30 d. Addition of gelatin to the medium had a slight negative effect on growth; biomass levels with 9.0 g L^{-1} gelatin were 23% lower than in the control cultures without gelatin. Results for antibody accumulation and secretion are shown in Figure 8.3a. Gelatin improved total antibody levels, which were increased by 14–68% compared with the control cultures. A marked increase in extracellular antibody levels also occurred; medium concentrations increased by a factor of between 4 and 8 compared with the control. The maximum antibody concentration in the medium at 9.0 g L^{-1} gelatin corresponds to 43% of the total antibody present. These results are similar to those observed previously with tobacco plant cell suspensions expressing heavy chain monoclonal antibody; addition of 5.0 g L^{-1} gelatin was found to improve extracellular antibody concentrations by a factor of 10 (LaCount et al., 1995).

Polyvinylpyrrolidone (PVP)

Addition of 0.2–2.0 g L^{-1} PVP (MW: 10,000) to B5 medium had no significant effect on biomass production by Clone 8 hairy roots compared with control cultures without PVP. The effect of PVP on antibody accumulation and secretion is shown in Figure 8.3b. The total antibody present after 30 d culture was enhanced by 20–92% in cultures with PVP compared with the control. Antibody concentration in the medium was also significantly increased by a factor of up to 9.3. The maximum

extracellular antibody level at 2.0 g L^{-1} PVP corresponds to secretion of 43% of the total antibody present, compared with ca. 8% for the control. Similar results have been reported previously for suspended tobacco cells expressing heavy chain antibody after treatment with 0.75 g L^{-1} PVP (LaCount et al., 1995).

Nitrate

The effect of added nitrate was tested using Clone 8 hairy roots. KNO$_3$ was added to B5 medium to give supplements of between 0 and 0.5% (w/v) in addition to the 0.25% (w/v) KNO$_3$ already present in B5 medium. Additions of up to 0.05% (w/v) had a negligible effect on growth during the 30-d culture period; higher concentrations reduced growth by up to 28% compared with the control cultures. Total antibody accumulation as a function of nitrate supplementation is shown in Figure 8.3c. Nitrate had a significant beneficial effect on antibody accumulation at an added level of 0.1% (w/v); antibody levels were increased by a maximum of 90% relative to the controls. Results for extracellular antibody show that 0.1% (w/v) added nitrate also had a marked effect on accumulation of antibody in the medium; the maximum antibody concentration in the treated cultures represents ca. 17% of the total antibody present.

Yeast Extract

Yeast extract was added to B5 medium with the aim of providing more of the building blocks for foreign protein synthesis than are usually present in plant nutrient media. Experiments were conducted using Clone 7 and yeast extract concentrations of between 0.25 and 2.0 g L^{-1}. Addition of yeast extract had no significant effect on growth measured after 30 d culture. As shown in Figure 8.4a, total antibody accumulation and extracellular antibody levels were either unchanged or reduced with increasing yeast extract concentration. With yeast extract at 2.0 g L^{-1}, total antibody levels were reduced by 20% compared with the control.

Vitamins

Experiments were conducted to determine the effect on hairy root Clone 2 of adding myo-inositol and thiamine.HCl to vitamin-free MS medium. The final concentrations of vitamins in the media tested are given in Table 8.1. For comparison, normal MS medium contains 100 mg L^{-1} myo-inositol and 0.1 mg L^{-1} thiamine.HCl; B5 medium contains 100 mg L^{-1} myo-inositol and 10 mg L^{-1} thiamine.HCl. Growth, antibody accumulation and antibody secretion were measured after 35 d culture. Neither inositol nor thiamine had any significant effect on growth. As shown in Figure 8.4b, accumulation and

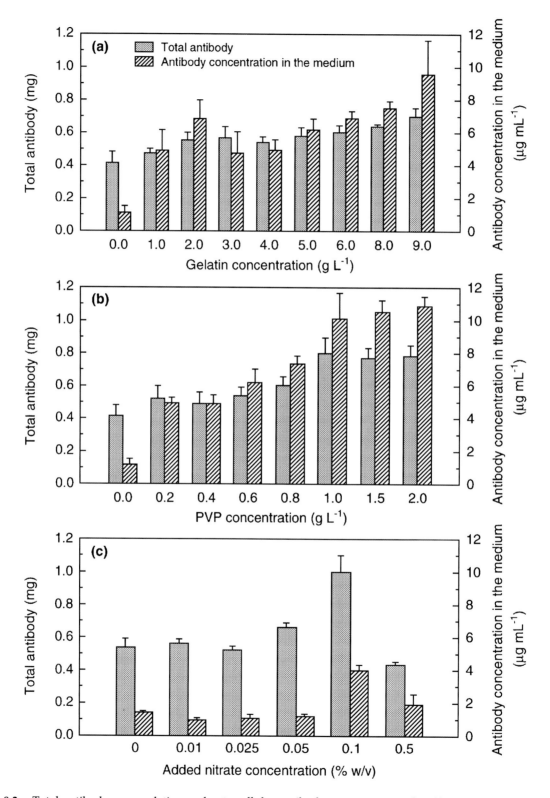

Figure 8.3. Total antibody accumulation and extracellular antibody concentrations for Clone 8 cultured for 30 d in B5 medium containing (a) gelatin, (b) polyvinylpyrrolidone (PVP), and (c) added nitrate. The initial liquid volume was 50 mL.

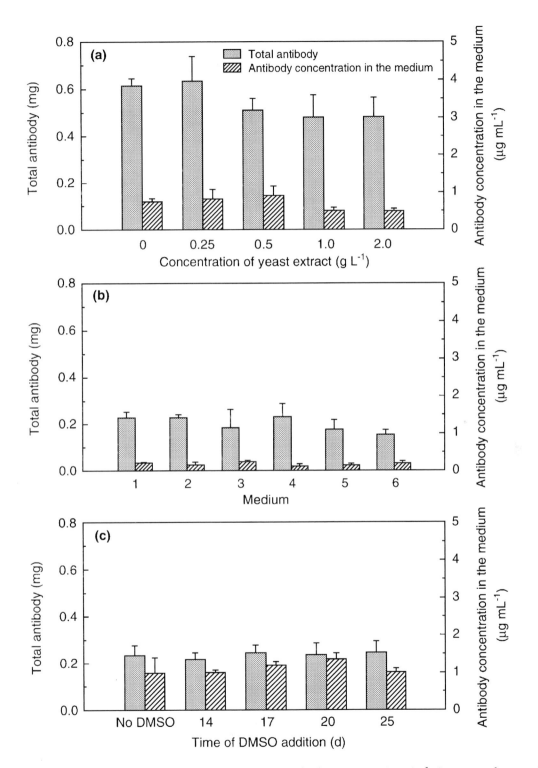

Figure 8.4. Total antibody accumulation and extracellular antibody concentrations in hairy root cultures with various medium additives. (a) Clone 7 cultured for 30 d in B5 medium containing yeast extract; (b) Clone 2 cultured for 35 d in vitamin-free MS medium containing various levels of myo-inositol and thiamine.HCl as listed in Table 8.1; (c) Clone 6 cultured for 30 d in B5 medium with and without DMSO. The time of DMSO addition to the roots was varied from 14 to 25 d. The initial liquid volume for all cultures was 50 mL.

Table 8.1. Concentrations of inositol and thiamine added to vitamin-free MS medium.

Medium	Concentration of myo-inositol (mg L^{-1})	Concentration of thiamine.HCl (mg L^{-1})
1	0	0
2	100	0
3	0	10
4	100	10
5	200	20
6	300	30

secretion of antibody were also largely insensitive to vitamin concentration, except for a possible reduction in total antibody levels at the highest vitamin concentrations tested (Medium 6).

Dimethylsulphoxide (DMSO)

DMSO is a membrane permeabilising agent, and was considered a potential effector for antibody secretion from hairy roots. DMSO at a concentration of 2.8% (v/v) was added to B5 medium and used for culture of Clone 6. The time of addition after inoculation was varied between 14 and 25 d; biomass and antibody levels were measured 30 d after inoculation. Biomass production was not affected by DMSO. As shown in Figure 8.4c, the amounts of antibody accumulated and secreted by the roots were also unaffected by this treatment. These results are in contrast to previous reports of enhanced heavy chain antibody production in plant suspensions after addition of 2.8% and 4.0% DMSO. Wahl *et al.* (1995) increased total antibody levels by up to 3-fold and significantly enhanced extracellular concentrations using DMSO.

Initial Medium pH

pH can have a significant effect on the functioning of membranes and secretion processes, and has been used in several studies as an elicitor of secondary metabolism and to enhance product excretion in plant cell and root cultures (Jardin *et al.*, 1991; Ermayanti *et al.*, 1994). To investigate the effect of medium pH on antibody accumulation and secretion, Clone 8 and wild-type tobacco hairy roots were cultured for 30 d in B5 media adjusted to pH values before autoclaving of between 3.0 and 11.0. Growth of the wild-type roots was relatively unaffected between pH 4.0 and 9.0; below and above these values at pH 3.0 and 11.0, growth was severely reduced by ca. 56 and 78%, respectively, compared with pH 4.0–9.0. The amount of biomass produced

by Clone 8 increased gradually as the initial pH was raised from 3.0 to 8.0, then declined at pH 9.0 and 11.0. The maximum biomass accumulated by Clone 8 at initial pH 8.0 was 0.60 ± 0.02 g, which is the same as that produced by wild-type roots in B5 medium at the same initial pH. Total antibody accumulation by Clone 8 declined steadily with increasing initial pH from an average of 0.47 ± 0.13 mg at initial pH 5.0–6.0, to 0.11 ± 0.06 mg at initial pH 11.0. Antibody levels below pH 5.0 were not measured because of the very low biomass levels produced by these cultures. Variation in the percentage of the antibody secreted into the medium was not significant between pH 5.0 and 11.0.

Conclusion

As a heterologous antibody expression system, hairy roots are capable of high levels of production of assembled antibody protein. In B5 medium, most of the antibody remains associated with the biomass rather than being secreted extracellularly; however, medium additives such as gelatin, polyvinylpyrrolidone and nitrate can be used to improve total antibody accumulation and allowed up to 43% of the product to be recovered from the medium. Antibody degradation appears to occur to a greater extent in hairy roots than in hybridoma systems; further work is needed to stabilise 'plantibody' production in hairy root culture.

Acknowledgments

Seeds of transgenic *Nicotiana tabacum* plants used in this work were kindly provided by Dr Andrew Hiatt and Scripps Research Institute, La Jolla, USA. P.M.D. acknowledges support from an Australian Research Council (ARC) Queen Elizabeth II Research Fellowship.

References

Berlin J, C Rügenhagen, N Greidziak, IN Kuzovkina, L Witte and V Wray (1993). Biosynthesis of serotonin and β-carboline alkaloids in hairy root cultures of *Peganum harmala*. *Phytochemistry* 33: 593–597.

Better M, CP Chang, RR Robinson and AH Horwitz (1988). *Escherichia coli* secretion of an active chimeric antibody fragment. *Science* 240: 1041–1043.

Bibila TA and MC Flickinger (1991). A model of inter-organelle monoclonal antibody transport and secretion in mouse hybridoma cells. *Biotechnol Bioeng* 38: 767–780.

De Neve M, M De Loose, A Jacobs, H Van Houdt, B Kaluza, U Weidle, M Van Montagu and A Depicker (1993). Assembly of an antibody and its derived antibody fragment in *Nicotiana* and *Arabidopsis*. *Transgenic Res* 2: 227–237.

Domansky N, P Ehsani, A-H Salmanian and T Medvedeva (1995). Organ-specific expression of hepatitis B surface antigen in potato. *Biotechnol Lett* 17: 863–866.

Düring K, S Hippe, F Kreuzaler and J Schell (1990). Synthesis and self-assembly of a functional monoclonal antibody in transgenic *Nicotiana tabacum*. *Plant Molec Biol* 15: 281–293.

Ermayanti TM, JA McComb and PA O'Brien (1994). Stimulation of synthesis and release of swainsonine from transformed roots of *Swainsona galegifolia*. *Phytochemistry* 36: 313–317.

Firek S, J Draper, MRL Owen, A Gandecha, B Cockburn and GC Whitelam (1993). Secretion of a functional single-chain Fv protein in transgenic tobacco plants and cell suspension cultures. *Plant Molec Biol* 23: 861–870.

Gao J, JM Lee and G An (1991). The stability of foreign protein production in genetically modified plant cells. *Plant Cell Rep* 10: 533–536.

Gao J and JM Lee (1992). Effect of oxygen supply on the suspension culture of genetically modified tobacco cells. *Biotechnol Prog* 8: 285–290.

Hamill JD, RJ Robins, AJ Parr, DM Evans, JM Furze and MJC Rhodes (1990). Over-expressing a yeast ornithine decarboxylase gene in transgenic roots of *Nicotiana rustica* can lead to enhanced nicotine accumulation. *Plant Molec Biol* 15: 27–38.

Hashimoto T, D-J Yun and Y Yamada (1993). Production of tropane alkaloids in genetically engineered root cultures. *Phytochemistry* 32: 713–718.

Hiatt A, R Cafferkey and K Bowdisk (1989). Production of antibodies in transgenic plants. *Nature* 342: 76–78.

Hiatt A (1990a). Antibodies produced in plants. *Nature* 344: 469–470.

Hiatt A (1990b). The potential of antibodies in plants. *Agbiotechnol News Informat* 2: 653–655.

Hogue RS, JM Lee and G An (1990). Production of a foreign protein product with genetically modified plant cells. *Enzyme Microb Technol* 12: 533–538.

Horwitz AH, CP Chang, M Better, KE Hellstrom and RR Robinson (1988). Secretion of a functional antibody and Fab fragment from yeast cells. *Proc Nat Acad Sci USA* 85: 8678–8682.

Jardin B, R Tom, C Chavarie, D Rho and J Archambault (1991). Stimulated indole alkaloid release from *Catharanthus roseus* immobilized cultures: initial studies. *J Biotechnol* 21: 43–62.

LaCount W, JR Ryland, PM Linzmaier and JM Lee (1995). The stabilization of heavy chain monoclonal antibody produced in plant suspension culture. *Abstr 209th ACS National Meeting*. Anaheim, USA, #084.

Ma JK-C, R Smith and T Lehner (1987). Use of monoclonal antibodies in local passive immunization to prevent colonization of human teeth by *Streptococcus mutans*. *Infect Immun* 55: 1274–1278.

Ma JK-C, T Lehner, P Stabila, CI Fux and A Hiatt (1994). Assembly of monoclonal antibodies with IgG_1 and IgA heavy chain domains in transgenic tobacco plants. *Eur J Immunol* 24: 131–138.

Miller WM, HW Blanch and CR Wilke (1988). A kinetic analysis of hybridoma growth and metabolism in batch and continuous suspension culture: effect of nutrient concentration, dilution rate, and pH. *Biotechnol Bioeng* 32: 947–965.

Owen M, A Gandecha, B Cockburn and G Whitelam (1992). Synthesis of a functional anti-phytochrome single-chain F_v protein in transgenic tobacco. *Bio/Technol* 10: 790–794.

Renard JM, R Spagnoli, C Mazier, MF Salles and E Mandine (1988). Evidence that monoclonal antibody production kinetics is related to the integral of the viable cells curve in batch systems. *Biotechnol Lett* 10: 91–96.

Tavladoraki P, E Benvenuto, S Trinca, D De Martinis, A Cattaneo and P Galeffi (1993). Transgenic plants expressing a functional single-chain Fv antibody are specifically protected from virus attack. *Nature* 366: 469–472.

Wahl MF, G An and JM Lee (1995). Effects of dimethyl sulfoxide on heavy chain monoclonal antibody production from plant cell culture. *Biotechnol Lett* 17: 463–468.

Weidle UH, A Borgya, R Mattes, H Lenz and P Buckel (1987). Reconstitution of functionally active antibody directed against creatine kinase from separately expressed heavy and light chains in non-lymphoid cells. *Gene* 51: 21–29.

Wongsamuth R and PM Doran (1997). Production of monoclonal antibodies by tobacco hairy roots. *Biotechnol Bioeng*, in press.

Transgenic Crop Plants Using *Agrobacterium rhizogenes-* Mediated Transformation

9

Mary C. Christey

Crop & Food Research, Private Bag 4704, Christchurch, New Zealand; email: ChristeyM@crop.cri.nz

Introduction

Agrobacterium rhizogenes is a soil bacterium responsible for the development of hairy root disease on a range of dicotyledonous plants (Tepfer, 1990). *In vitro*, hairy roots are easily distinguished by their rapid, highly branching growth on hormone-free media and plagiotropic root development (Tepfer, 1989). This phenotype is caused by genetic transformation in a manner similar to the development of crown gall disease by *A. tumefaciens*. Infection of wound sites by *A. rhizogenes* is followed by the transfer, integration and expression of T-DNA (T_R-DNA and T_L-DNA in agropine strains) from the Ri (root-inducing) plasmid and subsequent development of the hairy root phenotype (Grant *et al.*, 1991). Hairy roots can be induced on a wide range of plant species and many can be regenerated into plants, often spontaneously (Table 9.1). Plants regenerated from hairy roots often exhibit an altered phenotype due to expression of the *rol* loci. This phenotype is characterised by several morphological changes including wrinkled leaves, shortened internodes, reduced apical dominance, reduced fertility, altered flowering, and plagiotropic roots (Tepfer, 1989; Figure 9.1). The characteristic phenotypic changes are due predominantly to the T_L-DNA (Taylor *et al.*, 1985). Insertional mutagenesis in the T_L-DNA has identified four loci (*rol*A, *rol*B, *rol*C, *rol*D) involved in hairy root formation (White *et al.*, 1985). Transformation of individual *rol* genes into plants has provided information on the function of these genes, both individually and in combination. In tobacco, the combined expression of *rol*A, *rol*B and *rol*C loci confers the full hairy root phenotype, whereas plants transgenic for single *rol* genes or various combinations show distinct specific growth abnormalities (Schmülling *et al.*, 1988; Mariotti *et al.*, 1989).

A variety of dicotyledonous plants are susceptible to *A. rhizogenes*. Tepfer (1990) lists 116 dicotyledonous species for which stable, transformed hairy root cultures have been reported, with plants regenerated from 37 species. A recent literature survey indicates that transgenic plants have been regenerated from hairy roots of 62 different taxa, representing 53 species from 24 families (Table 9.1). In addition to the Ri plasmid, over half of these plants have been transformed with foreign genes. A diverse range of dicotyledonous plant families is represented, including gymnosperms, with ten or more examples from each of the Fabaceae, Brassicaceae

99

Table 9.1. Transgenic plants obtained from *Agrobacterium rhizogenes*-mediated transformation.

Family and species	Common name	Gene(s) introduced[a]	Shoots[b]	Reference[c]
Actinidiaceae				
Actinidia deliciosa	kiwifruit	WT	S	Yazawa *et al.* (1995)
Apiaceae (=Umbelliferae)				
Daucus carota	carrot	WT	H	David *et al.* (1984)
Foeniculum vulgare	fennel	WT	S	Mugnier (1988)
Apocynaceae				
Catharanthus roseus	periwinkle	WT	S	Brillanceau *et al.* (1989)
Vinca minor	lesser periwinkle	NPTII, GUS	H	Tanaka *et al.* (1994)
Asteraceae (=Compositae)				
Cichorium intybus	chicory	WT	S	Sun *et al.* (1991)
Brassicaceae (=Cruciferae)				
Arabidopsis thaliana	mouse ear cress	WT	S	Pavingerova and Ondrej (1986)
Armoracia lapathifolia	horseradish	WT	S	Noda *et al.* (1987)
Brassica campestris	Chinese cabbage	NPTII	H	Christey *et al.* (1996)
	turnip	GUS, NPTII, ALS	H	Christey and Sinclair (1992)
B. napus	forage rape	GUS, NPTII, ALS	H	Christey and Sinclair (1992)
	oilseed rape	NPTII	H	Boulter *et al.* (1990)
	rapid cycling	NPTII	H	Boulter *et al.* (1990)
B. oleracea	broccoli	NPTII	H	Christey *et al.* (1996)
	Brussels sprouts	NPTII	H	Christey *et al.* (1996)
	cabbage	NPTII, GUS	H	Christey *et al.* (1996)
	cauliflower	NPTII, GUS	H	Christey *et al.* (1996)
	forage kale	GUS, NPTII, ALS	H	Christey and Sinclair (1992)
	rapid cycling	NPTII, GUS	S	Christey *et al.* (1996)
	ornamental kale	WT	H	Hosoki *et al.* (1989)
Casuarinaceae				
Allocasuarina verticillata		WT	S, H	Phelep *et al.* (1991)
Convolvulaceae				
Convolvulus arvensis	morning glory	WT	S	Mugnier (1988)
Ipomoea batatas	sweet potato	NPTII, GUS	S	Otani *et al.* (1993)
Crassulaceae				
Kalanchoe daigremontiana		WT	S	White *et al.* (1985)
Cucurbitaceae				
Cucumis sativus	cucumber	NPTII	H	Trulson *et al.* (1986)
Ebenaceae				
Diospyros kaki	Japanese persimmon	WT	H	Tao *et al.* (1994)
Fabaceae (=Leguminosae)				
Anthyllis vulneraria	kidney vetch	NPTII, *ipt*	S	Stiller *et al.* (1992)
Glycine argyrea	wild soybean	NPTII	H	Kumar *et al.* (1991)
G. canescens	wild soybean	NPTII	H	Rech *et al.* (1989)
Lotus corniculatus	bird's-foot trefoil	WT	S	Webb *et al.* (1990)
Medicago arborea		HPT	H	Damiani and Arcioni (1991)
M. sativa	alfalfa/lucerne	WT	H	Golds *et al.* (1991)
M. truncatula		NPTII	H	Thomas *et al.* (1992)
Onobrychis viciifolia	sainfoin	WT	S	Golds *et al.* (1991)
Robinia pseudoacacia	black locust	NPTII	H	Han *et al.* (1993)
Stylosanthes humilis		NPTII	H	Manners and Way (1989)

Table 9.1. Continued.

Family and species	Common name	Gene(s) introduced[a]	Shoots[b]	Reference[c]
Gentianaceae				
Eustoma grandiflorum	prairie gentian	WT	S	Handa *et al.* (1995)
Gentiana scabra	gentian	WT	H	Suginuma and Akihama (1995)
Geraniaceae				
Pelargonium sp.	lemon geranium	WT	S	Pellegrineschi *et al.* (1994)
Labiatae				
Ajuga reptans		WT	S	Tanaka and Matsumoto (1993)
Linaceae				
Linum usitatissimum	flax	WT	H	Zhan *et al.* (1988)
Myrtaceae				
Verticordia grandis		NPTII, GUS	H	Stummer *et al.* (1995)
Papaveraceae				
Papaver somniferum	opium poppy	WT	S	Yoshimatsu and Shimomura (1992)
Pinaceae				
Larix decidua	European larch	NPTII, *aro*A, BT	S	Shin *et al.* (1994)
Primulaceae				
Anagallis arvensis	pimpernel	WT	S	Mugnier (1988)
Rosaceae				
Malus pumila	apple	WT	H	Lambert and Tepfer (1992)
Salicaceae				
Populus trichocarpa × *P. deltoides*	hybrid poplar	NPTII	H	Pythoud *et al.* (1987)
Scrophulariaceae				
Antirrhinum majus	snapdragon	WT	S	Handa (1992)
Solanaceae				
Atropa belladonna	deadly nightshade	*bar*	S	Saito *et al.* (1992)
Hyoscyamus muticus	Egyptian henbane	WT	S	Oksman-Caldentey *et al.* (1991)
Lycopersicon esculentum	tomato	NPTII	H	Shahin *et al.* (1986)
L. peruvianum		NPTII	S	Morgan *et al.* (1987)
Nicotiana debneyi		NPTII	H	Davey *et al.* (1987)
N. glauca		WT	S	Sinkar *et al.* (1988)
N. hesperis		WT	S	Walton and Belshaw (1988)
N. plumbaginifolia		NPTII	H	Davey *et al.* (1987)
N. tabacum	tobacco	NPTII	H	Hatamoto *et al.* (1990)
Petunia hybrida	petunia	WT	S	Ondrej and Biskova (1986)
Solanum dulcamara	bittersweet	NPTII, *rol*	S	McInnes *et al.* (1991)
S. nigrum	black nightshade	NPTII	S	Davey *et al.* (1987)
S. tuberosum	potato	NPTII, GUS	H	Visser *et al.* (1989b)
Vitaceae				
Vitis vinifera	grapevine	NPTII, GUS	S	Nakano *et al.* (1994)

[a] WT: a wild-type *A. rhizogenes* strain was used. Otherwise gene(s) introduced are listed. ALS: mutant acetolactate synthase; *aro*A: mutant 5-enolpyruvylshikimate-3-phosphate synthase; *bar*: phosphinothricin acetyltransferase; BT: *Bacillus thuringiensis* toxin; GUS: β-glucuronidase; HPT: hygromycin phosphotransferase; *ipt*: isopentenyl transferase; NPTII: neomycin phosphotransferase II; *rol*: root loci genes.

[b] Method of shoot regeneration. H: hormone(s) required; S: spontaneous.

[c] This is not a comprehensive list, but provides one key reference for each plant.

Figure 9.1. (a) A group of control and hairy root derived transgenic Giant rape plants showing the lack of hairy root phenotype, one year after transfer to soil. (b) Medium Stem kale plants showing differences between control (left) and hairy root derived transgenic (right) plants, one year after transfer to soil.

and Solanaceae families. There are no examples of transgenic monocotyledonous plants, but onion (Dommisse *et al.*, 1990) and asparagus (Hernalsteens *et al.*, 1993) have been reported as hosts for *A. rhizogenes*.

A. rhizogenes-derived hairy roots and plants have application for many areas of research. For example, hairy root cultures have been used extensively in root nodule research (Jensen *et al.*, 1986; Macknight *et al.*, 1995; Petit *et al.*, 1987), for production of secondary products (Hamill *et al.*, 1987; Flores *et al.*, 1987), as an experimental system to study responses to chemicals (Downs *et al.*, 1994), and to study interactions with other organisms such as nematodes (Verdejo *et al.*, 1988) and mycorrhizal fungi (Tepfer, 1989). In addition, over recent years there has been increased interest in the use of *A. rhizogenes* due to the effect of *rol* genes on plant morphology and development and the ability to introduce foreign genes via *A. rhizogenes*-mediated transformation. In this chapter, use of *A. rhizogenes* is confined to specific examples with potential applications for plant improvement including: increasing rooting of recalcitrant plants, altering plant phenotype, and the introduction of foreign genes. Emphasis is placed on research where plants have progressed to greenhouse and field testing stages to demonstrate the successes and potential for using *A. rhizogenes* for plant improvement.

Regeneration of Plants

Infection of plants with *A. rhizogenes* is conducted either by co-cultivating explants or by inoculating wounds with a bacterial suspension. Hairy roots that develop are maintained *in vitro* on hormone-free media. Hairy root morphology varies considerably between species with differences in root thickness, degree of branching and amount of root hair production (see Figure 1 in Mugnier, 1988). Production of hairy roots is the normal response after *A. rhizogenes* infection; however, some plants show other responses. In larch, swelling, adventitious roots and adventitious buds were induced on wound-inoculated hypocotyls (Shin *et al.*, 1994). In poppy, calli that spontaneously regenerated shoots were induced by infection with *A. rhizogenes* (Yoshimatsu and Shimomura, 1992). In *Verticordia grandis*, Stummer *et al.* (1995) obtained tumours on inoculated explants, although abnormal roots and shoots were also obtained. In rapid cycling cabbage, transformed roots had a normal phenotype (Berthomieu and Jouanin, 1992). These differences in response to *A. rhizogenes* infection could be due to a complex mix of genotype–strain interactions in response to *rol* gene products, combined with the effects of endogenous and exogenous hormones.

Plants can be regenerated from hairy root cultures either spontaneously, directly from the root cultures, or by transferring roots to hormone-containing media. An advantage of *A. rhizogenes* over other methods of gene transfer is that spontaneous shoot regeneration is often obtained, which avoids a callus phase and therefore the risk of somaclonal variation in the resulting plants. Table 9.1 indicates that in half of the plants regenerated from hairy roots, regeneration occurred spontaneously. In some cases, e.g. horseradish (Noda *et al.*, 1987) and trefoil (Petit *et al.*, 1987), transferring root cultures from the dark to light was sufficient to induce spontaneous shoot regeneration.

A. rhizogenes-mediated transformation offers several advantages over other methods of transformation, including *A. tumefaciens*. While the host range of *A. rhizogenes* is narrower than that of *A. tumefaciens* (De Cleene and De Ley, 1981), there are examples where *A. rhizogenes* transformation is an efficient alternative for genetic engineering. In flax, Zhan *et al.* (1988) readily regenerated transgenic plants from hairy roots induced by *A. rhizogenes*, whereas lack of regeneration from transgenic callus precluded *A. tumefaciens*-mediated transformation. In *Stylosanthes humilis*, *A. rhizogenes*-mediated transformation produced a far higher rate of transformation. In addition, the transformation and regeneration system was more rapid than previous *A. tumefaciens* methods (Manners and Way, 1989). *A. rhizogenes*-mediated transformation can be an extremely rapid method to produce transgenic plants. For example, in black locust, hairy roots appeared within a week of co-cultivation, with regenerated shoots obtained within four weeks (Han *et al.*, 1993).

In genotypes where low transformation efficiencies are obtained, methods to increase the virulence of *A. rhizogenes* have been applied. These studies demonstrate approaches that may be used to further widen the host range of *A. rhizogenes*. Pythoud *et al.* (1987) found the presence of the *vir* region from the supervirulent plasmid pTiBo542 in *trans* dramatically increased the *A. rhizogenes* transformation efficiency in a poplar hybrid. In flax, Zhan *et al.* (1990) demonstrated that the *tzs* gene of pTiC58 may promote transformation, since hairy root induction was markedly increased by co-inoculation with an *A. tumefaciens* strain containing the *tzs* gene.

An advantage of *A. rhizogenes*-mediated transformation is that transgenic plants can be obtained without using a selection agent, thus avoiding use of chemicals which may inhibit shoot regeneration. Hairy root morphology is used for the primary selection of transgenic cells and provides an efficient alternative to chemical selection. High rates of co-transfer of genes on the second binary vector can occur in the absence of selection. In tobacco and

Medicago arborea, 67 and 70%, respectively, of hairy roots were co-transformed with T-DNA from a binary vector (Hatamoto *et al.*, 1990; Damiani and Arcioni, 1991).

Transformation with *A. rhizogenes* should prevent regeneration of chimeric plants. With *A. tumefaciens*-mediated transformation, non-transformed cells can develop due to ineffective selection caused by the nurse effect of surrounding transformed cells, resulting in a high frequency of escapes and/or regeneration of chimeric plants. In contrast, with hairy root cultures transformation is confirmed by hairy root growth, and hairy root clones consisting of only transformed cells can be obtained after several cycles of root-tip subculture. In addition, the maintenance of transgenic hairy root cultures as organ cultures allows long-term culture and subsequent shoot regeneration without the problems associated with cytological abnormalities in long-term callus cultures.

Agricultural Applications of *A. rhizogenes*

To Increase Rooting

In some species, root initiation is a factor limiting vegetative propagation. Due to the highly branching root system induced by *A. rhizogenes*, intentional inoculation with *A. rhizogenes* has been used to improve the rooting of cuttings from some recalcitrant crops, particularly woody species. The following examples demonstrate success with the use of *A. rhizogenes* to improve rooting in a wide range of plants, both *in vivo* and *in vitro*. Plants were successfully transferred to soil, unless stated otherwise.

Roy (1989) demonstrated a 10–20% increase in the rooting percentage of softwood cuttings of several fruit trees and herbaceous species including peach, apple, cherry, olive, *Choisya ternata*, *Elaeagnus pungens*, *Magnolia soulangiana*, *Pieris japonica*, and *Viburnum tinus*. Only rooted fruit tree cuttings were grown further and after 3 months hydroponic growth, there was no visible difference in shoot growth between controls and inoculated cuttings.

McAfee *et al.* (1993) noted improved rooting of *Pinus monticola*, *P. banksiana* and *Larix laricina* from various *in vitro* and *in vivo* sources. After 6 months greenhouse growth, plants were similar to controls. In *Eucalyptus* species, Macrae and Van Staden (1993) obtained increased *in vitro* rooting. Root quality was also improved with extensive lateral root development when plants were hardened off.

In hazelnut, rooting of softwood cuttings was increased by both *A. rhizogenes* and IBA treatments (Bassil *et al.*, 1991). However, *A. rhizogenes* treatment was advantageous as less bud abscission occurred, a factor which limits hazelnut propagation.

In almond, Damiano *et al.* (1995) obtained 97% rooting with *A. rhizogenes* infection of *in vitro* microcuttings. Strobel and Nachmias (1985) demonstrated that application of *A. rhizogenes* to almond bare root stock at planting increased root number and mass. A significant increase in leaf number, stem diameter and shoot elongation was noted during the first growing season in the field.

Rugini and Mariotti (1991) demonstrated successful rooting of olive, apple, almond and pistachio apical cuttings. Untreated cuttings did not root. Plant survival after transplanting was similar to auxin-treated controls for olive, but very low (10–15%) for the others. Strobel *et al.* (1988) conducted a field experiment over 41 months that demonstrated the long-term benefits of *A. rhizogenes* inoculation of bare rootstock olive trees. Treated trees grew faster and produced significantly more flowers, fruit, and oil per tree.

Transformation with *rol* genes is another method to exploit the root-inducing effects of *A. rhizogenes*. Kiwifruit plants transformed with *rol*A, *rol*B and *rol*C in combination via *A. tumefaciens* showed the characteristic hairy root phenotype. In addition, *in vitro* transgenic microcuttings had higher rooting (100%) than the control (< 40%) (Rugini *et al.*, 1991). Rinallo and Mariotti (1993) developed a method for clonal propagation of adult chestnut trees by *A. tumefaciens* transformation with *rol*B. *Rol*B in combination with growth regulator treatment and basal etiolation induced rapid root formation that enabled rapid repotting, with good survival of shoots 11 months after potting.

These examples demonstrate an agricultural advantage to root induction either by *A. rhizogenes* or by specific *rol* gene(s). Such methods have the potential to increase the efficiency of plant propagation in crops where root formation is difficult. In addition, such methods may enable rooting at non-optimal times of the year. Host range analyses are important to conduct as different *A. rhizogenes* strains produce a marked difference in the rooting of different cuttings (Roy, 1989). In addition, further studies on the subsequent development of the cuttings or plants, both in the greenhouse and field, are required. It is important to ascertain that the *A. rhizogenes*-derived root system is fully functional in water–nutrient transport, therefore providing long-term benefits. Other applications of *A. rhizogenes* for altered root development include the production of increased root mass for increased water and nutrient uptake to enable better adaptation to environmental stress. In addition, *A. rhizogenes*-mediated transformation has the potential to introduce foreign genes specifically into the root system, e.g. resistance to root pathogens or pests; resistance to heavy metals.

To Alter Phenotypes

The characteristic altered phenotype induced by *A. rhizogenes* transformation is usually regarded as undesirable, but these altered phenotypic features can have potential applications for plant improvement. For example, in the horticultural industry some morphological alterations such as: dwarfing, altered flowering, wrinkled leaves and/or increased branching, may appeal to customers and be useful for ornamental potted plants. In addition, reduced fertility and inhibition of flowering may be of benefit for environmental release of transgenics.

Pellegrineschi *et al.* (1994) improved the ornamental quality of a scented *Pelargonium* species by *A. rhizogenes* transformation. This plant has a pleasant odour but is unattractive due to its long internodes and chaotic, ungainly growth. Hairy root regenerants were of shorter stature, with increased leaf and branch production. In addition to improving ornamental quality, other associated benefits included increased rooting of cuttings, altered root system architecture, inhibition of flowering and increased production of essential oils.

A dwarfing response due to reduced internode distances is often noted in hairy root regenerants, e.g. *Ajuga reptans* (Tanaka and Matsumoto, 1993), prairie gentian (Handa *et al.*, 1995) and gentian (Suginuma and Akihama, 1995). In snapdragon, flower number was increased dramatically due to increased branching (Handa, 1992). These phenotypes may be of particular benefit for ornamental potted plants. Other phenotype alterations include altered root systems. In *Medicago arborea*, transformation totally changed the structure of the root system, from tap-root to fasciculated, a characteristic of *Medicago* species growing in semiarid regions (Damiani and Arcioni, 1991).

In addition to alterations in vegetative morphology, plants transformed with *A. rhizogenes* often show alterations in their life cycle. Damiani and Arcioni (1991) noted that some perennial forage legumes became annual, while others remained perennial but lost or delayed the capacity to flower. Carrot and chicory, which are normally biennial, became annual when transformed with *A. rhizogenes* (Tepfer, 1984; Sun *et al.*, 1991). Other flowering alterations noted include inhibition of flowering, e.g. *Ajuga reptans* (Tanaka and Matsumoto, 1993) and *Pelargonium* (Pellegrineschi *et al.*, 1994), and in gentian accelerated flowering was noted (Suginuma and Akihama, 1995).

In addition to the full hairy root phenotype induced by *A. rhizogenes*, it is possible to obtain more specific alterations in morphology and/or development by introduction of specific *rol* genes. This may produce genetic variability useful for incorporation into breeding programmes. Of particular interest is the demonstration that *rol*C expression leads to male sterility (Schmülling *et al.*, 1993).

The horticultural characteristics of greenhouse grown tobacco (Scorza *et al.*, 1994) and potato (Fladung, 1990) plants transformed with the *rol*C gene via *A. tumefaciens* have been studied in detail. Tobacco plants were dwarf in stature, flowered earlier, with smaller leaves, flowers and seed capsules than controls. Potato *rol*C transformants showed severe abnormalities including dwarfism and increased tillering. Tuber number per plant increased dramatically, but the tubers were very small. In *Medicago sativa* plants transformed with various combinations of *rol* genes, stem number per plant was increased in all cases, with large increases in *rol*B and *rol*C transformants (Frugis *et al.*, 1995). Flowering was delayed by 10 days in *rol*B plants. *Rol*B, *rol*C and *rol*B+C transgenics showed significant increases in dry root weight.

In tomato, plants transgenic for *rol*A showed severe leaf wrinkling, reduced flower bud length, hyperstyly and reduced pollen viability. *Rol*B plants were characterised by reduced apical dominance and reduced internode length (Van Altvorst *et al.*, 1992). In tobacco, *rol*B action led to stimulation of root formation and affected flower and leaf morphology, whereas *rol*A and *rol*C modified shoot morphology (Schmülling *et al.*, 1988). Similar results were obtained in *Solanum dulcamara* (McInnes *et al.*, 1991) and by Mariotti *et al.* (1989) who conducted detailed field evaluation of transgenic tobacco.

The demonstration of specific functions associated with specific *rol* gene(s) and combinations provides useful information to help predict the altered phenotype. However, the severity varies between genotypes and most studies are limited to solanaceous crops. Promoter changes and antisense genes can be used to alter expression of the *rol* gene(s) (Schmülling *et al.*, 1988, 1993). In addition, greenhouse and field evaluations are important to determine the full phenotype of transgenic plants. Visser *et al.* (1989a) and Davey *et al.* (1987) noted that after transfer to soil, transgenic plants showed additional morphological variations that were not apparent during *in vitro* growth. In field studies, Mariotti *et al.* (1989) noted that some traits previously described as part of the "hairy root phenotype", e.g. early flowering and increased development of axillary branches, fell within the variability of normal plants.

To Introduce Foreign Genes

The demonstration that a wide range of crops can be regenerated from *A. rhizogenes*-induced hairy roots (Table 9.1) is an important prerequisite to introducing foreign genes for traits of agricultural interest. Of the transgenic plants obtained to date from *A. rhizogenes*-mediated transformation, half have had foreign gene(s) inserted in addition to the Ri T-DNA (Table 9.1). In addition to reporter genes, e.g. GUS, and selectable marker genes, e.g. NPTII, traits of potential agricultural use have been introduced into larch and brassica crops (Table 9.1).

A. rhizogenes-mediated transformation can be used to produce transgenic hairy root cultures, and subsequently plants, containing foreign genes through the use of either co-integrate or binary vectors. Co-integrate Ri plasmids, where the additional foreign genes are contained within the Ri T-DNA, have been used to produce transgenic plants of tobacco (Comai *et al.*, 1985), *Solanum* and *Nicotiana* species (Davey *et al.*, 1987), trefoil (Jensen *et al.*, 1986) and tomato (Morgan *et al.*, 1987). Binary vectors have the foreign gene(s) located on an additional plasmid between T-DNA border elements. As T-DNA from the second binary vector can be mobilised in *trans* by *vir* gene products of the Ri plasmid, co-transformation results in the introduction of both the foreign gene(s) and the Ri T-DNA. This enables the production of transgenic plants containing foreign genes after regeneration from hairy roots. Binary vectors have been used to produce transgenic plants containing foreign genes in numerous crop plants including *Stylosanthes humilis* (Manners and Way, 1989), cucumber (Trulson *et al.*, 1986), potato (Visser *et al.*, 1989b) and several brassica crops (Christey and Sinclair, 1992; Christey *et al.*, 1996).

One perceived disadvantage of *A. rhizogenes*-mediated transformation, probably limiting its use for the introduction of foreign genes, is the altered phenotypes often obtained. As discussed above these altered phenotypes can be advantageous. In addition, there are examples where transgenic plants show no or minimal Ri phenotype effects. Manners and Way (1989) obtained some transgenic *Stylosanthes humilis* plants with normal phenotype. Molecular analysis revealed all plants with normal phenotype contained the binary vector T-DNA, but one lacked the T_L-DNA. In *Nicotiana* sp., Sinkar *et al.* (1988) noted that after several months growth, normal shoots developed from basal axillary buds of plants with the Ri phenotype. Molecular analysis indicated transcriptional inactivation of T_L-DNA in revertant normal shoots. Christey *et al.* (1994) obtained two lines of forage rape which were barely distinguishable from the non-transgenic control, in both vegetative and reproductive characteristics (Figure 9.1). In transgenic *L. corniculatus*, only a few minor changes in plant morphology were noted, even though two or more copies of T_L-DNA were present (Webb *et al.*, 1990). In rapid cycling cabbage, Berthomieu and Jouanin (1992) obtained transgenic roots with no Ri phenotypic characteristics. Some plants were also normal, though with reduced male fertility. Molecular analysis demonstrated the presence of T_L- and

T_R-DNA. There are several possible explanations for these reduced Ri phenotypic effects. In some cases, molecular studies have shown lack of insertion or expression of some or all of the Ri T-DNA. In addition, genotype, copy number and position effects could be responsible for the variation in Ri phenotypic effects in some transgenic plants.

In plants where Ri phenotypic effects are present and undesirable, it is still possible to obtain phenotypically normal transgenic plants. When binary vectors are used, due to the independent insertion of the Ri T-DNA and binary vector T-DNA, segregation of the T-DNAs at meiosis can occur in subsequent generations. This allows the identification of phenotypically normal transgenic plants, transformed with only the binary vector T-DNA containing the foreign DNA of interest, but without the Ri genes and therefore the associated phenotypic changes. The independent segregation of the hairy root phenotype from the other transgenes has been demonstrated in tobacco (Hatamoto *et al.*, 1990) and oilseed rape (Boulter *et al.*, 1990). In both cases, transgenic plants were outcrossed with wild-type pollen but, due to multiple insertion events, the segregation ratios did not show the expected Mendelian patterns.

Field Evaluation

From 1986 to 1994, over 1000 field trials of transgenic plants were conducted world-wide, mainly in North America and Europe, with potato, oilseed rape, tobacco and maize accounting for over half the trials (Deshayes, 1994). While *A. rhizogenes*-mediated transformation has been used to produce transgenic plants in a large number of crop plants (Table 9.1), there are only rare examples of the field evaluation of *A. rhizogenes*-transformed plants with these limited mainly to potato. As with *A. tumefaciens*-derived transgenic plants, field testing of *A. rhizogenes*-derived transgenic plants normally requires detailed submission to the appropriate regulatory authority.

Since the early trials of Ooms *et al.* (1986) who field-tested potato cvs. Désirée and Maris Bard transformed with wild-type *A. rhizogenes*, there have been several field trials of potatoes transgenic for foreign genes. Kuipers *et al.* (1992) conducted a field trial in The Netherlands with potatoes transformed via *A. rhizogenes* with an antisense version of the granule-bound starch synthase (GBSS) gene. As expected, GBSS activity and amylose content were significantly reduced compared with the non-transformed control. Under field conditions, dry matter content of the four transgenic lines was comparable to the control. Transformants produced fewer tubers per plant, with high variation between and within lines, compared with the control. In two lines, yield and

Figure 9.2. Field trial of Kapeti kale 9 weeks after transplanting showing the hairy root derived plants (H) and control seed derived plants (C). (Reproduced from MC Christey and BK Sinclair, 1993, Field-testing of Kapeti kale regenerated from *Agrobacterium*-induced hairy roots. *NZ J Agricult Res* 36: 389–392, Figure 1; with permission from SIR Publishing, Wellington, New Zealand.)

dry matter were significantly reduced compared with the control. Fertility was not studied as flower buds were removed every two days, presumably due to regulatory conditions. Stable antisense inhibition was noted in one line and subsequent field experiments demonstrated that this was reproducible (Kuipers *et al.*, 1994).

In New Zealand, Conner *et al.* (1991) field-tested potato cv. Iwa transgenic for kanamycin resistance and thaumatin production. The two transgenic lines exhibited a slight Ri phenotype with slightly wrinkled leaves and a more erect habit. Transgenic lines had a lower number of tubers per plant than the control. Weight of individual tubers varied, with one line higher and one line lower than the control. Flowering and berry formation were allowed, which enabled studies of pollen dispersal. No pollen dispersal was detected on wild-type plants within the trial or 5.25 m east of the trial. However, there was only limited seed set due to late planting and/or somaclonal variation resulting in some non-flowering phenotypes.

Siffel *et al.* (1992) field-tested potato cv. Zvikov that were transgenic for kanamycin resistance. Stems and leaves were normal but the inflorescence was condensed with more flowers than the controls. Field studies indicated that chlorophyll content, dry matter and water content were similar in controls and transgenics.

Christey and Sinclair (1993) conducted a small-scale field trial in New Zealand of Kapeti kale plants regenerated from hairy roots from *A. rhizogenes* strain A4T. Regulatory approval was not necessary as these plants were regenerated following transformation with a wild-type *A. rhizogenes* strain with no genetic manipulation, so that the transformation events that resulted in the hairy roots were natural phenomena. At all stages in the development of these plants, both *in vivo* and *in vitro*, hairy root derived plants were clearly distinguishable from control plants. Hairy root derived plants were shorter with smaller, more wrinkled leaves (Figure 9.2). Plants from both sources showed the same phenology. All plants were fertile with seed set considerably reduced on the hairy root derived plants. A field trial of several vegetable brassicas containing foreign genes is currently in progress. Ri phenotypic effects vary considerably between lines, with some lines exhibiting minimal effects (Christey, unpublished observations).

Conclusions

Recent progress in *A. rhizogenes*-mediated transformation of plants from production of hairy roots, to shoot regeneration, to introduction of foreign genes, through to field-testing of transgenic plants, indicates the potential of *A. rhizogenes* for the production of transgenic crop plants. This progress means that already available genes encoding a wide range of desirable traits can be introduced into numerous plants via *A. rhizogenes*-mediated transformation, including genes for herbicide, pest and disease resistance, environmental stress and quality attributes. In addition, the many demonstrations of other uses of *A. rhizogenes* and *rol* genes, including increased rooting and altered phenotypes, indicates the potential application of *A. rhizogenes* for many areas of plant improvement.

Acknowledgments

I thank Tony Conner for useful discussions and for comments on the manuscript. The assistance of Michelle Johnston with typing the references is appreciated.

References

Bassil NV, WM Proebsting, LW Moore and DA Lightfoot (1991). Propagation of hazelnut stem cuttings using *Agrobacterium rhizogenes*. *Hort Sci* 26: 1058–1060.

Berthomieu P and L Jouanin (1992). Transformation of rapid cycling cabbage (*Brassica oleracea* var. *capitata*) with *Agrobacterium rhizogenes*. *Plant Cell Rep.* 11: 334–338.

Boulter ME, E Croy, P Simpson, R Shields, RRD Croy and AH Shirsat (1990). Transformation of *Brassica napus* L. (oilseed rape) using *Agrobacterium tumefaciens* and *Agrobacterium rhizogenes* — a comparison. *Plant Sci* 70: 91–99.

Brillanceau MH, C David and J Tempé (1989). Genetic transformation of *Catharanthus roseus* G. Don by *Agrobacterium rhizogenes*. *Plant Cell Rep* 8: 63–66.

Christey MC and BK Sinclair (1992). Regeneration of transgenic kale (*Brassica oleracea* var. *acephala*), rape (*B. napus*) and turnip (*B. campestris* var. *rapifera*) plants via *Agrobacterium rhizogenes* mediated transformation. *Plant Sci* 87: 161–169.

Christey MC and BK Sinclair (1993). Field-testing of Kapeti kale regenerated from *Agrobacterium*-induced hairy roots. *NZ J Agricult Res* 36: 389–392.

Christey MC, BK Sinclair and RH Braun (1994). Phenotype of transgenic *Brassica napus* and *B. oleracea* plants obtained from *Agrobacterium rhizogenes* mediated transformation. *Abstr VIIIth Internat Congr Plant Tissue and Cell Culture*, Florence, Italy, p 157.

Christey MC, BK Sinclair, RH Braun and L Wyke (1996). Regeneration of transgenic vegetable brassicas (*Brassica oleracea* and *B. campestris*) via *Ri*-mediated transformation. *Plant Cell Rep.* Submitted for publication.

Comai L, D Facciotti, WR Hiatt, G Thompson, RE Rose and DM Stalker (1985). Expression in plants of a mutant *aro*A gene from *Salmonella typhimurium* confers tolerance to glyphosate. *Nature* 317: 741–744.

Conner AJ, MK Williams and MC Christey (1991). Report on field testing of genetically modified potatoes, asparagus and broccoli. Ministry for the Environment, Wellington, New Zealand.

Damiani F and S Arcioni (1991). Transformation of *Medicago arborea* L. with an *Agrobacterium rhizogenes* binary vector carrying the hygromycin resistance gene. *Plant Cell Rep* 10: 300–303.

Damiano C, T Archilletti, E Caboni, P Lauri, G Falasca, D Mariotti and G Ferraiolo (1995). *Agrobacterium* mediated transformation of almond: *in vitro* rooting through localised infection of *A. rhizogenes*. w.t. *Acta Hort* 392: 161–169.

Davey MR, BJ Mulligan, KMA Gartland, E Peel, AW Sargent and AJ Morgan (1987). Transformation of *Solanum* and *Nicotiana* species using an Ri plasmid vector. *J Exp Bot* 38: 1507–1516.

David C, MD Chilton and J Tempé (1984). Conservation of T-DNA in plants regenerated from hairy root cultures. *Bio/Technol* 2: 73–76.

De Cleene M and J De Ley (1981). The host range of infectious hairy-root. *Bot Rev* 47: 147–194.

Deshayes AF (1994). Environmental and social impacts of GMO's: what have we learned from the past few years? In *Proc 3rd Internat Symp on Biosafety Results of Field Tests of Genetically Modified Plants and Microorganisms*, edited by DD Jones, pp. 5–19. University of California.

Dommisse EM, DWM Leung, ML Shaw and AJ Conner (1990). Onion is a monocotyledonous host for *Agrobacterium*. *Plant Sci* 69: 249–257.

Downs CG, MC Christey, KM Davies, GA King, BK Sinclair and DG Stevenson (1994). Hairy roots of *Brassica napus*: II. Glutamine synthetase overexpression alters ammonia assimilation and the response to phosphinothricin. *Plant Cell Rep* 14: 41–46.

Fladung M (1990). Transformation of diploid and tetraploid potato clones with the *rol*C gene of *Agrobacterium rhizogenes* and characterization of transgenic plants. *Plant Breeding* 104: 295–304.

Flores HE, MW Hoy and JJ Pickard (1987). Secondary metabolites from root cultures. *Trends in Biotechnol* 5: 64–69.

Frugis G, S Caretto, L Santini and D Mariotti (1995). *Agrobacterium rhizogenes rol* genes induce productivity-related phenotypical modifications in "creeping-rooted" alfalfa types. *Plant Cell Rep* 14: 488–492.

Golds TJ, JY Lee, T Husnain, TK Ghose and MR Davey (1991). *Agrobacterium rhizogenes* mediated transformation of the forage legumes *Medicago sativa* and *Onobrychis viciifolia*. *J Exp Bot* 42: 1147–1157.

Grant JE, EM Dommisse, MC Christey and AJ Conner (1991). Gene transfer to plants using *Agrobacterium*. In *Advanced Methods in Plant Breeding and Biotechnology*, edited by DR Murray, pp. 50–73. CAB International.

Hamill JD, AJ Parr, MJC Rhodes, RJ Robins and NJ Walton (1987). New routes to plant secondary products. *Bio/Technol* 5: 800–804.

Han KH, DE Keathley, JM Davis and MP Gordon (1993). Regeneration of a transgenic woody legume (*Robinia pseudoacacia* L., black locust) and morphological alterations induced by *Agrobacterium rhizogenes*-mediated transformation. *Plant Sci* 88: 149–157.

Handa T (1992). Genetic transformation of *Antirrhinum majus* L and inheritance of altered phenotype induced by Ri T-DNA. *Plant Sci* 81: 199–206.

Handa T, T Sugimura, E Kato, H Kamada and K Takayanagi (1995). Genetic transformation of *Eustoma grandiflorum* with *rol* genes. *Acta Hort* 392: 209–218.

Hatamoto H, ME Boulter, AH Shirsat, EJ Croy and JR Ellis (1990). Recovery of morphologically normal transgenic tobacco from hairy roots co-transformed with *Agrobacterium rhizogenes* and a binary vector plasmid. *Plant Cell Rep* 9: 88–92.

Hernalsteens JP, B Bytebier and M Van Montagu (1993). Transgenic *Asparagus*. In *Transgenic Plants*, edited by SD Kung and R Wu, vol 2, pp. 35–46. *Present Status and Social and Economic Impacts*. Academic Press, San Diego.

Hosoki T, K Shiraishi, T Kigo and M Ando (1989). Transformation and regeneration of ornamental kale (*Brassica oleracea* var. *acephala* DC) mediated by *Agrobacterium rhizogenes*. *Scientia Hort* 40: 259–266.

Jensen JS, KA Marcker, L Otten and J Schell (1986). Nodule-specific expression of a chimaeric soybean leghaemoglobin gene in transgenic *Lotus corniculatus*. *Nature* 321: 669–674.

Kuipers GJ, JTM Vreem, H Meyer, E Jacobsen, WJ Feenstra and RGF Visser (1992). Field evaluation of antisense RNA mediated inhibition of GBSS gene expression in potato. *Euphytica* 59: 83–91.

Kuipers AGJ, E Jacobsen and RGF Visser (1994). Formation and deposition of amylose in the potato tuber starch granule are affected by the reduction of granule-bound starch synthase gene expression. *Plant Cell* 6: 43–52.

Kumar V, B Jones and MR Davey (1991). Transformation by *Agrobacterium rhizogenes* and regeneration of transgenic shoots of the wild soybean *Glycine argyrea*. *Plant Cell Rep* 10: 135–138.

Lambert C and D Tepfer (1992). Use of *Agrobacterium rhizogenes* to create transgenic apple trees having an altered organogenic response to hormones. *Theor Appl Genet* 85: 105–109.

Macknight RC, PHS Reynolds and KJF Farnden (1995). Analysis of the lupin *Nodulin-45* promoter: conserved regulatory sequences are important for promoter activity. *Plant Mol Biol* 27: 457–466.

Macrae S and J Van Staden (1993). *Agrobacterium rhizogenes*-mediated transformation to improve rooting ability of eucalypts. *Tree Physiol* 12: 411–418.

Manners JM and H Way (1989). Efficient transformation with regeneration of the tropical pasture legume *Stylosanthes humilis* using *Agrobacterium rhizogenes* and a Ti plasmid–binary vector system. *Plant Cell Rep* 8: 341–345.

Mariotti D, GS Fontana, L Santini and P Costantino (1989). Evaluation under field conditions of the morphological alterations ("hairy root phenotype") induced on *Nicotiana tabacum* by different Ri plasmid T-DNA genes. *J Genet Breed* 43: 157–164.

McAfee BJ, EE White, LE Pelcher and MS Lapp (1993). Root induction in pine (*Pinus*) and larch (*Larix*) spp. using *Agrobacterium rhizogenes*. *Plant Cell Tiss Organ Cult* 34: 53–62.

McInnes E, AJ Morgan, BJ Mulligan and MR Davey (1991). Phenotypic effects of isolated pRiA4 TL-DNA *rol* genes in the presence of intact TR-DNA in transgenic plants of *Solanum dulcamara* L. *J Exp Bot* 42: 1279–1286.

Morgan AJ, PN Cox, DA Turner, E Peel, MR Davey, KMA Gartland and BJ Mulligan (1987). Transformation of tomato using an Ri plasmid vector. *Plant Sci* 49: 37–49.

Mugnier J (1988). Establishment of new axenic hairy root lines by inoculation with *Agrobacterium rhizogenes*. *Plant Cell Rep* 7: 9–12.

Nakano M, Y Hoshino and M Mii (1994). Regeneration of transgenic plants of grapevine (*Vitis vinifera* L.) via *Agrobacterium rhizogenes*-mediated transformation of embryogenic calli. *J Exp Bot* 45: 649–656.

Noda T, N Tanaka, Y Mano, S Nabeshima, H Ohkawa and C Matsui (1987). Regeneration of horseradish hairy roots incited by *Agrobacterium rhizogenes* infection. *Plant Cell Rep* 6: 283–286.

Oksman-Caldentey KM, O Kivelä and R Hiltunen (1991). Spontaneous shoot organogenesis and plant regeneration from hairy root cultures of *Hyoscyamus muticus*. *Plant Sci* 78: 129–136.

Ondrej M and R Biskova (1986). Differentiation of *Petunia hybrida* tissues transformed by *Agrobacterium rhizogenes* and *Agrobacterium tumefaciens*. *Biol Plant* 28: 152–155.

Ooms G, ME Bossen, MM Burrell and A Karp (1986). Genetic manipulation in potato with *Agrobacterium rhizogenes*. *Potato Res* 29: 367–379.

Otani M, M Mii, T Handa, H Kamada and T Shimada (1993). Transformation of sweet potato (*Ipomoea batatas* (L.) Lam.) plants by *Agrobacterium rhizogenes*. *Plant Sci* 94: 151–159.

Pavingerova D and M Ondrej (1986). Comparison of hairy root and crown gall tumors of *Arabidopsis thaliana*. *Biol Plant* 28: 149–151.

Pellegrineschi A, JP Damon, N Valtorta, N Paillard and D Tepfer (1994). Improvement of ornamental characters and fragrance production in lemon-scented geranium through genetic transformation by *Agrobacterium rhizogenes*. *Bio/Technol* 12: 64–68.

Petit A, J Stougaard, A Kühle, KA Marcker and J Tempé (1987). Transformation and regeneration of the legume *Lotus corniculatus*: a system for molecular studies of symbiotic nitrogen fixation. *Mol Gen Genet* 207: 245–250.

Phelep M, A Petit, L Martin, E Duhoux and J Tempé (1991). Transformation and regeneration of a nitrogen-fixing tree, *Allocasuarina verticillata* Lam. *Bio/Technol* 9: 461–466.

Pythoud F, VP Sinkar, EW Nester and MP Gordon (1987). Increased virulence of *Agrobacterium rhizogenes* conferred by the *vir* region of pTiBo542: application to genetic engineering of poplar. *Bio/Technol* 5: 1323–1327.

Rech EL, TJ Golds, T Husnain, MH Vainstein, B Jones, N Hammatt, BJ Mulligan and MR Davey (1989). Expression of a chimaeric kanamycin resistance gene introduced into the wild soybean *Glycine canescens* using a cointegrate Ri plasmid vector. *Plant Cell Rep.* 8: 33–36.

Rinallo C and D Mariotti (1993). Rooting of *Castanea sativa* Mill. shoots: effect of *Agrobacterium rhizogenes* T-DNA genes. *J Hort Sci* 68: 399–407.

Roy MC (1989). *Plant Growth Response to Agrobacterium rhizogenes*. MApplSc thesis, Lincoln College, University of Canterbury, New Zealand.

Rugini E and D Mariotti (1991). *Agrobacterium rhizogenes* T-DNA genes and rooting in woody species. *Acta Hort* 300: 301–308.

Rugini E, A Pellegrineschi, M Mencuccini and D Mariotti (1991). Increase of rooting ability in the woody species kiwi (*Actinidia deliciosa* A Chev) by transformation with *Agrobacterium rhizogenes rol* genes. *Plant Cell Rep* 10: 291–295.

Saito K, M Yamazaki, H Anzai, K Yoneyama and I Murakoshi (1992). Transgenic herbicide-resistant *Atropa belladonna* using an Ri binary vector and inheritance of the transgenic trait. *Plant Cell Rep* 11: 219–224.

Schmülling T, J Schell and A Spena (1988). Single genes from *Agrobacterium rhizogenes* influence plant development. *EMBO J* 7: 2621–2629.

Schmülling T, H Röhrig, S Pilz, R Walden and J Schell (1993). Restoration of fertility by antisense RNA in genetically engineered male sterile tobacco plants. *Mol Gen Genet* 237: 385–394.

Scorza R, TW Zimmerman, JM Cordts, KJ Footen and M Ravelonandro (1994). Horticultural characteristics of transgenic tobacco expressing the *rolC* gene from *Agrobacterium rhizogenes*. *J Am Soc Hort Sci* 119: 1091–1098.

Shahin EA, K Sukhapinda, RB Simpson and R Spivey (1986). Transformation of cultivated tomato by a binary vector in *Agrobacterium rhizogenes*: transgenic plants with normal phenotypes harbor binary vector T-DNA, but no Ri-plasmid T-DNA. *Theor Appl Genet* 72: 770–777.

Shin DI, GK Podila, Y Huang and DF Karnosky (1994). Transgenic larch expressing genes for herbicide and insect resistance. *Can J For Res* 24: 2059–2067.

Siffel P, E Sindelkova, M Durchan and M Zajicova (1992). Photosynthetic characteristics of *Solanum tuberosum* L. plants transformed by *Agrobacterium* strains. I. Pigment apparatus. *Photosynthetica* 27: 441–447.

Sinkar VP, FF White, IJ Furner, M Abrahamsen, F Pythoud and MP Gordon (1988). Reversion of aberrant plants transformed with *Agrobacterium rhizogenes* is associated with the transcriptional inactivation of the T_L-DNA genes. *Plant Physiol* 86: 584–590.

Stiller J, V Nasinec, S Svoboda, B Nemcova and I Machackova (1992). Effects of agrobacterial oncogenes in kidney vetch (*Anthyllis vulneraria* L.). *Plant Cell Rep* 11: 363–367.

Strobel GA and A Nachmias (1985). *Agrobacterium rhizogenes* promotes the initial growth of bare root stock almond. *J Gen Microbiol* 131: 1245–1249.

Strobel GA, A Nachmias and WM Hess (1988). Improvements in the growth and yield of olive trees by transformation with the Ri plasmid of *Agrobacterium rhizogenes*. *Can J Bot* 66: 2581–2585.

Stummer BE, SE Smith and P Langridge (1995). Genetic transformation of *Verticordia grandis* (Myrtaceae) using wild-type *Agrobacterium rhizogenes* and binary *Agrobacterium* vectors. *Plant Sci* 111: 51–62.

Suginuma C and T Akihama (1995). Transformation of gentian with *Agrobacterium rhizogenes*. *Acta Hort* 392: 153–160.

Sun LY, G Touraud, C Charbonnier and D Tepfer (1991). Modification of phenotype in Belgian endive (*Cichorium intybus*) through genetic transformation by *Agrobacterium rhizogenes*: conversion from biennial to annual flowering. *Transgenic Res* 1: 14–22.

Tanaka N and T Matsumoto (1993). Regenerants from *Ajuga* hairy roots with high productivity of 20-hydroxyecdysone. *Plant Cell Rep* 13: 87–90.

Tanaka N, M Takao and T Matsumoto (1994). *Agrobacterium rhizogenes*-mediated transformation and regeneration of *Vinca minor* L. *Plant Tiss Cult Lett* 11: 191–198.

Tao R, T Handa, M Tamura and A Sugiura (1994). Genetic transformation of Japanese persimmon (*Diospyros kaki* L.) by *Agrobacterium rhizogenes* wild type strain A4. *J Jap Soc Hort Sci* 63: 283–289.

Taylor BH, RM Amasino, FF White, EW Nester and MP Gordon (1985). T-DNA analysis of plants regenerated from hairy root tumors. *Mol Gen Genet* 201: 554–557.

Tepfer D (1984). Transformation of several species of higher plants by *Agrobacterium rhizogenes*: sexual transmission of the transformed genotype and phenotype. *Cell* 37: 959–967.

Tepfer D (1989). Ri T-DNA from *Agrobacterium rhizogenes*: a source of genes having applications in rhizosphere biology and plant development, ecology, and evolution. In *Plant–Microbe Interactions: Molecular and Genetic Perspectives*, edited by T Kosuge and EW Nester, vol. 3, pp. 294–342. McGraw-Hill, USA.

Tepfer D (1990). Genetic transformation using *Agrobacterium rhizogenes*. *Physiol Plant* 79: 140–146.

Thomas MR, RJ Rose and KE Nolan (1992). Genetic transformation of *Medicago truncatula* using *Agrobacterium* with genetically modified Ri and disarmed Ti plasmids. *Plant Cell Rep* 11: 113–117.

Trulson AJ, RB Simpson and EA Shahin (1986). Transformation of cucumber (*Cucumis sativus* L.) plants with *Agrobacterium rhizogenes*. *Theor Appl Genet* 73: 11–15.

Van Altvorst AC, RJ Bino, AJ van Dijk, AMJ Lamers, WH Lindhout, F van der Mark and JJM Dons (1992). Effects of the introduction of *Agrobacterium rhizogenes rol* genes on tomato plant and flower development. *Plant Sci* 83: 77–85.

Verdejo S, BA Jaffee and R Mankau (1988). Reproduction of *Meloidogyne javanica* on plant roots genetically transformed by *Agrobacterium rhizogenes*. *J Nematol* 20: 599–604.

Visser RGF, A Hesseling-Meinders, E Jacobsen, H Nijdam, B Witholt and WJ Feenstra (1989a). Expression and inheritance of inserted markers in binary vector carrying *Agrobacterium rhizogenes*-transformed potato (*Solanum tuberosum* L.). *Theor Appl Genet* 78: 705–714.

Visser RGF, E Jacobsen, B Witholt and WJ Feenstra (1989b). Efficient transformation of potato (*Solanum tuberosum* L.) using a binary vector in *Agrobacterium rhizogenes*. *Theor Appl Genet* 78: 594–600.

Walton NJ and NJ Belshaw (1988). The effect of cadaverine on the formation of anabasine from lysine in hairy root cultures of *Nicotiana hesperis*. *Plant Cell Rep* 7: 115–118.

Webb KJ, S Jones, MP Robbins and FR Minchin (1990). Characterization of transgenic root cultures of *Trifolium repens*, *Trifolium pratense* and *Lotus corniculatus* and transgenic plants of *Lotus corniculatus*. *Plant Sci* 70: 243–254.

White FF, BH Taylor, GA Huffman, MP Gordon and EW Nester (1985). Molecular and genetic analysis of the transferred DNA regions of the root-inducing plasmid of *Agrobacterium rhizogenes*. *J Bacteriol* 164: 33–44.

Yazawa M, C Suginuma, K Ichikawa, H Kamada and T Akihama (1995). Regeneration of transgenic plants from hairy root of kiwi fruit (*Actinidia deliciosa*) induced by *Agrobacterium rhizogenes*. *Breeding Sci* 45: 241–244.

Yoshimatsu K and K Shimomura (1992). Transformation of opium poppy (*Papaver somniferum* L.) with *Agrobacterium rhizogenes* MAFF 03-01724. *Plant Cell Rep* 11: 132–136.

Zhan XC, DA Jones and A Kerr (1988). Regeneration of flax plants transformed by *Agrobacterium rhizogenes*. *Plant Mol Biol* 11: 551–559.

Zhan X, DA Jones and A Kerr (1990). The pTiC58 *tzs* gene promotes high-efficiency root induction by agropine strain 1855 of *Agrobacterium rhizogenes*. *Plant Mol Biol* 14: 785–792.

Artificial Seed Production Through Hairy Root Regeneration

10

Nobuyuki Uozumi and Takeshi Kobayashi

Department of Biotechnology, Faculty of Engineering, Nagoya University, Chikusa-ku,
Nagoya 464-01, Japan; email: takeshi@proc.nubio.nagoya-u.ac.jp

General Introduction

Agrobacterium rhizogenes is responsible for hairy root induction in infected sensitive plants. The phenomenon is due to the transfer, integration, and expression in the plant cell genome of DNA (T-DNA) originating from large plasmids called Ri (root inducing) plasmids (Tepfer, 1990). Hairy roots are similar to roots of the parent plant in shape. Genetic modification using *A. rhizogenes* plasmids as vectors is believed to be feasible for the improvement of plant properties and for the production of transgenic plants. Induced hairy roots have several superior properties such as higher inherent genetic stability and growth rate increment compared with tissue induced by growth regulators.

Since tissue-specific behaviour and synthesis are maintained in hairy roots, hairy root cultures are suitable for large-scale production of secondary metabolites and other chemicals. So far, application of hairy roots has been focussed most consistently in this area. However, higher levels of some target metabolites have been found in the leaves of plants regenerated from hairy roots, so that plant regeneration is necessary for production of these chemicals (Uozumi and Kobayashi, 1994). Therefore, as shown in Figure 10.1, there are two main strategies for obtaining biochemical products from hairy roots:

(1) high density culture of hairy roots using bioreactors in combination with effective control techniques; and
(2) micropropagation of plants regenerated from hairy roots.

Considering the above properties, transformed plant organs in the form of hairy roots provide a promising means for the biotechnological exploitation of plant cells.

Murashige (1978) has proposed a concept of "artificial seed" as an approach to the mass propagation of elite plant varieties. Artificial seed is prepared by encapsulating somatic embryos or regenerated tissue in polysaccharide gels such as alginate and coating with a membrane to prevent water evaporation. Artificial seeds are expected to be a reliable delivery system for clonal propagation of elite plants (Kitto and Janick, 1985; Redenbaugh *et al.*, 1986, 1987; Redenbaugh and Walker, 1990) with the potential for genetic uniformity, high yield and low cost of production. Plant cells used for artificial seeds must have a good ability to regenerate and a high resist-

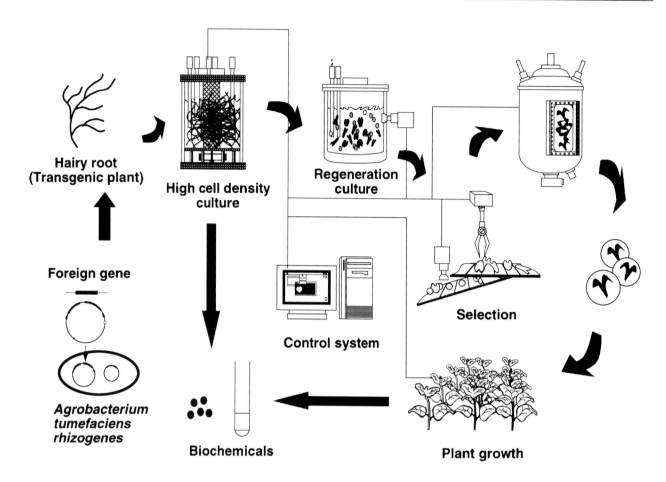

Figure 10.1. Process for production of biochemical products and plants from hairy roots.

ance to disease and mutation. Successful regeneration of whole plants from Ri-transformed cells has been obtained with various species (Tepfer, 1984; Ooms *et al.*, 1985; Noda *et al.*, 1987; Brillanceau and Tempé, 1989; Lambert and Tepfer, 1991; Phelep *et al.*, 1991). For some species of hairy root, it has also been observed that regeneration occurs in the light. Generally, genetic improvement of plants using conventional breeding and selection methods takes a long time. However, new gene transfer technologies offer the opportunity to produce plants easily with desirable traits such as disease or herbicide resistance within an acceptable time period. Reports of elite transgenic plant cells and their advantageous properties have stimulated interest in developing regeneration and delivery systems using hairy roots. Appropriate systems of plant regeneration are necessary for efficient production of transgenic plants from hairy roots. In particular, the production process should be constructed and improved to generate artificial seed from plants on an industrial scale.

Hairy roots must be cut to produce artificial seeds. The candidate tissue derived from hairy roots and used as explant material for artificial seeds can be classified as either root fragments, adventitious shoots, or plantlets. Figure 10.2 is a schematic diagram showing the different pathways for plant regeneration from these three kinds of explant. In this study, we focus on the bioengineering aspects of micropropagation from hairy roots using artificial seed systems (Uozumi and Kobayashi, 1995). The hairy root cells used to produce artificial seeds were classified as:

(1) hairy roots or root fragments;
(2) adventitious shoot primordia formed in the dark; or
(3) plantlets produced from hairy roots in the light.

In this study, regeneration frequencies associated with the three routes of plant regeneration from hairy roots were evaluated after excision, encapsulation and supplementation with growth regulators to develop manipulation procedures for artificial seeds. Horseradish (*Armoracia rusticana*) and *Ajuga reptans* hairy roots were used as model species.

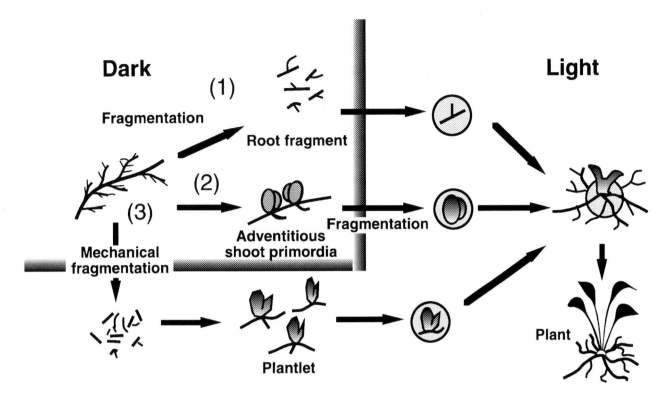

Figure 10.2. Three routes for plant regeneration from hairy roots.

Regeneration from Encapsulated Root Fragments

This method is the easiest of the three procedures for plant regeneration from hairy roots. Regeneration frequency is strongly correlated with the portion of the root encapsulated in the artificial seeds. This section describes how to increase the regeneration frequency (Uozumi *et al.*, 1992).

Dependence of Shoot Formation Frequency on the Portion of Hairy Root Encapsulated

Horseradish hairy roots were excised as fragments varying from 1.0 to 10 mm in length and encapsulated in alginate. If the root fragment did not have lateral roots, shoots appeared outside the beads after elongation of the root. If the encapsulated root fragment possessed lateral roots, shoots formed in the beads, and the shoot formation frequency was comparable with that of root fragments with only one apical meristem. The shoot formation frequency from root fragments without either apical meristem or lateral roots was significantly lower than from fragments with apical meristems or laterals.

The minimum root length enabling shoot formation was evaluated using encapsulated horseradish root fragments with apical meristems but no lateral roots. The shoot formation frequency increased with increasing root length up to 5 mm. Fragments of length greater than 5 mm produced shoots with almost equal frequency. These results indicated that fragments of 5 mm length were suitable for encapsulation.

Stimulation of Shoot Formation by Auxin

Auxin stimulated the emergence of root apical meristems and lateral roots. To harvest a large quantity of root fragments with apical meristems or lateral roots from a whole root culture, hairy roots can be treated with various concentrations of auxin, then randomly-picked root fragments can be encapsulated and transferred to Murashige and Skoog (MS) liquid medium in the light. Horseradish hairy roots treated with 1-naphthaleneacetic acid (NAA) or indole-3-butyric acid (IBA) showed a markedly higher shoot formation frequency compared with those without auxin treatment. In particular, the highest shoot formation frequency was obtained when roots were placed on medium containing 0.1 mg L^{-1} NAA. On the other hand, treatment with 2,4-dichlorophenoxyacetic acid (2,4-D) and indole-3-acetic acid (IAA) resulted in a decrease in shoot formation frequency.

Since NAA (0–1 mg L^{-1}) or IBA (1 mg L^{-1}) treatment apparently increased the number of lateral roots, the number of root fragments with lateral roots increased. Consequently, the shoot formation frequency of the root fragments with NAA or IBA treatment was higher than that without NAA or IBA treatment.

Healthy plantlets were formed in the light on the agar medium from horseradish hairy roots cultured with 0.1 mg L^{-1} and 1 mg L^{-1} NAA, and appeared larger than plantlets from non-treated roots. However, abnormal plantlet morphologies were observed in liquid culture with 5 mg L^{-1} NAA. From the frequencies of plantlet development and the morphological observations, the optimum NAA concentration for root preculture before encapsulation was thereby determined to be 0.1 mg L^{-1}. After encapsulation of roots, NAA inhibited plantlet formation extensively. Hence, plantlet development required removal of NAA from the medium in the light after preculture of hairy roots with NAA in the dark to enhance lateral root emergence.

For application of artificial seeds, carbohydrate should be contained only in the beads. Various concentrations of sucrose in the beads were tested. High shoot formation frequencies from horseradish hairy roots were observed at sucrose concentrations in the beads above 3%. Once leaves emerge from the roots, energy for differentiation and proliferation can be supplied by photosynthesis. Thus supplementation of the beads with carbohydrate is necessary only to produce plantlets from the root fragments.

Encapsulation of Adventitious Shoot Primordia

Horseradish and *Ajuga* hairy roots give rise to adventitious shoot primordia in dark culture. These adventitious primordia readily become green and grow into plantlets (Uozumi *et al.*, 1994a).

Morphology of Adventitious Shoot Primordia

When horseradish or *Ajuga* hairy roots were cultured in liquid in the dark, neoplasm appeared on the roots after a few weeks. One day after being transferred to light, the neoplasm turned green, and shoots emerged a few days later. From this observation, the neoplasm was considered to be adventitious shoot primordia. At the beginning of culture, adventitious shoot primordia developed in shake flasks only on the centre portions of whole roots which, being apart from the root apical meristems, were composed of older cells. Over the course of the culture, primordia also emerged close to the apical meristems. Although some of the primordia on horseradish hairy roots developed into etiolated plantlets in the dark, most remained 0.5–3 mm in size during the subculture period (3 weeks) in the dark. Taking the hairy root morphology into ac-

count, excision of roots carrying adventitious shoot primordia facilitated handling of the primordia for encapsulation. Adventitious shoot primordia formed in the dark were excised, encapsulated, and placed on agar medium in the light, and plantlets arose out of the beads after a few weeks. Small clusters of primordia on basal portions of lateral roots could be observed using scanning electron microscopy. Each adventitious shoot primordia consisted of two leaves.

Relationship Between Formation of Adventitious Shoot Primordia and Culture Time Before Encapsulation

The number of primordia and the size of individual primordia increased as the hairy root cultures proceeded. The number increased with culture times up to ca. 20 d and then reached a plateau. The maximum frequency of plantlet development occurred from primordia formed in 26 d cultures. After 40 d, ca. 43% of the adventitious shoot primordia on horseradish hairy roots formed multiple shoots which were difficult to grow into healthy plants, while the number of adventitious shoot primordia and the frequency of plantlet development in 8 d and 18 d cultures were low. These results suggest that primordia harvested early from the cultures were generally too premature to form plantlets, whereas the 40 d culture was too long resulting in a decreased frequency of plantlet development. Primordia harvested after 26–31 d culture were most suitable for use in artificial seeds, both in terms of number and plantlet development frequency. Hence these primordia were used in subsequent experiments.

To test the effect of auxin and cytokinin on primordia formation from horseradish hairy roots, roots were cultured in liquid MS medium containing varying concentrations of NAA (0–5 mg L^{-1}) or benzyladenine (BA; 0–5 mg L^{-1}) in the dark. NAA supplementation only stimulated the formation of callus which had a low capacity for plantlet development. Neither auxin nor cytokinin improved the number of adventitious shoot primordia or their regeneration frequency in the dark, although treatment with cytokinin increased the size of the primordia.

For embryogenesis from somatic adventitious shoots, root growth must be carried out *in vitro* under a high auxin concentration. In this work, use of hairy root adventitious shoot primordia overcame the difficulties of root emergence and elongation in a natural manner as root elongation occurred after encapsulation. We consider this an important advantage associated with root regeneration from hairy roots compared with the use of adventitious shoots derived from somatic embryos.

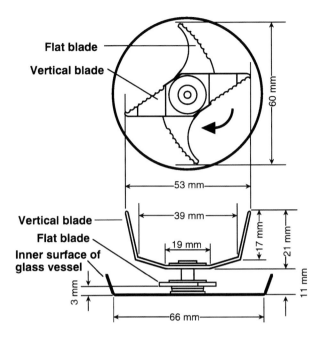

Figure 10.3. Schematic drawing of blades of the blender used for hairy root fragmentation.

Encapsulation of Plantlets Regenerated from Hairy Roots

Encapsulating plantlets seems to be an efficient method for growing up to the whole plant (Naka-shimada *et al.*, 1995). Procedures for production of large numbers of plantlets are examined in this section.

Fragmentation of Hairy Roots Using a Blender

Fragmentation of roots before plantlet formation is required to obtain plantlets suitable for use in artificial seeds. Mechanical fragmentation is preferred for large-scale production systems. Fragmentation was carried out using a commercial blender with blades shown diagrammatically in Figure 10.3. Horseradish hairy roots cultured for 35 d without addition of plant growth regulators were fragmented in the blender for 10, 20, 30 or 60 s. As the duration of fragmentation increased, the mean fragment size decreased. Root fragments obtained after fragmentation for 10 or 20 s were 1625 μm or less in length, while those obtained after fragmentation for 30 or 60 s were 1000 μm or less in length. Approximately 54% of the root fragments ($\geq$ 100 μm) on a fresh weight basis was recovered after fragmentation for 30 s. After fragmentation, the roots were transferred to plantlet formation medium without plant growth regulators. Shoots appeared after 4 d culture, and secondary roots began to elongate from the root fragments.

The number of plantlets formed from root fragments cut manually with a razor blade was 1.5 times higher than that produced from roots which were fragmented mechanically in the blender for 30 s. However, it took ca. 10 min to cut the roots manually to achieve a degree of fragmentation comparable to that achieved by fragmentation in the blender for 30 s. Although use of the blender for fragmentation resulted in a decrease in plantlet formation frequency, use of the blender is advantageous for rapid and large-scale root fragmentation.

Effect on Plantlet Formation of NAA Supplementation During Hairy Root Growth

As described above, auxin supplementation stimulated the emergence of root apical meristems on horseradish hairy roots and led to an increase in growth rate. In addition, NAA supplementation affected the frequency of plantlet formation from horseradish hairy roots. Hairy roots were cultured until stationary phase in liquid MS medium supplemented with various concentrations of NAA, fragmented in a blender for 30 s, and then cultured in liquid medium without plant growth regulators for 10 d. NAA supplementation (0.1 mg L^{-1} or 1.0 mg L^{-1}) resulted in an increase not only in the root growth rate but also in the plantlet formation frequency compared with non-treated roots. The frequency of plantlet formation from roots treated with 1.0 mg L^{-1} NAA increased about 1.2-fold compared with non-treated roots. However, supplementation with 5 mg L^{-1} NAA led to a decrease in plantlet formation frequency. The plantlet productivity of roots treated with 1.0 mg L^{-1} NAA was 79 L^{-1} d^{-1}, or 1.8 times that of non-treated roots. Horseradish hairy roots should therefore be cultured in medium supplemented with 1.0 mg L^{-1} NAA for 20 d at the hairy root growth stage.

Effect of Kinetin Supplementation on Plantlet Formation

The effects of kinetin supplementation at the plantlet formation stage were tested in an effort to stimulate development of plantlets from hairy roots. Addition of kinetin (0.01–1.0 mg L^{-1}) to horseradish root fragment culture led to an increase in the number of plantlets compared with that obtained without kinetin supplementation. The highest frequency of plantlet formation was achieved in medium supplemented with 0.1 mg L^{-1} kinetin. On the other hand, addition of more than 5 mg L^{-1} kinetin resulted in inhibition of plantlet formation, while 10 mg L^{-1} caused callus formation.

Kinetin supplementation also resulted in an increase in plantlet size. The optimal plantlet size for encapsulation is considered to be 2–4 mm, and control of plantlet size by adjusting the kinetin

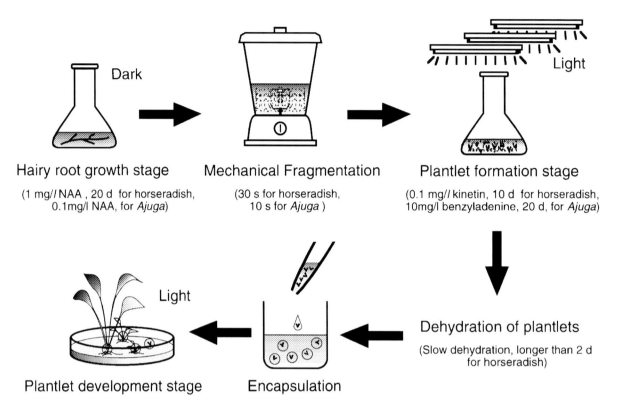

Figure 10.4. Artificial seed production system using plantlets derived from horseradish and *Ajuga* hairy roots fragmented in a blender. Optimal conditions determined in this study are indicated in parentheses.

concentration is necessary for efficient production of artificial seeds. Moreover, since plantlets with multiple shoots do not develop into healthy plants, plantlets with single shoots are needed for production of artificial seeds. Supplementation with 0.1 mg L^{-1} kinetin resulted in the largest number of horseradish plantlets 2–4 mm in size, and 91% of these plantlets had single shoots.

Morphological changes were observed during plantlet development from horseradish root fragments when the medium was supplemented with 0.1 mg L^{-1} kinetin. Four days after root fragmentation, development of two leaves from bud primordia was observed and chlorophyll was detected. At 6 d of culture, more extensive leaf development from bud primordia was observed.

The above results indicate that horseradish hairy roots should be cultured with 1.0 mg L^{-1} NAA before root fragmentation, and that the fragmented roots should be grown in medium supplemented with 0.1 mg L^{-1} kinetin to achieve optimal plantlet formation frequency (Figure 10.4). Fragmentation for 30 s resulted in formation of the highest number of plantlets 2–4 mm in size which had a single shoot, compared with that obtained after fragmentation for 10, 20 or 60 s.

Effect of Plantlet Dehydration on Frequency of Plantlet Development

At the plantlet development stage, roots emerge and two new leaves appear. The frequency of plantlet development from encapsulated plantlets was only 10% after 15 d culture and 58% after 45 d. On the other hand, non-encapsulated plantlets developed into healthy plants with high frequency (92% after 30 d), indicating that encapsulation decreases plantlet development frequency. When plants derived from encapsulated plantlets were observed under a microscope, their leaves were found to be water-soaked, translucent and glassy. These abnormal morphologies are characteristic of hyperhydricity. We considered that the inhibition of plantlet development from encapsulated plantlets was due to the prevention of dehydration after encapsulation.

Horseradish plantlets were encapsulated after dehydration by exposure to air flow in a laminar flow cabinet (rapid dehydration), and the frequency of plantlet development measured. After 3 h of dehydration in the air flow, the weight of the plantlets had decreased by 81% compared with the initial weight. Dehydration improved plantlet development frequency if the weight after dehydration was

Figure 10.5. (A) Artificial seeds containing encapsulated horseradish plantlets developed from hairy roots excised with a blender. The Ca-alginate gel beads shown are ca. 5 mm in diameter. (B) Plant regenerated from GUS-transformed *Ajuga* hairy root.

35% or less of the initial weight. Dehydration treatments longer than 2 h caused serious damage to the plantlets.

Dehydration treatment was also performed on horseradish plantlets cultured in sealed Petri dishes under light conditions (slow dehydration). After dehydration for 7 d resulting in a 23% decrease in plantlet weight, the frequency of plantlet development after 15 d culture reached its highest level of 98%, higher than that for plantlets subjected to rapid dehydration in the laminar flow cabinet. Slow dehydration was suitable for achieving high plantlet development frequency. In summary, the optimal conditions determined in this study are shown in parentheses in Figure 10.4. Figure 10.5A shows encapsulated plantlets.

Application of the Regeneration Procedure to *Ajuga* Hairy Roots and to Hairy Roots Containing a Foreign Gene

This section shows that the procedures for artificial seed production proposed above can be applied to other types of hairy root (Uozumi *et al.*, 1996), such as *Ajuga reptans* hairy roots and GUS-transformed hairy roots. *Ajuga* hairy roots produce an insect-moulting hormone, 20-hydroxyecdysone, at levels about four times higher than those from the original plant (Matsumoto and Tanaka, 1991).

Regeneration Procedure for *Ajuga* Hairy Roots

Rates of growth of *Ajuga* hairy roots in liquid MS medium supplemented with IBA, NAA and IAA were compared with those obtained without auxin addition. Branching was vigorous in cultures with IBA, NAA and IAA. Supplementation with NAA resulted in the highest shoot formation frequency, ca. 1.2 times that for non-treated roots. The optimum concentration of NAA for both hairy root growth and plantlet formation was 0.1 mg L^{-1}. Supplementation of hairy root growth medium with 0.1 mg L^{-1} NAA enhanced the rate of cell growth and the frequency of regeneration at the next plantlet formation stage.

Mechanical fragmentation for 10 s in the commercial blender illustrated in Figure 10.3 allowed the most efficient production of a large number of root fragments compared with treatments for 5, 15 and 20 s.

Shoots seldom emerged from fragments of *Ajuga* hairy roots in MS medium without growth regulators. To enhance shoot formation from root fragments, BA was added to the medium. This caused a dramatic increase in the number of plantlets obtained from root fragments produced by the blender. BA supplementation at 10 mg L^{-1} resulted in the highest number of plantlets.

After the *Ajuga* plantlets were encapsulated in Ca-alginate containing MS medium and glucose and placed on agar, they turned brown and did not grow. The dehydration treatment should be examined for improvement of regeneration frequency. On the other hand, non-encapsulated plantlets were able to grow up into plants. Plantlets of different size were transferred onto solid MS medium for further growth and development. Not only the size of the plantlet but also root emergence were key factors affecting proper plant development. As the original length of the plantlets increased, the number of plantlets formed with roots also increased. Plantlets more than 4 mm in length had roots and grew into plants with relatively high efficiency (86%). Figure 10.4 shows the procedures developed in this study for plant regeneration from *Ajuga* hairy roots.

Application of Transgenic Plants for Artificial Seed Production

Transgenic *Ajuga* hairy roots were obtained by *A. rhizogenes*-mediated co-transformation of the GUS gene under the control of the light-inducible promoter, ribulose-1,5-bisphosphate carboxylase/oxygenase small subunit, *rbc*S (Uozumi *et al.*, 1994b), and the root-inducing gene on the Ri plasmid. Seven transformants produced mikimopine, and one of them designated as PTR-4 was used in this study. *Ajuga* hairy roots capable of growing on solid medium containing kanamycin were selected and tested for opine production.

One hairy root clone showing GUS activity was cultured in liquid medium and used to test the plant regeneration procedure developed in this study. The final hairy root biomass obtained after addition of 0.1 mg L^{-1} NAA to the medium was 3.1 times higher than in control cultures without NAA. The hairy roots were subjected to fragmentation in a blender and the root fragments transferred to medium containing 10 mg L^{-1} BA. The number of root fragments after treatment with NAA was ca. 1.6 times that without NAA. NAA also increased the cell mass and number of plantlets. GUS activity was detected in leaves of the regenerated plants by histochemical reaction. There was no significant difference in appearance between plants regenerated from *Ajuga* hairy roots and GUS-positive hairy roots (Figure 10.5B). GUS-transformed *Ajuga* hairy roots responded well to the regeneration procedures determined in the above section.

Prospects

Growth of the transgenic plants produced from hairy roots in this study has been restricted to test pots inside the laboratory. However, some transgenic plants have recently been tested in the USA outside of the laboratory. Transgenic plants will be accepted for field growth in the near future, and artificial seed systems using hairy roots will become more useful.

In this study, application of a systematic regeneration procedure to *Ajuga* hairy roots shows the usefulness of artificial seed production in industrial plant production. Induction of hairy roots by *A. rhizogenes* can generate plant cells with high levels of secondary metabolites and with extremely stable regeneration ability after long-term culture (Ooms *et al.*, 1985; Phelep *et al.*, 1991). Since many hairy roots with properties superior to the original plant have been induced so far, artificial production systems are strongly required for mass propagation of hairy roots. This study has shown that, like somatic embryos, hairy roots have the potential

for use as artificial seed. Three regeneration routes from hairy roots have been proposed as basic procedures. The use of root fragments is the most convenient method for developing encapsulated hairy roots. Adventitious shoot primordia offer the advantages of high regeneration potential and relatively uniform greening under light conditions compared with the other cell materials tested. It is necessary to increase the number of shoots formed in the artificial seed system. Encapsulated plantlets have the highest regeneration frequency although the success of this procedure depends on the properties of the hairy roots. The procedures can be made more sophisticated after further extensive studies.

Process engineering including bioreactor design, high density cell culture, image analysis with computers and robotic systems can improve artificial seed production. High density culture of hairy roots could be performed in turbine reactors (Uozumi *et al.*, 1991, 1993a, 1995). On the other hand, machine vision analysis has been developed for celery somatic embryogenesis (Uozumi *et al.*, 1993b). These techniques are applicable to hairy root regeneration systems and improve several time-consuming and labour-intensive processes involved in conventional micropropagation.

References

Brillanceau MH and J Tempé (1989). Genetic transformation of *Catharanthus roseus* G. Don by *Agrobacterium rhizogenes*. *Plant Cell Rep* 8: 63–66.

Kitto SL and J Janick (1985). Production of synthetic seeds by encapsulating asexual embryos of carrot. *J Amer Soc Hort Sci* 110: 277–282.

Lambert C and D Tepfer (1991). Use of *Agrobacterium rhizogenes* to create chimeric apple trees through genetic grafting. *Bio/Technol* 9: 80–83.

Matsumoto T and N Tanaka (1991). Production of phytoecdysteroids by hairy root cultures of *Ajuga reptans* var. *atropurpurea*. *Agric Biol Chem* 55: 1019–1025.

Murashige T (1978). Plant growth substances in commercial uses of tissue culture. In *Frontiers of Plant Tissue Culture*, edited by T Thorpe, pp. 15–26. Calgary: International Association of Plant Tissue Culture.

Nakashimada Y, N Uozumi and T Kobayashi (1995). Production of plantlets for use as artificial seeds from horseradish hairy roots fragmented in a blender. *J Ferment Bioeng* 79: 458–464.

Noda T, N Tanaka, Y Mano, S Nabeshima, H Ohkawa and C Matsui (1987). Regeneration of horseradish hairy roots incited by *Agrobacterium rhizogenes* infection. *Plant Cell Rep* 6: 283–286.

Ooms G, A Karp, MM Burrell, D Twell and J Roberts (1985). Genetic modification of potato development using Ri T-DNA. *Theor Appl Genet* 70: 440–446.

Phelep M, A Petit, L Martin, E Duhoux and J Tempé (1991). Transformation and regeneration of a nitrogen-fixing tree, *Allocasuarina verticillata* Lam. *Bio/Technol* 9: 461–466.

Redenbaugh K and K Walker (1990). Role of artificial seeds in alfalfa breeding. In *Plant Tissue Culture: Applications and Limitations*, edited by SS Bhojwani, pp. 102–135. Amsterdam: Elsevier.

Redenbaugh K, BD Paasch, JW Nichol, ME Kossler, PR Viss and KA Walker (1986). Somatic seeds: encapsulation of asexual plant embryos. *Bio/Technol* 4: 797–801.

Redenbaugh K, D Slade, P Viss and J Fujii (1987). Encapsulation of somatic embryos in synthetic seed coats. *HortScience* 22: 803–809.

Tepfer D (1984). Genetic transformation of special species of higher plants by *Agrobacterium rhizogenes*: phenotypic consequences and sexual transmission of transformed genotype and phenotype. *Cell* 37: 959–967.

Tepfer D (1990). Genetic transformation using *Agrobacterium rhizogenes*. *Physiol Plant* 79: 140–146.

Uozumi N and T Kobayashi (1994). Application of hairy root and bioreactors. In *Advances in Plant Biotechnology*, edited by DDY Ryu and S Furusaki, pp. 307–338. Amsterdam: Elsevier.

Uozumi N and T Kobayashi (1995). Artificial seed production through encapsulation of hairy root and shoot tips. In *Biotechnology in Agriculture and Forestry: Somatic Embryogenesis and Synthetic Seed I*, edited by YPS Bajaj, pp. 170–180. Berlin: Springer-Verlag.

Uozumi N, K Kohketsu, O Kondo, H Honda and T Kobayashi (1991). Fed-batch culture of hairy root using fructose as a carbon source. *J Ferment Bioeng* 72: 457–460.

Uozumi N, Y Nakashimada, Y Kato and T Kobayashi (1992). Production of artificial seed from horseradish hairy root. *J Ferment Bioeng* 74: 21–26.

Uozumi N, K Kohketsu and T Kobayashi (1993a). Growth and kinetic parameters of *Ajuga* hairy root in fed-batch culture on monosaccharide medium. *J Chem Tech Biotechnol* 57: 155–161.

Uozumi N, T Yoshino, S Shiotani, K Suehara, F Arai, T Fukuda and T Kobayashi (1993b). Application of image analysis with neural network for plant somatic embryo culture. *J Ferment Bioeng* 76: 505–509.

Uozumi N, Y Asano and T Kobayashi (1994a). Micropropagation of horseradish hairy root by means of adventitious shoot primordia. *Plant Cell Tiss Organ Cult* 36: 183–190.

Uozumi N, Y Inoue, K Yamazaki and T Kobayashi (1994b). Light activation of expression associated with the tomato *rbcS* promoter in transformed tobacco cell line BY-2. *J Biotechnol* 36: 55–62.

Uozumi N, S Makino and T Kobayashi (1995). 20-Hydroxyecdysone production in *Ajuga* hairy root controlling intracellular phosphate content based on kinetic model. *J Ferment Bioeng* 80: 362–368.

Uozumi N, Y Ohtake, Y Nakashimada, Y Morikawa, N Tanaka and T Kobayashi (1996). Efficient regeneration from *GUS*-transformed *Ajuga* hairy roots. *J Ferment Bioeng* 81: 374–378.

Mycorrhizal Interactions and the Effects of Fungicides, Nematicides and Herbicides on Hairy Root Cultures

11

Jacques Mugnier

Rhône–Poulenc Agrochimie, 69263 Lyon Cedex 09, France; email: jacques.mugnier@rp.fr

Introduction

In vitro experiments with root organ cultures grown in Petri dishes offer the means to simplify the study of complicated root–organism interactions. Hairy roots are used here to establish cultures of obligate organisms (the vesicular-arbuscular mycorrhizae *Glomus mosseae* and *Gigaspora margarita*, and the Myxomycetes *Plasmodiophora brassicae* and *Polymyxa betae*), and to study the interactions with non-obligate parasites (*Rhizoctonia solani, Fusarium oxysporum, Cercospora beticola, Thielaviopsis basicola, Sclerotinia sclerotiorum*, and *Pythium* spp.). These organisms traditionally studied by mycologists include fungi, algae and plasmodial myxomycetes. In phylogenetic trees based on ribosomal DNA sequences (Taylor *et al.*, 1994), the Plasmodiophorales diverged from the Fungi. This is consistent with the many differences between the life cycle of Myxomycetes and those of fungi. Similarly, ribosomal sequence analysis of Oomycetes (*Pythium, Phytophthora*) shows these organisms to be part of the clade including brown algae, again consistent with morphologic characters (biflagellate zoospores). *Rhizoctonia solani* (teleomorph *Thanatephorus cucumeris*) is placed in the Corticiaceae (Basidiomycetes). The teleomorph of *Fusarium oxysporum* is *Gibberella* (Hypocreales, Ascomycetes) and *Cercospora betae* known exclusively as asexual forms is related to *Mycosphaerella* (Ascomycetes). *Sclerotinia* is placed in the Heliotiales (Ascomycetes). The teleomorphic position of *Thielaviopsis* is unknown. *Glomus mosseae* and *Gigaspora margarita* belong to the Endogonales (Zygomycetes). Hairy roots have also been used to investigate plant–nematode interactions. The effect of various fungicides, nematicides and herbicides was studied in this work through application of hairy root cultures.

Vesicular-Arbuscular Mycorrhizal Fungi

Vesicular-arbuscular (VA) mycorrhizae engage in mutually beneficial associations with land plants. The study and exploitation of this relationship have been hampered by the inability to culture the fungi. Mugnier *et al.* (1982) report that hairy root cultures are a suitable substrate for the symbiotic culture of *Glomus mosseae* Gerd. & Trappe and *Gigaspora margarita* Beker & Hall, two fungi capable of associating with numerous crop species. A bicompartmental Petri dish culture system (Mugnier and Mosse, 1987) was employed so that factors influencing root physiology and the process of infection

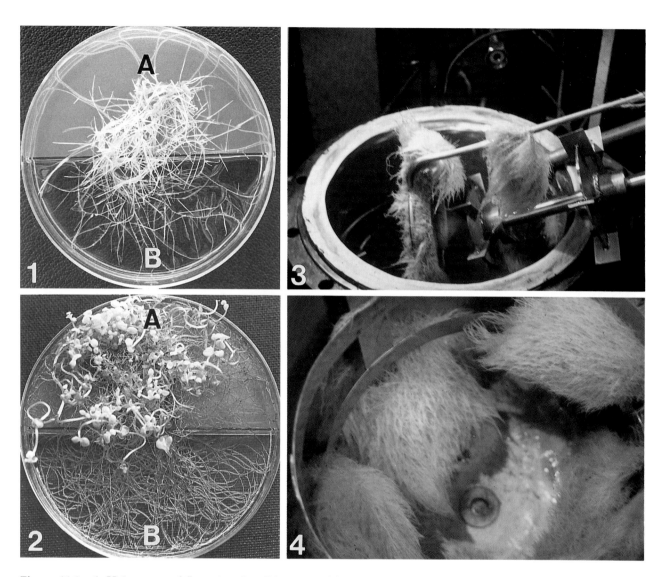

Figure 11.1.   1. Hairy roots of flowering flax (*Linum grandiflorum*) after 10 d growth in a 9-cm diameter bicompartmental Petri dish culture system. Compartment A contains Murashige and Skoog agar medium; compartment B contains water. 2. Hairy roots of pimpernel (*Anagalis arvensis*) after 10 d growth in a 9-cm diameter bicompartmental Petri dish. Compartment A contains Murashige and Skoog agar medium; compartment B contains water and the bleaching herbicide, norflurazon. Shoots regenerated only in compartment A. Note the bleaching effect on the regenerated shoots. 3, 4. Hairy roots of morning glory (*Convolvulus sepium*) after 14 d growth in an industrial 50-L fermenter.

could be regulated independently. The root culture was placed in compartment A (Figure 11.1) containing root growth medium. Root tips grew into compartment B where the medium consisted of distilled water, fragments of peat neutralised by $CaCO_3$, and agar. Compartment A simulated the aerial parts of the plant, and compartment B the soil deprived of nutrients. Spores were placed in the proximity of the root tips in compartment B. The composition of the nutrient medium was varied and the development of hyphae and subsequent infection were monitored. Although it was known that application of phosphate to whole plants inhibits VA mycorrhiza–plant interactions, phosphate concentrations did not affect hyphal development and root infection in this system. We found, however, that the essential determinant in the infection process was the concentration of nitrogen (NH_4NO_3 and KNO_3). At greater than 2 mM total reduced nitrogen, hyphal growth from spores was completely inhibited. In the range 0.2–1 mM the hyphae grew but were non-infective, passing through the medium without infecting the root. When the medium contained less than 0.2 mM reduced nitrogen, the hyphae formed highly branched, septate structures (Day 3), penetrated the roots (Day 5), and formed arbuscules (Day 10). In the final stage, the hyphae colonised the exterior of the root and the surrounding medium. Interestingly, all roots were not infected by the fungus; those which produced root hairs appeared to be immune. Use of such a system for studies of radioactive nutrient exchange (sucrose, phosphate) between hairy roots and fungus is discussed by Mugnier and Mosse (1987).

Other workers have described typical infections of hairy roots and production of spores by *Gigaspora margarita* (Bécard and Fortin, 1988; Bécard and Piché, 1990, 1992; Diop *et al.*, 1992; Balaji *et al.*, 1994). Chabot *et al.* (1992) showed that hairy root culture is also suitable for growing *Glomus intraradix*. However, consistent infection of a large-scale hairy root culture, a prerequisite for producing a commercially useful inoculum, was not achieved. Kilogram quantities of hairy roots can be readily obtained over 2–3 weeks of culture in fermenters (Figures 11.1.3 and 11.1.4). Systemic infection of such a culture will depend on defining the conditions in which *in vitro* symbiosis can occur on a large scale. The infection process is dependent on the physiology and morphology of the hairy roots: they must be subjected to nitrogen starvation and be devoid of root hairs. In Petri dishes, however, we believe that the co-culture system we have described can be used to study plant–microorganism interactions.

Pythium aphanidermatum

Responses of Zoospores to Hairy Roots

Zoospores play important roles in the life cycle of *Pythium* and *Phytophthora*. The possibility that root exudates increase their activity around roots has long been recognised. Zoospores normally accumulate at the site of maximum root exudation, namely, immediately behind the root tip, and at the point where adventitious roots erupt through the epidermis. We have investigated the taxic responses of *Pythium aphanidermatum* (Edson) Fitzp. to hairy roots of snapdragon (*Antirrhinum majus* L.) and flowering flax (*Linum grandiflorum* Desf.). Hairy root cultures were also used to investigate other aspects of root diseases, in particular, movement and selective action of fungicides. To evaluate zoospore–root interactions without allowing contact between zoospores and the high-salt Murashige and Skoog medium, the bicompartmental culture system (Figure 11.1) was used. Compartment A contained Murashige and Skoog medium and compartment B contained distilled water. A freshly-made zoospore suspension

containing 500–1000 zoospores was gently poured into compartment B. The progress of infection by zoospores was followed under a light microscope over a 0–10 min period. For long-term experiments, the cultures were incubated at 28°C.

Immediately after their introduction into compartment B, the zoospores moved towards the hairy roots with frequent changes in the direction of movement. In the immediate vicinity of the roots, zoospores swam smoothly following a helical rotating path. This was followed a few seconds later by immobilisation on the root surface and encystment of spores settling on the roots. Zoospores clustered just behind the root tips, at points where adventitious roots erupted, and on the root hairs (Figure 11.2.1); others parts of the root were non-attractive. However, when hairy roots were punctured or damaged in otherwise non-attractive regions, immediate massive accumulation of zoospores followed wounding (Figures 11.2.2 and 11.2.3). After aggregation and encystment the spores germinated: the emerging germ tubes grew towards the root surface and commonly produced dichotomous branched hyphae (Figure 11.2.4), their tips forming small appressoria by which *Pythium* penetrated the epidermal cells. Invasion of roots through either cracks or wounds was never observed. Hyphae which had penetrated snapdragon root hair grew in both directions, towards and away from the root tip (Figure 11.2.4). The hyphae were constricted at the point where root cells had penetrated. Thereafter, the penetration peg immediately enlarged to the normal diameter of the hyphae and this process was repeated as the hyphae passed through successive cell walls. The mycelium appeared to colonise the snapdragon tissues without apparent discoloration or death beneath the infection process. After 2 d there was an extensive external mycelium and the thoroughly colonised roots developed a brownish discoloration. In flax, when appressoria were formed, the root cells in these areas turned red and the invading hyphal growth stopped completely. Presumably, the resistance mechanism was due to hypersensitive reactions and/or inhibitory phenolic compounds produced by flax hairy roots. The behaviour of zoospores of *Phytophthora parasitica* to tomato hairy roots and that of zoospores of *Aphanamyces cochlioides* to sugar beet hairy roots were essentially the same.

Effect of Fungicides

Experiments were carried out on snapdragon hairy roots highly susceptible to *Pythium* infection. Well-established root cultures were treated in compartment A of the bicompartmental culture system with fungicides. After a 4 d uptake period, zoospores were placed in compartment B. Tests were duplicated in

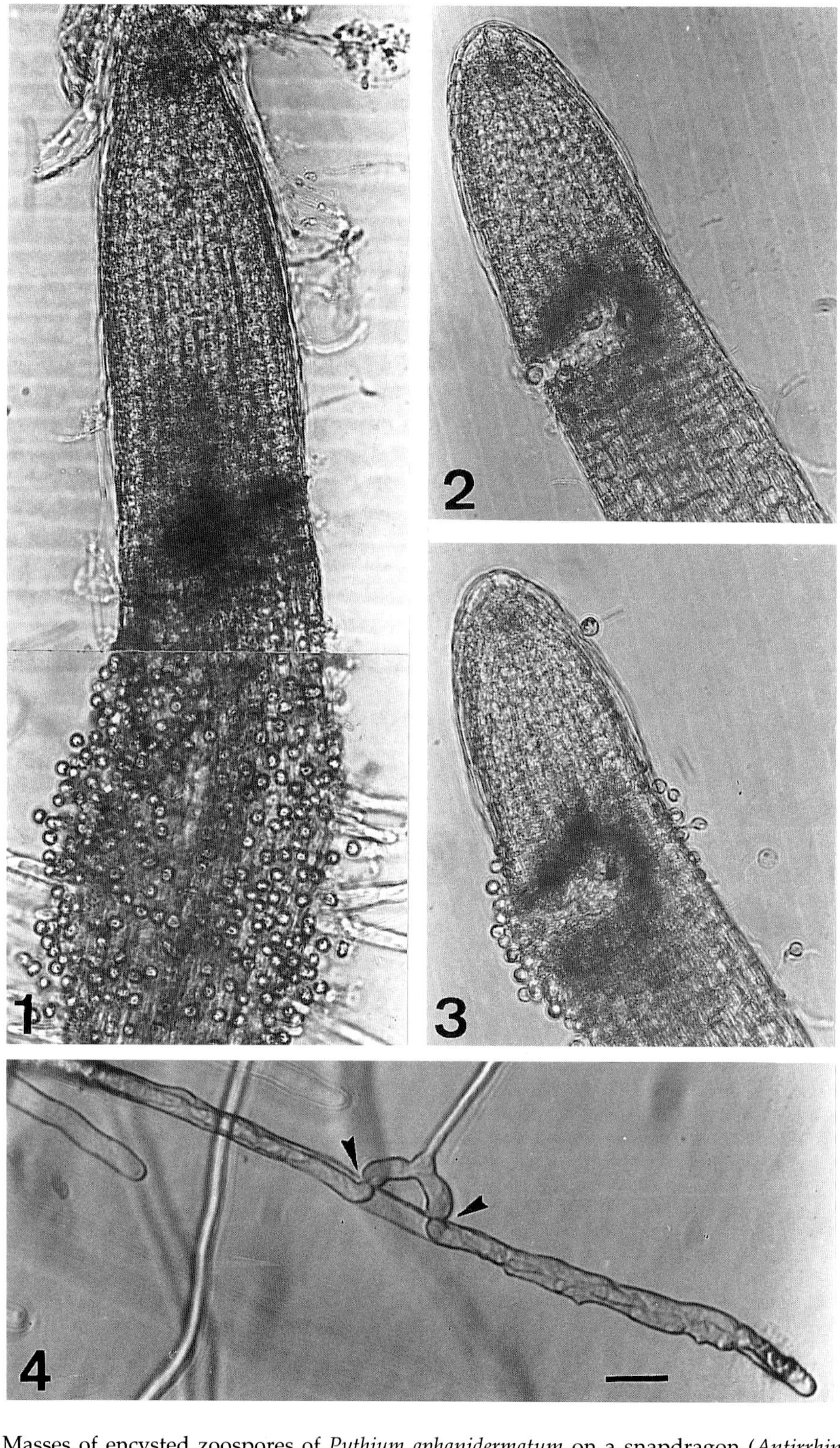

Figure 11.2.   1. Masses of encysted zoospores of *Pythium aphanidermatum* on a snapdragon (*Antirrhinum majus*) hairy root in the region of root elongation. Scale bar (shown in 4) = 40 μm. 2. Damaged snapdragon hairy root. Scale bar (shown in 4) = 40 μm. 3. Accumulation of zoospores following wounding. Scale bar (shown in 4) = 40 μm. 4. Branched appressorial pegs (arrowed) from a germ tube of *Pythium aphanidermatum* to a snapdragon root hair. Scale bar = 160 μm.

bicompartmental Petri dishes without roots to evaluate the effect of fungicides suspected of being volatile. The anti-oomycetes fungicides tested were: captan, chlorotalonil, fenaminosulf and mancozeb (protectant fungicides), and systemic fungicides propamocarb, prothiocarb (carbamates), cymoxamyl (cyanoacetamide-oxime), metalaxyl, furalaxyl, milfuran (acylalanines), oxadixyl (oxazolinidone), hymexazol (isoxazole), and fosetyl (phosphanate). As may be expected, the protectant fungicides were not active in our system; they were not taken up by the roots in compartment A and thus not translocated within the roots in compartment B. The acylalanine fungicides have high systemic activity on the development of the mycelium, but the motility and aggregation of zoospores around the roots were not inhibited. Oxadixyl exhibited a much stronger fungicidal action, out-performing all other fungicides in our tests. Prothiocarb and propamocarb have apparently no action against *P. aphanidermatum*; these compounds have been reported to be systemic, but are known to be weakly fungitoxic. Cymoxamyl was ineffective; the deficiency of this fungicide can be related to its local systemic activity in the entire plant. Hymexazol has a systemic activity.

In general, all these compounds were translocated apoplastically in the xylem. In the bicompartmental hairy root culture, these systemic fungicides moved in the direction of the sucrose gradient, i.e. sucrose in compartment A to water in compartment B. The compounds, after having traversed the endodermal cells, were loaded within the vascular system with sucrose. However, it remains difficult to judge their exact partitioning process in the xylem vessels or in the sieve-tube companion cells in the phloem of the hairy roots. Fosetyl is an unique example of a fungicide that is truly systemic, i.e. apoplastic and symplastic. In our tests, fosetyl retarded primary infection and inhibited the external mycelial growth of *P. aphanidermatum*.

The present work has shown that no systemic fungicide was able to prevent the invasion by masses of zoospores, nor the initiation of the first appressoria. It is tempting to look for new systemic fungicides which can act on these sites that are unique to the Oomycetes.

Rhizoctonia solani

Rhizoctonia solani Kühn exists as a complex species: the fungus is divided into fourteen anastomosis groups (AGs) that differ in culture appearance and plant specificity. AG1 includes aerial pathogens that cause web blight and leaf blight. AG2-1 is specific to plants belonging to the Cruciferae. AG2-2 is divided into two groups: rice sheath blight caused by C-96 (AG2-2 IIIB) and root rot of sugar beet

(AG2-2 IV). AG3 is specific to the Solanaceae. AG4 is not host specific. AG5, 6, 7, 8, 9, 10 and BI are considered to be weak pathogens. Here we present a study on the specificity of interaction between a wide range of hairy root cultures and *R. solani* isolates belonging to different AGs. The hairy roots tested were: *Anagallis arvensis* L. (pimpernel), *Antirrhinum majus* L. (snapdragon), *Arachis hypogea* L. (peanut), *Brassica hirta* L. (mustard), *Brassica napus* L. (rapeseed), *Convolvulus sepium* L. (bindweed), *Cichorium endivia* L. (endive), *Cichorium intybus* L. (chicory), *Coriander sativum* L. (coriander), *Daucus carota* L. (carrot), *Dianthus caryophyllus* L. (carnation), *Foeniculum vulgare* Mill. (fennel), *Ipomea purpurea* L. (morning glory), *Linum grandiflorum* Desf. (flowering flax), *Lupinus albus* L. (white lupine), *Lycopersicon esculentum* Mill. (tomato), *Polygonum aviculare* L. (knotweed), *Sinapis alba* L. (white mustard), *Solanum tuberosum* L. (potato), *Tagetes erecta* L. (marigold), and *Valerianella locusta* L. (valerian). Other hairy root cultures (Mugnier, 1988d) have been tested but they were not included in this study because they turned brown when inoculated with *R. solani* (e.g. red beet, sugar beet, sunflower). The hairy root cultures were cultured in bicompartmental Petri dishes as described previously. Roots in compartment B (Figure 11.1) were inoculated with small pieces of potato dextrose agar fungal cultures. After 2–4 d, the progress of infection was followed under a light microscope and anatomical details were observed in roots stained with Trypan blue and lactophenol.

Of the 23 hairy root species tested, only the roots of Cruciferae, *Brassica* spp. and *Sinapis alba* became infected by isolates AG2-1. During the infection process, the fungus produced long hyphae that could extend over several millimetres of root surface and intersecting hyphae were frequently anastomosed. Lobate branches gave rise to appressoria and in some areas the fungus formed infection cushions (Figure 11.3.1) and the hyphae penetrated the root cells (Figure 11.3.3). Extensively infected roots continued to grow, except the beets roots as mentioned above. A detailed study was made of the relationship between AG2-1 and a non-susceptible root such as flax. The fungus was present on the root surface without penetration. Figure 11.3.2 shows that hyphae, closely adpressed to the epidermis, grew along the line of junction of cells, producing side branches at right angles. There was no shortening or swelling of hyphal tips which never gave rise to appressorial pegs or infection cushions. The isolates AG2-2 IV were pathogenic only on mustard roots and could also infect hairy roots of sugar beet and red beet, but browning reactions limited the fungal infection. Isolate AG2-2 IIIB infected all the hairy root cultures tested. The isolates AG1 (rice web and

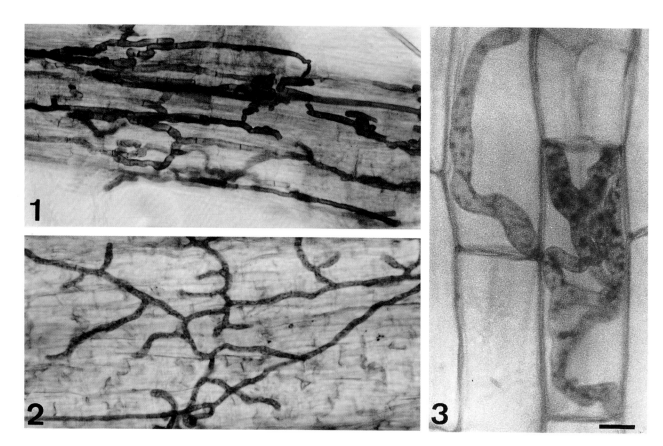

Figure 11.3. 1. Relationship between *Rhizoctonia solani* AG2-1 and a mustard (*Brassica hirta*) hairy root. Scale bar (shown in 3) = 160 μm. 2. Relationship between *R. solani* AG2-1 and a flax hairy root. Scale bar (shown in 3) = 160 μm. 3. Intracellular infection of a flax (*Linum grandiflorum*) hairy root by *R. solani* AG4. Scale bar = 40 μm. The roots were stained with Trypan blue.

sheath blight types) are aerial pathogens, but apparently the mode of infection of hairy roots was similar to that described for root-infecting isolates. Certain AG1 isolates that were not virulent in greenhouse tests did not infect the hairy roots. AG4 isolates were pathogenic to all the hairy root species tested. The isolates AG3 were pathogenic only on potato hairy roots which were particularly susceptible to infection. The other AGs tested (AG5, 6, 7 and BI) were weakly pathogenic on certain hairy roots.

The mode of infection of hairy roots by *R. solani* was not morphologically different from that found in natural infection, and infection of hairy roots was host specific.

Fusarium oxysporum

Tests of pathogenicity of vascular wilt *Fusarium oxysporum* were carried out with hairy roots of *Dianthus caryophillus*, *Lycopersicon esculentum*, *Cucumis sativus* and *Tagetes erecta*. The cultures were inoculated in compartment B (Figure 11.1) with spores of their potential pathogens: respectively, spores of *F. oxysporum* f. sp. *dianthi*, *F. oxysporum* f. sp. *lycopersici*, *F. oxysporum* f. sp. *melonis* (race 0), and *F. oxysporum* f. sp. *callistephi*. We studied the development of germ tubes and hyphae near the hairy roots. The details of this experiment are described elsewhere (Mugnier, 1989). Unfortunately, none of the isolates of the different formae speciales was capable of infecting the root hairs or the epidermal cells of their potential host, and therefore colonisation of the xylem vessels could not be observed. Various modifications of the original solid Murashige and Skoog medium were tested (e.g. nitrogen concentration and form, NH_4^+ vs NO_3^-), but even when the nutrient medium contained lower nitrogen and sugar concentrations, the infection process was not observed. This result emphasises the possibly central role of root nutrition in the development of vascular wilt infection, and suggests that the bicompartmental root system which can supply separate nutrition for the roots and the vascular wilt fungus may be an effective tool for further study. In contrast to *Rhizoctonia solani* which depends little on hairy root nutrition but is much more dependent on hairy root species (which must be that of a susceptible plant), *Fusarium oxysporum* f. sp. *chrysanthemi*, the vascular wilt agent of chrysanthemums, has been found to infect mycorrhizal hairy roots of carrot. Benhamou *et al.* (1994) have investigated the cytological modifications which occur during this infection.

Screening of Molecules in Hairy Root Cultures

Sterol biosynthesis inhibitors (SBIs) are systemic fungicides with strong fungitoxicity to fungi such as *Fusarium* sp., *Cercospora beticola* and *Rhizoctonia solani*. SBIs inhibit sterol metabolism in fungi and also have plant growth retardant or phytotoxic activity. Various SBIs were applied to different hairy root cultures to determine any phytotoxic effects; the roots were incubated in liquid Murashige and Skoog medium containing the fungicides at different concentrations. Root growth inhibition was estimated visually after 5 d incubation at 25°C in the dark. The fungicides tested were: fenarimol, propiconazole, flutriafol, flusilazol, prochloraz, triadimefon, triadimenol, tridemorph, bitertanol and tebuconazole. Root growth was reduced in the presence of the SBIs at concentrations less than 1 mg L^{-1}. Treatment with higher concentrations of fungicides altered the root morphology, which was marked by swelling of the root tips in the region of cell division.

The fungicide transport process was tested using the bicompartmental Petri dish system as described previously. Fungicides in liquid agar at 40°C were placed in compartment A and the fungal inoculum was placed in compartment B. Controls of Petri dishes without roots were used to investigate vapour phase activity. In general, the SBIs applied to compartment A moved into the roots, were translocated through the root tissues into compartment B, and then exuded into the water. The fungicides present in the water inhibited spore germination and hyphal growth. In cultures not treated with fungicide, *Fusarium oxysporum* invaded the Petri dish.

We are continuing to develop the bicompartmental culture system for use as a tool in the screening of agrichemicals for useful biological activities, i.e. fungicidal, non-phytotoxic and systemic activities. Mugnier *et al.* (1991) reported a new SBI fungicide, triticonazole (REAL®), which is unique among SBIs in its high plant selectivity for root and seed uptake.

Cercospora beticola

Cercospora beticola Sacc. causes the development of spots surrounded by reddish borders on the leaves of sugar beet; the reddening is caused by a toxin, cercosporin. We have studied the regulation of toxin synthesis using bicompartmental hairy root cultures. The fungus was inoculated into compartment B and the Petri dishes were incubated in the light (1500 lux) because the toxin is a photosensitising compound, i.e. a compound that sensitises plant cells to visible light. Toxin synthesis was greatly influenced by the hairy root species. In the presence of Cruciferae (*Brassica* spp., *Sinapis alba*) and Chenopodiaceae (*Beta vulgaris*), the fungus produced cercosporin which coloured the medium red and accumulated in the root tissue resulting in rapid death of the roots. The other hairy roots tested did not induce toxin synthesis. The exact conditions prevailing in the hairy root cultures which regulate synthesis of the toxin remain unknown.

Thielaviopsis basicola

Thielaviopsis basicola (Berk. & Br.) Ferr. is a destructive parasite found on roots of many crops. When grown together with hairy roots it penetrated the root hairs and epidermis without forming any special structures. Hyphae emerged from the infected roots and formed large, dark, thick-walled chlamydospores. As the infection advanced, the roots became deep black.

Sclerotinia sclerotiorum

Sclerotinia sclerotiorum (Lib.) de Bary has an extremely wide host range. Hairy roots of sunflower and rapeseed growing in the bicompartmental culture system were infected with sclerotia. Germination rapidly produced a cotton-like mycelial mass and the roots turned brown in advance of the mycelial front. Necrosis and disintegration of hairy roots resulted apparently from enzymes and/or toxins produced by the fungus, since sterile filtrates from *S. sclerotiorum* cultures also produced these symptoms. Sclerotia formed in the Petri dishes sometimes produced one or more stipes, each with an apothecium.

Polymyxa betae and *Plasmodiophora brassicae*

Polymyxa betae Keskin is associated with rhizomania disease in sugar beet, and *Plasmodiophora brassicae* (Woron.) is the causative agent of club root disease in rape and other *Brassica* species. Rhizomania is reported to be caused by the combined effects of *P. betae* and beet necrotic yellow vein virus, which is believed to be vectored by *P. betae*. The obligate nature of these Plasmodiophorales and the difficulty of obtaining inocula have significantly retarded study of the host–parasite interactions, development of anti-Myxomycetes compounds, and the breeding of resistant plants. Typical infections by *Polymyxa betae* and *Plasmodiophora brassicae* of hairy root cultures of host species have been reported by Mugnier (1987). However the practical problem of culturing these organisms remains; many cultures were co-contaminated with bacteria and other organisms such as *Olpidium brassicae* (Chytridiales), *Phialophora* sp. (Ascomycetes) and Protozoa. *O. brassicae* is a "true" fungus (Taylor *et al.*, 1994) which is an intracellular, non-mycelial, zoosporic, biotrophic organism. The preparations contaminated by *O. brassicae* were invaded by masses of their uniflagellate zoospores. Although the hairy root cultures were usually contaminated, this technique is potentially valuable for studying the puzzling aspects of the Plasmodiophorales.

Nematodes

Phytopathogenic nematodes can be divided into migratory ectoparasites and endoparasites, and sedentary endoparasites. Species of migratory ectoparasites and endoparasites (*Trichodorus* spp., *Tylenchorhynchus* spp., *Ditylenchus* spp., *Pratylenchus* spp.) can be maintained *in vitro* on suitable plant callus tissues. In contrast, sedentary endoparasites (*Heterodera* spp., *Meloidogyne* spp., *Globodera* spp.) cannot be maintained on undifferentiated callus. Hairy roots have been used to propagate the root-knot nematodes *Meloidogyne incognita* and *M. javanica* (Mugnier, 1988c; Verdejo and Jaffe, 1988; Adachi, 1992), the cyst nematode *Heterodera* (Paul *et al.*, 1987; Mugnier, 1988c; Savka *et al.*, 1990; Paul *et al.*, 1990). Such cultures are being used to study the parasitism of *M. javanica* by the bacteria *Pasteuria penetrans* (Verdejo *et al.*, 1988).

We have obtained carrot hairy roots showing root-knot galls of *M. incognita*. Although galls were obtained, we have not established lines of infected hairy roots because the nematodes did not reproduce well and their development was retarded. The effect of the systemic nematicide oxamyl on the behaviour of *M. incognita* and *H. schachtii* during the early stages of their development was studied in the bicompartmental culture system. Juvenile nematodes that thrust their stylets into the cells of oxamyl-treated roots died.

Effects of Herbicides on Hairy Root Cultures

The effect of various herbicides on the growth of hairy roots was studied. When a compound affected the growth of different root species differentially, the difference might be attributed to root uptake and metabolism of the herbicide. In general, metabolism of the herbicides tested led to inactivation

(clopyralid, linuron, phenmedipham), but in certain instances herbicide treatment resulted in activation (quizalofop-ethyl). Visible effects on root morphology were observed; dinitroanilines and certain carbamates led to remarkable swelling of the root tips; norflurazon and diflufenican were effective bleaching agents in greening cultures in Murashige and Skoog medium; and the presence of sucrose in the medium antagonised the effect of triazines herbicides. Growth inhibition by sulfonylureas was antagonised by addition of valine and leucine, asulam inhibition was antagonised by addition of folic acid, but glyphosate inhibition was not significantly reversed by aromatic amino acids. With certain exceptions, bipyridinium and diphenyl ether herbicides had rapid and devastating phytotoxic effects on root growth. Movement of the symplastic herbicide 14C-glyphosate in hairy roots was similar to that of sucrose. The phytotoxic effects of herbicides on hairy root cultures is discussed in more detail by Mugnier (1988a) with particular reference to their mode of action in intact plants.

Regenerated plants from hairy roots of pimpernel (*Anagallis arvensis*) were propagated in bicompartmental Petri dishes (Figure 11.1.2). Shoots were regenerated from roots established in compartment A but never from roots growing in water in compartment B. We studied the effect on the shoots of uptake and translocation of herbicides by the roots placed in compartment B. Aminotriazole, glyphosate and dichlorophenoxyacetic acid (2,4-D) had strong bleaching and wilting effects. The performance of S-triazines, triazinones and substituted ureas was apparently limited by the sucrose and moisture conditions in the Petri dish. We have obtained autoradiographs showing the distribution of radioactive atrazine throughout regenerated shoots of *Atropa belladonna* (Mugnier, 1988b). Problems of restricted gas exchange in the environment of a Petri dish and the fact that plants grew under a low evaporative water demand suggest that artefacts totally unrelated to normal plant conditions may have affected the results.

In conclusion, hairy root culture can be used to establish cultures *in vitro* of soil organisms which require roots for growth. The system offers the opportunity to culture and study biotrophic fungi and nematodes, root-colonising bacteria such as *Rhizobia* or *Frankia* (Diouf *et al.*, 1995), root-feeding insect pests, and other soil organisms which interact with roots but which are not necessarily biotrophic. Study of host–parasite interactions, screening of agrichemicals for fungicide, herbicide and nematicide activity, and research into crop disease resistance have been enhanced through the application of hairy root cultures.

References

Adachi H (1992). Culture of *Meloidogyne incognita* on oriental melon roots genetically transformed by *Agrobacterium rhizogenes*. *Japanese J Appl Entomol Zool* 36: 225–230.

Balaji B, AM Ba, TA LaRue, D Tepfer and Y Piché (1994). *Pisum sativum* mutants insensitive to nodulation are also insensitive to invasion *in vitro* by the mycorrhizal fungus, *Gigaspora margarita*. *Plant Sci* 102: 195–203.

Bécard G and JA Fortin (1988). Early events of vesicular-arbuscular mycorrhiza formation on RI-TDNA transformed roots. *New Phytol* 108: 211–218.

Bécard G and Y Piché (1990). Physiological factors determining vesicular arbuscular mycorrhizal formation in host and non-host Ri T-DNA transformed roots. *Can J Bot* 68: 1260–1264.

Bécard G and Y Piché (1992). Establishment of VA mycorrhizae in root organ culture: review and proposed methodology. In *Methods in Microbiology: Experiments with Mycorrhizae*, edited by A Varma, JR Dorris and DJ Read, pp. 89–108. New York: Academic Press.

Benhamou N, JA Fortin, C Hamel, M St-Arnaud and A Shatilla (1994). Resistance responses of mycorrhizal Ri T-DNA-transformed carrot roots to infection by *Fusarium oxysporum* f. sp. *chrysanthemi*. *Phytopathology* 84: 958–968.

Chabot S, G Bécard and Y Piché (1992). Life cycle of *Glomus intraradix* in root organ culture. *Mycologia* 84: 315–321.

Diop TA, G Bécard and Y Piché (1992). Long-term *in vitro* culture of an endomycorrhizal fungus, *Gigaspora margarita*, on Ri T-DNA transformed roots of carrot. *Symbiosis* 12: 249–259.

Diouf D, H Gherbi, Y Prin, C Franche, E Duhoux and D Bogusz (1995). Hairy root nodulation of *Casuarina glauca*: a system for the study of symbiotic gene expression in an actinorhizal tree. *Molecular Plant–Microbe Interactions* 8: 532–537

Mugnier J (1987). Infection by *Polymyxa betae* and *Plasmodiophora brassicae* of roots containing root-inducing transferred DNA of *Agrobacterium rhizogenes*. *Phytopathology* 77: 539–542.

Mugnier J (1988a). Behavior of herbicides in dicotyledonous roots transformed by *Agrobacterium rhizogenes*. I. Selectivity. *J Exp Bot* 39: 1045–1056.

Mugnier J (1988b). Behavior of herbicides in dicotyledonous roots transformed by *Agrobacterium rhizogenes*. II. Transport to regenerated shoots. *J Exp Bot* 39: 1057–1064.

Mugnier J (1988c). Transport of the nematicide oxamyl in roots transformed with *Agrobacterium rhizogenes*. *Ann Appl Nematol* 2: 29–33.

Mugnier J (1988d). Establishment of new axenic hairy root lines by inoculation with *Agrobacterium rhizogenes*. *Plant Cell Rep* 7: 9–12.

Mugnier J (1989). Studies on selectivity and systemicity of sterol biosynthesis inhibitors in transformed roots inoculated with *Fusarium oxysporum*. In *Vascular Wilt Diseases of Plants*, edited by EC Tjamos and C Beckman, pp. 529–535. NATO ASI Series. Berlin: Springer-Verlag.

Mugnier J and B Mosse (1987). Vesicular-arbuscular mycorrhizal infection in transformed root-inducing T-DNA roots grown axenically. *Phytopathology* 77: 1045–1050.

Mugnier J, G Jung and JL Prioul (1982). Method of producing endomycorrhizian fungi with arbuscules and vesicles *in vitro*. Patent FR 82,10768; US 4,599,312.

Mugnier J, M Chazalet and F Gatineau (1991). *RPA 400727: A New Fungicide For Cereal Seed Treatment*, ANPP, Bordeaux.

Paul, H, C Zijlstra, JE Leeuwangh, FA Krens and HJ Huizing (1987). Reproduction of the beet cyst nematode *Heterodera schachtii* Schm. on transformed root cultures of *Beta vulgaris* L. *Plant Cell Rep* 6: 379–381.

Paul H, JEM van Deelen, B Henken, TSM de Bock, W Lange and FA Krens (1990). Expression *in vitro* of resistance to *Heterodera schachtii* in hairy roots of an alien monotelosomic addition plant of *Beta vulgaris*, transformed by *Agrobacterium rhizogenes*. *Euphytica* 48: 153–157.

Savka MA, B Ravillon, GR Noel and SK Farrand (1990). Induction of hairy roots on cultivated soybean genotypes and their use to propagate the soybean cyst nematode. *Phytopathology* 80: 503–508.

Taylor JW, EC Swann and ML Berbee (1994). Molecular evolution of ascomycete fungi: phylogeny and conflict. In *Ascomycete Systematics: Problems and Perspectives in the Nineties*, edited by DL Hawksworth, pp. 201–212. NATO ASI Series. New York: Plenum.

Verdejo S and BA Jaffe (1988). Reproduction of *Pasteuria penetrans* in a tissue-culture system containing *Meloidogyne javanica* and *Agrobacterium rhizogenes*-transformed roots. *Phytopathology* 78: 1284–1286.

Verdejo S, A Jaffee and R Mankau (1988). Reproduction of *Meloidogyne javanica* on plant roots genetically transformed by *Agrobacterium rhizogenes*. *J Nematol* 20: 599–604.

Accumulation of Cadmium Ions by Hairy Root Cultures

12

Tomáš E. Macek[1], Pavel Kotrba[2], Tomáš Ruml[2],
František Skácel[3] and Martina Macková[2]

[1] *Institute of Organic Chemistry and Biochemistry, Academy of Sciences of the Czech Republic,*
   Flemingovo n. 2, 166 10 Prague 6, Czech Republic; email: tom.macek@uochb.cas.cz
[2] *Department of Biochemistry and Microbiology, Faculty of Food and Biochemical Technology,*
   ICT Prague, Technická 3, 166 28 Prague 6, Czech Republic
[3] *Department of Analytical Chemistry, Faculty of Chemical Engineering, ICT Prague, Technická 5,*
   166 28 Prague 6, Czech Republic

Introduction

Environmental contamination with heavy metals comes from a variety of sources. Such pollution may result from many industrial, military and agricultural activities. Some metals such as Zn, Cu, Ni, Co and Cr are essential or beneficial micronutrients for microorganisms, plants and animals, whereas others such as Cd, Hg or Pb have no known biological and/or physiological functions. All of these metals are toxic at higher concentrations and their accumulation in the environment is a health hazard.

Many systems, employing living or non-living microorganisms rather than plants, have been developed for removal of cadmium and other heavy metals from the environment. Although relatively little information is available about the mechanisms of accumulation and detoxification of heavy metals by plants, it is now clear that use of vegetation for *in situ* treatment of contaminated soils and sediments is an emerging technology promising effective and inexpensive clean-up of certain hazardous waste sites and contaminated areas. This technology, phytoremediation, has already been shown effective in a number of full-scale and pilot studies (Schnoor *et al.*, 1995). Because phytoremedia-tion is still under development, the technology is not yet commonly used. Phytoremediation may take longer than traditional approaches to reach clean-up goals or may be limited by soil toxicity. However, as a rule, plants will survive higher concentrations of hazardous wastes than will most microorganisms used for bioremediation. Conventional treatment of soils contaminated with toxic metals costs as much as US$1000 per ton of soil, while treating sites using phytoremediation will cost an estimated US$15–50 per ton (Jones, 1994). Furthermore, use of plants for cleaning up contaminated sites has an almost certain prospect of public acceptance.

Potential applications of phytoremediation include bioremediation of petrochemical spills, ammunition wastes, chlorinated solvents, contaminated chemical storage areas, landfill leachates, etc.

Organic pollutants can be removed by many mechanisms, including degradation to non- or less toxic compounds, mineralisation to carbon dioxide and water, or transformation to non-phytotoxic metabolites such as lignin. Heavy metals present a different situation in that they will not disappear after biodegradation. They, however, can be removed from the environment by accumulation in plants, after being concentrated in different plant parts. Heavy metals are bound in the form of complexes with many types of protein.

Plant cells are able to form metallothioneins of class II, and peptides similar to metallothioneins, bearing two cysteine-rich domains. The actual role of these peptides is still not known, but they are most likely involved in homeostasis of essential heavy metals (Kotrba et al., 1994a). The genes for these peptides have been cloned and sequenced (Robinson et al., 1993). The resistance of plant cells to many of the heavy metals results from metal binding to phytochelatins ([γ-Glu-Cys]$_n$Gly peptides, where n is 2–11, characterised as class III metallothioneins, accumulated in response to the presence of the metal, especially Cd^{2+} or Cu^{2+} ions). The phytochelatin synthase from Silene cucubalus (γ-glutamylcysteine dipeptidyltranspeptidase) was described by Grill et al. (1989). Phytochelatin synthase is expressed constitutively as an enzyme consisting of four identical subunits, each of molecular weight 25 kDa. Its activity depends on the concentration of metal ions (Robinson et al., 1993). The crucial step determining phytochelatin mediated heavy metal tolerance at the cellular level is probably the transport of the metal–phytochelatin complex into the vacuole (Ornitz et al., 1992; Vogeli-Lange and Wagner, 1992).

In general, plant species show a very wide variation in accumulation of and tolerance to heavy metals. Both low and high accumulation of heavy metals in tolerant cultivars of different species have been reported (Harmens et al., 1993). Many plant species show little or no genetically-based variation in heavy metal tolerance (e.g. Lolium perenne, Poa pratensis, Trifolium repens), while others show greater variability (e.g. Agrostis capillaris, A. stonifera, Festuca rubra) (Symeonidis et al., 1985). In studies with Salyx, Landberg and Greger (1994) showed that tolerant clones, but also sensitive ones, exhibited high or low accumulation; thus net uptake and accumulation in Salyx do not seem to be correlated with tolerance. Clones with very high or low accumulation of heavy metals in roots, shoots or whole plants were found. Since both sensitive and tolerant clones of Salyx have a similar ability to accumulate or exclude heavy metals, it would seem that tolerance to heavy metals is not the only important aspect to study.

The physiological effects of heavy metals, particularly of Cd on plants, are being studied in many laboratories. These studies include histological and cytological changes, growth, etc., as well as the mechanisms of accumulation and detoxification of heavy metals. In vitro cultured plant cells can serve as a very useful tool in such studies. This type of experimental material allows relatively rapid and reproducible evaluation of bioaccumulation.

Roots are the plant parts which facilitate the primary contact between a plant and soil-originating xenobiotics. Uptake and transport start here, and thus a better understanding and description of phenomena related to heavy metal accumulation in roots is needed.

Cultures of transformed hairy roots grown in vitro are most suitable models for such studies. Their many advantages include a similarity with normal roots and a high growth rate. Resistant clones can be selected, new features can be introduced using molecular biology techniques, and whole plants can be regenerated.

Despite the above-mentioned advantages, very limited data are available about bioaccumulation or biosorption of heavy metals by transformed root cultures. Metzger et al. (1992) described transformed root cultures of tobacco, sugar beet and morning glory as models for evaluating the bioavailability of Cd from sewage sludge. Recently we attempted to study cadmium bioaccumulation using Solanum nigrum and S. aviculare hairy root cultures. (Macek et al., 1994; Kotrba et al., 1994b, 1995a, 1995b). These cultures were successfully used also in studies of biotransformation of polychlorinated biphenyls (Macek et al., 1995). The aim of this chapter is to summarise our results on hairy root–Cd interactions.

Experimental Materials and Methods

Strains and Cultivation

All our experiments described here were performed using hairy root clones of Solanum nigrum or Solanum aviculare, transformed by Agrobacterium rhizogenes bearing the Ri plasmid C58C1 (Macek et al., 1988). The clones were cultivated in the dark at 27°C in liquid medium according to Linsmeier and Skoog (LS) without any exogenous regulators, as described for S. aviculare by Macek (1989).

Accumulation Experiments

Cadmium was added as $Cd(NO_3)_2$. Cd content was measured by atomic absorption spectrophotometry (Varian Spectra A300) after mineralisation of the roots (HNO_3, 4 h, 170°C, under pressure).

Cd accumulation by hairy roots was followed over a period of up to 50 h using media containing Cd inoculated with 14-d-old roots in the ratio of 1 g fresh weight to 50 mL of liquid medium. All accumulation studies were performed in triplicate.

To study the effect of Cd on root elongation, root tips 7–10 mm long were cultivated for 24 d as described by Macek and Green (1993) on LS agar plates containing a range of Cd concentrations.

Cell walls were prepared by the method of Sentenac and Grignan (1981). Desorption of physically bound Cd was estimated using roots (1 g fresh weight) or cell walls (0.11 g fresh weight) incubated at room temperature in 50 mL of 0.06 M $CaCl_2$ (15 min), 0.1 M EDTA (5 min) or 0.1 M HCl (5 min, cell walls only).

Analyses

Viability of hairy roots was evaluated following treatment with $CaCl_2$ or EDTA for desorption of Cd. The roots were washed in water, transferred into liquid LS medium and cultivated for 10 d under standard conditions. The obtained biomass was than dried to constant weight at 90°C and compared with the biomass of untreated roots cultured in the same way. Reduction of triphenyltetrazolium chloride yielding red formazan was used as an indicator of respiratory chain activity and was performed according to Towill and Mazur (1975).

Total thiols were determined as described by Florjin et al. (1993).

Results and Discussion

The accumulation of cadmium from liquid medium by hairy root cultures was, after preliminary selection (10 clones of *Solanum aviculare* and *S. nigrum* were compared on the basis of Cd uptake ability, biomass yield, root elongation and branching), studied mostly using *S. nigrum* clone SNC-9O. This clone grew well with a high branching frequency.

The effect of different compositions of test solution on cadmium uptake by roots was examined using 0.1 M MES, 0.1 M acetate, 0.1 M citrate and 0.1 M phosphate buffers, and a 0.05 M sodium borate solution with pH adjusted by HNO_3. These experiments showed that MES buffer and borate ions create the optimal environment for Cd uptake studies (Macek *et al.*, 1994). The use of phosphate buffer resulted in a lower accumulation of Cd, which was probably caused by precipitation of Cd with phosphate ions, as described also by Gadd and White (1993). The highest cadmium removal efficiencies were found over the pH range 5–7 from an initial Cd content of 2 mg L^{-1} (Kotrba *et al.*, 1994b). For further experiments, pH 5.6 was chosen because it is also the pH value of normal growth medium. Cadmium uptake over an exposure period of 5 h from borate solutions containing 1 mg L^{-1} and 2 mg L^{-1} Cd showed first a rapid stage of accumulation indicating physical adsorption. This was followed by a slower uptake suggesting that accumulation was in part metabolically sponsored. The dependence of Cd uptake on initial concentration was followed in the borate solution using initial Cd levels in the range 0.2 to 2000 mg L^{-1}. The increased content of Cd in the solution was reflected by an increase in cadmium accumulated by the plant

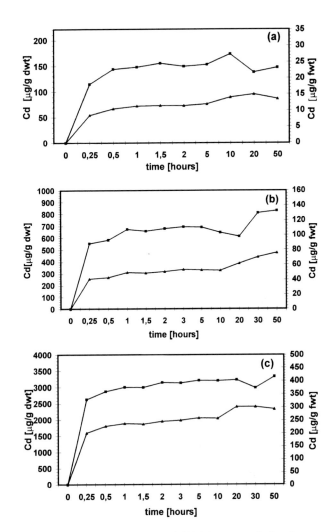

Figure 12.1. Time course of Cd bioaccumulation by *in vitro* cultured hairy roots of *S. nigrum* from liquid LS media with initial Cd concentrations (a) 2 mg L^{-1}, (b) 10 mg L^{-1}, and (c) 50 mg L^{-1}. (■) Cd in dry weight; (▲) Cd in fresh weight. The data are plotted using a non-linear time scale.

tissue. Cadmium content reached from 0.1 mg g^{-1} to 24 mg g^{-1} dry weight after 2 h exposure to solutions containing the above-mentioned initial concentrations. Nevertheless the efficiency of Cd accumulation, expressed as the percentage of cadmium accumulated relative to the initial cadmium content in the borate solution, was significantly decreased at concentrations of cadmium over 2 mg L^{-1}. Temperature also influenced the process; the percentage cadmium removal after 2 h exposure to the borate solution initially containing 2 mg L^{-1} cadmium decreased to 29% at 4°C compared with 60% at 26°C.

Accumulation of Cd from nutrient medium over time (15 min–50 h) was followed using Cd concentrations of 2, 10 and 50 mg L^{-1}. Figure 12.1 shows the

Table 12.1. Amount of Cd removed from 0.11 g fresh weight of cell walls of *S. nigrum* hairy roots.

Cd in LS medium (mg L⁻¹)	Cd (μg g⁻¹ fresh weight) removed by		
	HCl	CaCl₂	EDTA
2	45.5	51.8	38.2
10	186	218	197
50	855	764	768

Removal by 0.1 M HCl: 5 min; 0.06 M CaCl₂: 15 min; 0.1 M EDTA: 5 min.

time course of Cd uptake. Taking into account biomass production, no difference in the total amount (μg) of accumulated Cd was observed between 5 and 20 h. However, after desorption of Cd from the root surface, roots contained a higher amount of Cd after longer cultivation (20 h) (Kotrba *et al.*, 1995a).

To distinguish between Cd accumulated inside the cells and Cd bound physically to cell walls, the amount of cadmium removed from cells exposed previously to 2, 10 and 50 mg L⁻¹ Cd was measured using treatment with CaCl₂ and EDTA. The amounts of Cd removed from 1 g fresh weight of isolated cell walls are summarised in Table 12.1. The efficiency of removal by these methods was compared with the use of HCl (0.1 M) for desorption. The effect of treatment with CaCl₂ and EDTA on the viability of hairy root cells is shown in Figure 12.2. EDTA at concentrations 0.1 M and 0.06 M resulted in total arrest of cell growth due to lysis. The cadmium remaining in the hairy roots cultivated for 5 and 20 h in liquid LS medium with Cd, after different

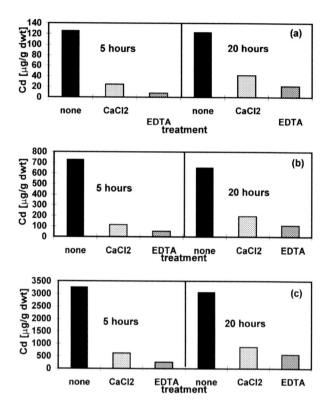

Figure 12.3. Residual Cd concentrations in roots after removal of adsorbed Cd by CaCl₂ or EDTA from *S. nigrum* hairy roots cultivated for 5 and 20 h in liquid LS media with initial Cd concentrations of (a) 2 mg L⁻¹, (b) 10 mg L⁻¹ and (c) 50 mg L⁻¹.

treatments to remove the surface-bound Cd, are shown in Figure 12.3.

Inhibition of *S. nigrum* main root elongation on agar medium by 2 and 10 mg L⁻¹ Cd reached 20 and 50%, respectively, after 24 d. Nearly full inhibition of elongation was observed on agar media at 50 mg L⁻¹ (Figure 12.4). When calculating the relative rate of elongation, the major inhibitory effect of 10 mg L⁻¹ Cd appeared in the first 10 d of cultivation. Root branching was observed at 0 mg L⁻¹ Cd (branching angle = 52 ± 2°) and 2 and 10 mg L⁻¹ (branching angle = 60 ± 2°). Branching was suppressed at 50 mg L⁻¹ Cd; the roots became dark with callus-like objects on their surface (Kotrba *et al.*, 1995b).

The reduction of triphenyltetrazolium chloride (TTC) by the aa₃ complex of the plant respiratory chain was followed. Comparing to control, an increase in reduced TTC was observed with roots cultivated for 5 h in media containing 2 and 10 mg L⁻¹ Cd, indicating the impact of Cd on the respiratory chain in the initial phase of cultivation (Figure 12.5). No significant effect was observed with 50 mg L⁻¹ Cd.

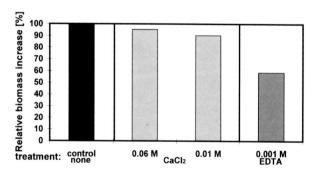

Figure 12.2. Effect of treatment with CaCl₂ and EDTA on the viability of hairy roots of *S. nigrum*. Roots were treated with CaCl₂ for 15 min, or with EDTA for 5 min. Biomass yields were evaluated after treatment and further cultivation, and are shown relative to the control culture without treatment.

ELONGATION OF THE MAIN ROOT

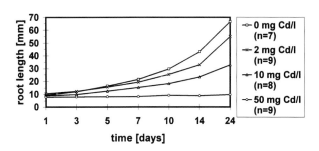

Figure 12.4. Elongation of the main root of *S. nigrum* on agar plates with LS media containing Cd at concentrations of (□) 0, (×) 2, (△) 10 and (○) 50 mg L⁻¹. The data shown for each concentration are averages from 7, 9, 8 and 9 root length measurements, respectively. The data are plotted using a non-linear time scale. n = number of repeats.

In plants, cadmium is detoxified by phytochelatins which are synthesised from glutathione. We measured the total thiols (cysteine, glutathione, phytochelatins) in *S. nigrum* hairy roots cultivated in liquid LS medium. The amount of total thiols increased during cultivation in medium containing Cd as shown in Figure 12.6.

Conclusions

The absorption equilibrium between the *S. nigrum* root surface and Cd in LS medium containing 2, 10 and 50 mg L⁻¹ Cd was reached between the first and third hour of cultivation. The initial rapid phase of accumulation (sorption) was followed by a slower

REDUCTION OF TTC

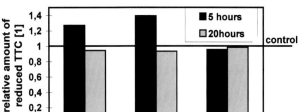

Figure 12.5. Reduction of triphenyltetrazolium chloride (TTC) by *S. nigrum* hairy roots cultivated for 5 h (black bars) and 20 h (grey bars) in the presence of 2, 10 and 50 mg L⁻¹ Cd in liquid LS medium. Values are shown relative to the control culture without added Cd.

TOTAL THIOL LEVELS

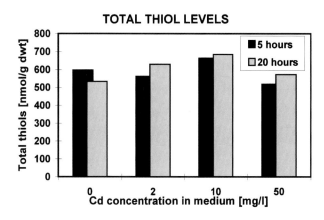

Figure 12.6. Differences in total thiol levels in *S. nigrum* hairy roots cultivated for 5 and 20 h in LS medium with various Cd concentrations.

continuous uptake of cadmium at all Cd concentrations in LS medium. This uptake is more readily interpreted relative to the fresh weight of biomass because of the increase in dry mass content during cultivation in media containing Cd.

The amount of Cd accumulated between the 5th and 20th hour of cultivation by mechanisms other than sorption (as determined by residual Cd concentrations after desorption) was nearly double that seen after the first 5 h. CaCl₂ treatment proved to be the most efficient means for the desorption measurements because EDTA could lead to breakage of the cell membranes during treatment. Additionally, CaCl₂ treatment did not affect *S. nigrum* viability.

The influence of Cd on growth, morphology and physiological aspects of *S. nigrum* hairy root clone SNC-9O was evaluated in the presence of cadmium at concentrations ranging from 2 to 50 mg L⁻¹. The presence of low concentrations of Cd in solid medium did not cause a dramatic decrease in root elongation and branching. However, no or very slight growth was observed in the presence of higher Cd concentration (50 mg L⁻¹). Roots became dark and callus-like objects were formed on the root surface. Low Cd concentrations (2 mg L⁻¹) did not decrease the biomass yield within 21 d of cultivation. Cadmium had an impact on the respiratory chain, as seen from TTC reduction. Our preliminary experiments concerning the content of thiols revealed that exposure of hairy roots to Cd causes an increase in the concentration of total thiols in the roots. The total thiol pattern could indicate synthesis and accumulation of phytochelatins in roots exposed to Cd at 2 and 10 mg L⁻¹.

Altogether, the results mentioned above prove the usefulness of *Agrobacterium* transformed root cultures in studies of heavy metal uptake and as a starting point for genetic manipulations to enhance bioaccumulation.

Acknowledgments

This work was supported by grant No. 204/93/0836 of the Grant Agency of the Czech Republic.

References

Florjin PJ, JA De Knecht and ML Van Beusichem (1993). Phytochelatin concentrations and binding state of Cd in roots of maize genotypes differing in shoot/root Cd partitioning. *J Plant Physiol* 142: 537–542.

Gadd GM and C White (1993). Microbial treatment of metal pollution — a working biotechnology. *Trends in Biotechnol* 11: 353–359.

Grill E, S Loffler, E-L Winnacker and MH Zenk (1989). Phytochelatins, the heavy-metal-binding peptides of plants, are synthetised from gluthathione by a specific γ-glutamylcysteine dipeptidyl transpeptidase (phytochelatin synthase). *Proc Nat Acad Sci USA* 84: 6838–6846.

Harmens H, NGCPB Gusmao, PR Den Hartog, JAC Verkleij and WHO Ernst (1993). Uptake and transport of zinc in Zn-sensitive and Zn-tolerant *Silene vulgaris*. *J Plant Physiol* 141: 309–315.

Jones RL (1994). ASPP recommends hazardous waste remediation technologies to DOE. *Am Soc Plant Physiol Newslett* 21(6): 12–13.

Kotrba P, T Ruml and T Macek (1994a). Cadmium binding by microbial and plant cells. *Chemické listy* 88: 642–649.

Kotrba P, T Ruml, M Macková-Šůchová, F Skácel, K Demnerová and T Macek (1994b). Cadmium accumulation by transformed root cultures of *Solanum nigrum*. In *Plant Cell, Tissue and Organ Cultures in Liquid Media*, edited by T Macek and T Vanek, pp. 111–113. Prague: Abstr. Int. Symp, CSBMB.

Kotrba P, T Macek and T Ruml (1995a). The effect of cadmium on the growth of hairy-root culture of *Solanum nigrum*. In *Biosorption and Bioremediation*, edited by T Macek, K Demnerová and M. Macková, 1–7. Prague: Abstr Int Symp, Merín, CSBMB.

Kotrba P, T Macek, F Skácel and T Ruml (1995b) Accumulation of cadmium by hairy-root culture of *Solanum nigrum* from nutrient medium. In *Biosorption and Bioremediation*, edited by T Macek, K Demnerová and M Macková, 1–8. Prague: Abstr Int Symp, Merín, CSBMB.

Landberg T and M Greger (1994). Can heavy metal tolerant clones of *Salyx* be used as vegetation filters on heavy metal contaminated land? In *Willow Vegetation Filters For Municipal Wastewater and Sludges*, edited by P Aronsson and K Perttu, pp. 133–144. Uppsala: Proc Conf, Swedish University of Agricultural Science.

Macek T (1989). Poroporo, *Solanum aviculare* and *S laciniatum: in vitro* culture and the production of solasodine. In *Biotechnology in Agriculture and Forestry 7*, edited by YPS Bajaj, pp. 443–467. Berlin: Springer-Verlag.

Macek T and KD Green (1993). Growth studies using transformed roots of *Solanum aviculare* and *S nigrum*. *Planta Med* 59 (Suppl.): A659.

Macek T, M Czakó, T Vanek and L Márton (1988). *Agrobacterium* transformed plant cell lines: *Solanum aviculare* — opine content and the production of betulinic acid. Abstr 14th Congr Internat. Union Biochemistry, Prague, 5: p. 191.

Macek T, P Kotrba, M Šůchová, F Skácel, K Demnerová and T Ruml (1994). Accumulation of cadmium by hairy root cultures of *Solanum nigrum*. *Biotechnol Lett* 16: 621–624.

Macek T, M Macková, A Holubková, J Burkhard, K Demnerová and J Pazlarová (1995). Biodegradation of polychlorinated biphenyls by plant cells and the effect of peroxidase activity. In *Biosorption and Bioremediation*, edited by T Macek, K Demnerová and M Mackova, L 2–7. Prague: Abstr Int Symp, Merín, CSBMB.

Metzger L, I Fouchault, Ch Glad, R Prost and D Tepfer (1992). Estimation of cadmium availability using transformed roots. *Plant Soil* 143: 249–257.

Ornitz DF, L Krepper and DM Speiser (1992). Heavy metal tolerance in the fission yeast requires an ATP-binding cassette-type vacuolar membrane transporter. *EMBO J* 11: 3491–3499.

Robinson NJ, AM Tommey, Ch Kuske and PJ Jackson (1993). Plant metallothioneins. *Biochem J* 295: 1–10.

Schnoor JL, LA Licht, SC McCutcheon, NL Wolfe and LH Carreira (1995). Phytoremediation of organic and nutrient contaminants. *Env Sci Technol* 29: 318A–326A.

Sentenac H and C Grignan (1981). A model for predicting ionic equilibrium concentrations in cell walls. *Plant Physiol* 68: 415–419.

Symeonidis L, T McNeilly and AD Bradshaw (1985). Interpopulation variation in tolerance to cadmium, copper, lead, nickel and zinc in nine populations of *Agrostis capillaris* L. *New Phytol* 101: 317–324.

Towill LE and P Mazur (1975). Studies on the reduction of 2,3,5-triphenyltetrazolium chloride as a viability assay for plant tissue cultures. *Can J Bot* 53: 1097–1102.

Vogeli-Lange R and GJ Wagner (1992). Subcellular localization of cadmium and cadmium-binding peptides in tobacco leaves. Implication of transport function for cadmium binding peptides. *Plant Physiol* 92: 1086–1093.

Oxygen Effects in Hairy Root Culture

13

Shaoxiong Yu, M.G.P. Mahagamasekera,
Gary R.C. Williams, Kanokwan Kanokwaree and
Pauline M. Doran

Department of Biotechnology, University of New South Wales, Sydney NSW 2052, Australia;
email: p.doran@unsw.edu.au

Introduction

Culture conditions which affect the performance of hairy roots in shake flasks and bioreactors have been studied intensively for about a decade. Nutrient levels (Hilton and Rhodes, 1990), carbon source (Inomata *et al.*, 1993), hormone regime (Yoshikawa and Furuya, 1987; Robins *et al.*, 1991; Repunte *et al.*, 1993), medium osmolality (Yu *et al.*, 1996), medium pH (Mukundan and Hjortso, 1991; Ermayanti *et al.*, 1994), gas-phase carbon dioxide concentration (DiIorio *et al.*, 1992a), incubation temperature (Hilton and Rhodes, 1990; Yu *et al.*, 1996), and other parameters have been addressed in the literature. Hairy root cultures have thus been investigated using approaches which have proven useful in the past with other types of cell culture. However, because hairy roots are a solid-phase biocatalyst and not dispersed in liquid like suspended cells, the effect of internal tissue and external boundary layer concentration gradients on transport of dissolved components to the root cells must also be considered. Despite oxygen being one of the least soluble substances necessary for hairy root culture and therefore very likely to be subject to mass transfer restrictions, relatively little attention has been focused on

mass transfer conditions and the oxygen requirements of hairy roots (Kondo *et al.*, 1989; Yu and Doran, 1994; Ramakrishnan and Curtis, 1995). Hairy roots are heterotrophic, respiratory organisms which rely on oxygen for energy generation and other metabolic functions; inadequate oxygen supply can be expected to have a direct influence on growth rate and may also affect the synthesis of certain secondary metabolites. It is often argued that because the metabolic rate of plant cells is low compared with microorganisms, oxygen demand is concomitantly low, and therefore oxygen transfer is of little concern. While this may be reasonable logic for plant cell suspensions, the solid-phase nature of hairy roots introduces important additional considerations which affect the balance between oxygen supply and oxygen demand.

The Oxygen Transfer Situation in Hairy Root Cultures

Most hairy roots are grown submerged in liquid medium in shake flasks or air-sparged bioreactors; this is usually the simplest and easiest form of root culture. Other types of reactor such as the nutrient

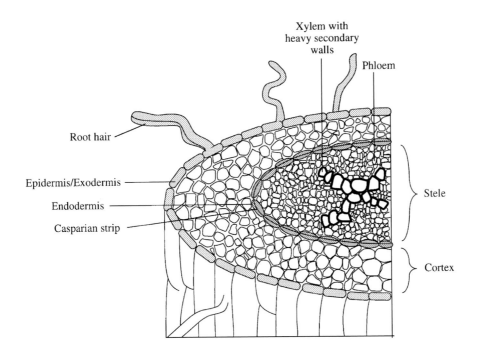

Figure 13.1. A simplified representation of the structural components of roots. (Adapted from Taiz and Zeiger, 1991.)

Xylem with heavy secondary walls

Phloem

Root hair

Epidermis/Exodermis

Endodermis

Casparian strip

Stele

Cortex

mist (Wilson *et al.*, 1990; DiIorio *et al.*, 1992b), trickle bed (Flores and Curtis, 1992) and rotating drum (Kondo *et al.*, 1989) have also been applied. Irrespective of the vessel configuration, hairy roots in shake flasks and reactors tend to form dense root mats or clumps which grow to a diameter about the same as that of the culture container. This characteristic presents considerable difficulties in terms of mass transfer, producing a hierarchy of mass transfer resistances affecting oxygen delivery to the root cells.

Mass Transfer Resistance Within the Root Tissue

First, there is the resistance to mass transfer posed by the root tissue itself. Mechanisms must be available for transport of nutrients including oxygen from the external surface of the root to its interior. Diffusion is considered to account for the bulk of gas transport in roots, although other processes such as cytoplasmic streaming, water throughflow and pressure gradients caused by the differential solubilities of oxygen, nitrogen and carbon dioxide may also contribute (Armstrong *et al.*, 1991).

Submerged culture of hairy roots in liquid medium can be considered an unnatural condition, as relatively few plant species can tolerate continuous waterlogging. In whole plants, aeration of root tissue occurs by a combination of radial diffusion from the surrounding soil and internal longitudinal oxygen transfer from the leaves; survival in flooded conditions depends largely on oxygen transport to the roots from aerial tissues. Longitudinal diffusion is facilitated by aerenchyma, which are continuous

intercellular gas spaces or pores in the cortex of the root that may develop in response to certain levels of anoxia. In hairy root cultures, supplementation of oxygen by transfer from aerial structures such as leaves is impossible as the roots are excised from other plant organs. In the absence of evidence of additional oxygen transfer mechanisms, radial diffusion can therefore be considered the dominant mechanism of oxygen transfer upon which survival of the biomass and performance of the culture depend.

Diffusion of oxygen in roots has been the subject of considerable experimental investigation and modelling. As shown in Figure 13.1, root tissues are not uniform in composition, and contain various differentiated structures and non-porous layers which can impede radial diffusion. Armstrong and Beckett (1985, 1987) have modelled this heterogenous root structure as a series of three concentric cylindrical shells corresponding to the stele, the cortex and the wall layers. The cortex is usually porous with channels available between the individual cells; the diffusion properties of cortical tissues depends largely on the amount of intercellular gas space and the extent to which this space is connected. The stele and wall layers are generally non-porous or have extremely low porosity. The value of the oxygen diffusion coefficient for stelar and other non-porous zones has been estimated to be ca. 30 times lower than the radial diffusivity in the cortex (Armstrong and Beckett, 1985). Oxygen demand in root tissues is also not spatially uniform across the root radius.

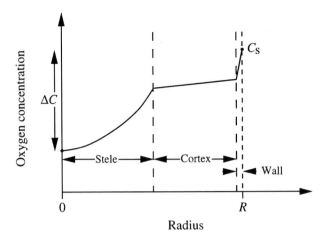

Figure 13.2. Oxygen concentration profile in a root of radius R with porous cortex. C_s is the oxygen concentration at the root surface; ΔC is the total oxygen deficit across the root radius.

Stelar and exodermal respiration rates per unit volume of tissue can be more than twice those in the cortex, most likely because the cortex comprises vacuolated and expanded cells (Armstrong *et al.*, 1991). Several models of oxygen diffusion in roots have been developed with varying degrees of complexity to account for the non-uniformity of root tissues and, to some extent, metabolic responses to anoxia. Solution of the model equations can be used to determine the shape of the oxygen concentration profile through the root tissue: for example, as shown in Figure 13.2 (Armstrong *et al.*, 1991). With a deterioration in the rate of oxygen supply, the stele is the first tissue in the root to experience anoxia. The total oxygen concentration difference across the root radius can be expected to vary considerably depending on the root thickness, the thicknesses of the component structures, the tortuosity of the diffusion path, and the oxygen demand of the various tissues. These parameters can change significantly down the length of the root: for example, the oxygen requirements of root tips are up to 10 times those of more mature tissues (Luxmoore *et al.*, 1970; Ramakrishnan and Curtis, 1995), while the diameters of the root and its component tissues also change with distance from the apex (Armstrong *et al.*, 1991).

In terms of reactor engineering, once a particular species of hairy root has been selected, the resistance to internal mass transfer offered by the root itself may not be easily altered.

Liquid–Solid Boundary Layer Mass Transfer Resistance

A second zone of mass transfer resistance in hairy root cultures is the liquid boundary layer at the surface of each root inside the root clump. For hairy roots submerged in liquid, the thickness of this boundary layer (and therefore the extent to which it reduces the rate of oxygen transfer) depends on several factors, the most amenable to manipulation from an engineering standpoint being the liquid velocity in the vicinity of the root. To minimise mass transfer resistance external to the root surface, high local liquid velocities inside the root clump must be maintained. The actual velocity required will depend on the oxygen demand of the roots, the external dissolved oxygen concentration, the root diameter and surface characteristics, and liquid properties.

Although roots in some reactors such as the nutrient mist, trickle bed and rotating drum may not be continuously submerged in liquid, a liquid boundary layer will be present nonetheless as the individual roots must be prevented from drying out. Factors affecting the thickness of boundary layers in these reactors are diverse and perhaps more difficult to analyse theoretically, but include the rate of liquid drainage through the root bed. Development of stagnant pockets of liquid and high levels of liquid entrainment in the root clump (Flores and Curtis, 1992) reduce the effectiveness of external oxygen transfer. In vessels without mechanisms for direct agitation, liquid boundary layers around individual roots may not be easily reduced. In any case, the need to eliminate external mass transfer restrictions must be balanced against the risk of shear damage; vigorous stirring can induce callus formation on hairy roots resulting in an abrupt decline in secondary metabolite production (Hamill *et al.*, 1987).

Other Mass Transfer Effects

A third consideration is the resistance to convective mass transfer posed by the entire root clump in the reactor. In submerged systems, the presence of the roots reduces bulk liquid velocities and mixing in the vessel outside the clump, and also affects the dispersion of air bubbles. If the level of turbulence becomes inadequate, gas–liquid boundary layer resistances will also contribute to a reduction in oxygen transfer. If the sparger is located under or inside the root ball, gas pockets readily develop around the roots causing channelling and poor overall oxygen distribution. To minimise these effects, convective currents in the bulk liquid must remain strong with sufficient energy to penetrate the root clump.

The relationship between the liquid velocity outside and conditions inside root clumps has been investigated by Prince *et al.* (1991) using untransformed onion roots, and by Yu and Doran (1994) using *Atropa belladonna* hairy roots. For a 3.6-cm root ball at a biomass density of 0.32 g mL^{-1} fresh weight, Prince *et al.* (1991) found that external oxygen transfer restrictions were eliminated at a linear velocity outside the clump of about 1 cm s^{-1}. In other systems, this value will vary depending on the oxygen demand of the roots, the clump size and porosity, the biomass density, and the bulk flow configuration (Yu and Doran, 1994). In reactors with considerably larger root clumps and/or ill-defined bulk circulation currents, it can be expected that much higher extraclump velocities are required.

Intraclump Dissolved Oxygen Levels in Bioreactors

The question arises as to what range of dissolved oxygen concentrations is experienced by hairy roots within root clumps in reactors or shake flasks. To investigate this, the stirred and bubble column reactors shown in Figure 13.3 were used to measure intraclump oxygen tensions, and to examine the effects of clump size, root density and external flow conditions (Yu and Doran, 1994). Fresh *Atropa belladonna* hairy roots were packed into spherical wire mesh 'tea balls' (as used for domestic tea making: Prince *et al.*, 1991) to densities between 15.7 and 44.2 g L^{-1} dry weight, and placed in Murashige and Skoog (MS) medium in the reactors. Use of the tea balls suspended in the bulk flow field of the reactors allowed convenient control over the size and position of the root clumps and the biomass density they contained. Dissolved oxygen electrodes were positioned outside and at the centre of the root balls. These electrodes had sufficiently small cathodes so that the difference in response between stationary and well-mixed environments was less than 2–4% air saturation; this meant that measurements inside the root ball were essentially unaffected by the lack of fluid movement at the electrode membrane. Care was taken to prevent the roots pressing directly against the probe surface. The stirred tank reactor was operated at stirrer speeds between 50 and 190 rpm at a constant air flow rate of 300 cm^3 min^{-1}; the bubble column was operated at varying air flow rates between 60 and 500 cm^3 min^{-1}. Steady state measurements of dissolved oxygen tension were made at 25°C using tea balls with diameters 3.5, 4.0 and 5.0 cm.

Results for intraclump dissolved oxygen levels are shown in Figure 13.4. In all cases, the extraclump oxygen tension remained high at 97–100% air saturation. As indicated in Figure 13.4a, intraclump oxygen levels in the stirred vessel increased as the stirrer speed was raised from 50 rpm; at the lower biomass densities, 100 rpm was sufficient to achieve above 90% air saturation inside the root ball. Oxygen concentration was considerably lower at the higher root densities at constant clump diameter. Clump size within the relatively small range tested appeared to have the least effect on oxygen levels; only at the low stirrer speeds did oxygen tension decrease significantly with increasing clump size at approximately the same root density. Figure 13.4b shows results for the 5.0-cm root ball in the bubble column. Internal dissolved oxygen tensions decreased as the fluid flow rate in the reactor was reduced; however, increases in air flow above 200 cm^3 min^{-1} produced relatively little improvement. At less than 200 cm^3 min^{-1}, oxygen levels were significantly lower at the higher root densities.

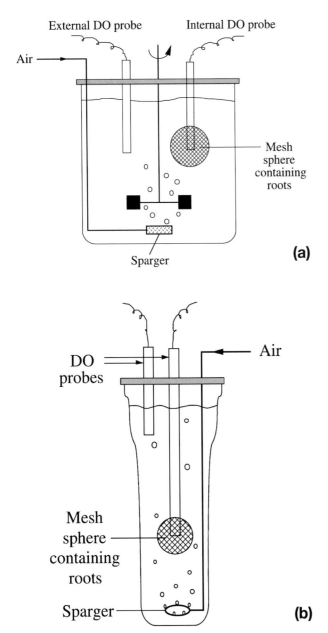

Figure 13.3. Experimental apparatus for measuring intraclump dissolved oxygen (DO) concentrations under reactor conditions (Yu and Doran, 1994). (a) 1.5-L stirred tank reactor; (b) 2.5-L bubble column reactor.

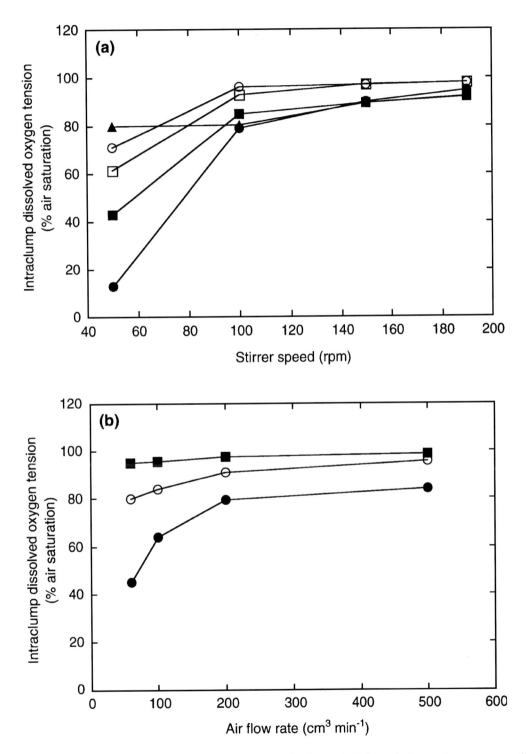

Figure 13.4. Typical results for dissolved oxygen tension inside *Atropa belladonna* hairy root clumps as a function of reactor operating conditions. (a) Stirred tank reactor: (▲) 3.5-cm ball containing 44.2 g L^{-1} dry weight roots; (○) 4.0-cm ball containing 29.7 g L^{-1} roots; (●) 4.0-cm ball containing 42.8 g L^{-1} roots; (□) 5.0-cm ball containing 25.9 g L^{-1} roots; and (■) 5.0-cm ball containing 32.1 g L^{-1} roots. (b) Bubble column reactor with 5.0-cm root ball containing (■) 15.7, (○) 29.6, and (●) 41.7 g L^{-1} dry weight roots.

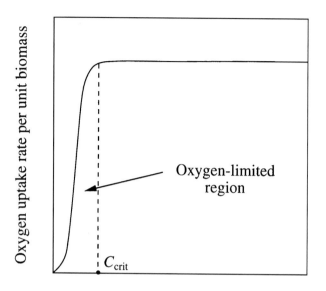

Figure 13.5. General relationship between dissolved oxygen concentration and rate of oxygen uptake per unit biomass.

Intraclump dissolved oxygen concentrations measured in these experiments varied widely with operating conditions from 13 to 98% air saturation. As the diameters of all the root balls used were much smaller than those usually formed in reactors of comparable size, these results should be considered optimistic indications of actual oxygen levels likely to be found in hairy root cultures. To evaluate whether these oxygen concentrations are adequate for hairy root growth and metabolism, some idea of the oxygen requirements of hairy roots is needed.

Critical Oxygen Levels for Hairy Roots

The relationship between rate of oxygen uptake and dissolved oxygen concentration in cell cultures generally conforms to the model illustrated in Figure 13.5. Above a certain critical oxygen level C_{crit}, rate of oxygen uptake is independent of oxygen concentration; below this critical level, the rate declines as dissolved oxygen tension is reduced. For optimal culture performance at the fastest rates possible, dissolved oxygen concentrations must be kept above the critical. The value of C_{crit} depends on the organism: for microbial cultures under average growth conditions, C_{crit} usually falls between 4×10^{-4} and 8×10^{-4} g L^{-1}, or about 5–10% of air saturation at ambient conditions. Values for suspended plant cells range higher, and have been reported as 1.3–1.6×10^{-3} g L^{-1}, or 16–20% air satu-

ration under average culture conditions (Kessell and Carr, 1972; Payne *et al.*, 1987).

For the rate of oxygen uptake by solid, three-dimensional tissue such as hairy roots to be independent of oxygen concentration, all respiring cells in the root must be exposed to oxygen levels above the critical. However, to maintain cells in the interior of the root above, e.g. 16–20% air saturation, the external dissolved oxygen tension must be greater than this value because of the concentration gradient through the tissue as illustrated in Figure 13.2. The 'apparent' or 'whole root' critical oxygen tension measured in the bulk liquid in the absence of boundary layer effects at the root surface can be estimated according to the equation:

$$C_{crit,wr} = C_{crit,true} + \Delta C \qquad (1)$$

where $C_{crit,wr}$ is the measured 'whole root' critical oxygen tension, $C_{crit,true}$ is the 'true' or 'biological' critical oxygen tension (e.g. 16–20% air saturation), and ΔC is the total oxygen deficit across the root radius from wall to inner stele as indicated in Figure 13.2. As ΔC is likely to change with location on the root, the above equation represents an average over the entire root length. We expect therefore that critical oxygen levels for hairy roots measured in the bulk liquid will be greater than the true critical oxygen concentration for plant cells, even in the absence of external boundary layer effects.

Experiments were performed using *Atropa belladonna* hairy roots to estimate the whole root critical oxygen tension. To eliminate liquid boundary layers around individual roots, a small gradientless closed-loop recirculation reactor was operated at flow velocities high enough so that the rate of oxygen uptake was independent of external flow conditions (Williams and Doran, 1995). Oxygen uptake rates were measured at $25.0 \pm 0.1°C$ at dissolved oxygen levels between 21 and 212% air saturation (i.e. about 1.8–18×10^{-3} g L^{-1}). As shown in Figure 13.6a, the results indicate that the whole root critical oxygen tension is 110–120% air saturation, considerably higher than C_{crit} for suspended plant cells. Consideration of the oxygen gradients existing within the roots (e.g. Figure 13.2) suggests that the observed reductions in respiration rate at lower dissolved oxygen tensions reflect the development of an anoxic core within the stele rather than a general reduction in respiratory activity throughout the root tissue.

Critical oxygen levels can also be defined for other metabolic activities such as growth and product synthesis; these values may be somewhat different from C_{crit} for oxygen uptake. For example, as oxygen levels are lowered, reductions in stelar respiration may occur to signal C_{crit} for oxygen uptake,

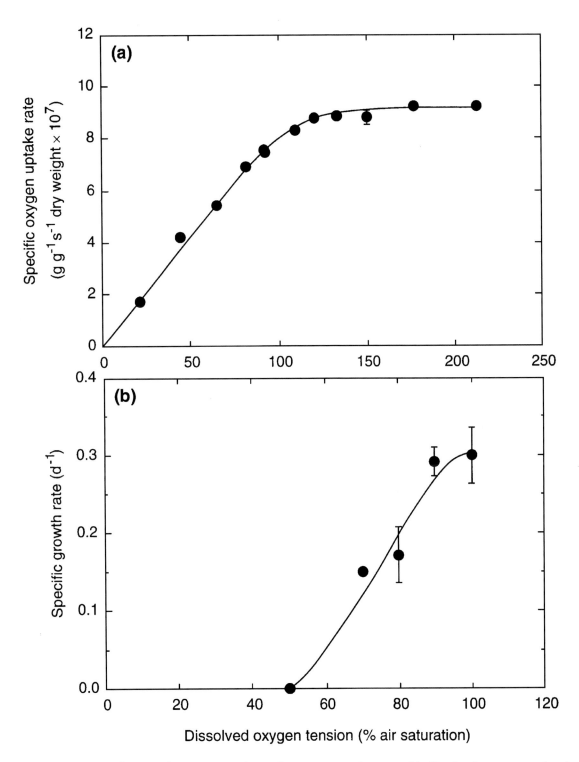

Figure 13.6. (a) Results showing the variation of specific oxygen uptake rate with dissolved oxygen tension for *Atropa belladonna* hairy roots. The critical oxygen tension for oxygen uptake is 110–120% air saturation. (b) Results for the specific growth rate of single *A. belladonna* hairy roots in liquid culture as a function of dissolved oxygen tension. Growth was measured in terms of total root length. The critical oxygen tension for growth is 90–100% air saturation (Yu and Doran, 1994).

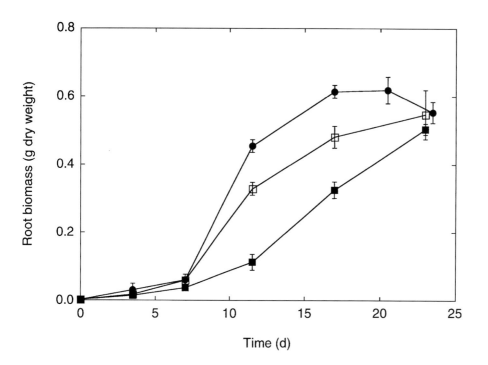

Figure 13.7. Growth of *A. belladonna* roots in shake flasks provided with an atmosphere outside the flask of (□) air; (●) oxygen-enriched air at an oxygen level of 150% that in air; and (■) 100% oxygen.

without oxygen levels or growth rate at the root tip meristem being affected (Armstrong and Webb, 1985). If this happens, the critical oxygen tension measured for growth will be lower than that for oxygen uptake. Localisation of secondary metabolite synthesis within the root tissue might also cause critical oxygen levels for product formation to be different from those for growth and respiration. The critical oxygen tension for growth of *A. belladonna* hairy roots was investigated using shake flask cultures in the absence of external boundary layer effects, with growth measured in terms of root length from photographs taken from the bottom of the flasks (Yu and Doran, 1994). As indicated in Figure 13.6b, growth was negligible at 50% air saturation, and C_{crit} for growth can be estimated as close to 100% air saturation. This result for C_{crit} has been confirmed in additional experiments with dissolved oxygen tensions up to 400% air saturation (Williams and Doran, unpublished).

The results of Figure 13.6 make very clear the tendency of *A. belladonna* hairy root cultures to become oxygen-limited at dissolved oxygen tensions around 100% air saturation. When considered together with the results of Figure 13.4 which show that oxygen levels inside hairy root clumps are readily reduced below 100% air saturation, it must be concluded that hairy root cultures are very likely to be oxygen-limited unless the culture is aerated with oxygen-enriched air or some other measures are taken to improve local dissolved oxygen concentrations. Some variations may be expected in other

root cultures due to differences in root diameter, structural component thicknesses and oxygen demand; however the results obtained with *A. belladonna* mean that oxygen effects must at least be considered when analysing kinetic data from hairy root systems.

Oxygen Limitations in Shake Flasks

Shake flasks have been applied extensively for culturing hairy roots; to date, most of the kinetic and stoichiometric information available about hairy roots has been obtained from shake flask experiments. Shake flasks can be an effective device for gas–liquid mass transfer: besides the type of flask closure (Lee and Shuler, 1991), the important factors affecting aeration of shake flasks are the shaking speed, the size of the flask, and the volume of the liquid. Gas–liquid mass transfer occurs primarily by absorption in the thin liquid film thrown up onto the flask walls by the action of the shaker (Gaden, 1962). The presence of hairy roots in flasks tends to dampen the motion of the liquid and the efficiency of this process. Supply of oxygen to hairy roots depends on the rate of liquid–solid as well as gas–liquid mass transfer; however, relatively little can be predicted about the adequacy of liquid–solid transfer in shake flasks. Investigations by Rhodes and Gaden (1957) have suggested that the performance of shake flasks in this area may be generally poor, with the liquid–solid mass transfer coefficient only weakly dependent on operating parameters.

To establish whether oxygen limitations affect shake flask cultures of hairy roots, experiments were performed using 250-mL shake flasks with silicone sponge stoppers. Each flask containing 100 mL MS medium was inoculated with *Atropa belladonna* hairy roots and placed in $59 \times 27 \times 19$-cm perspex boxes. The boxes used in the experiment had tight-fitting lids, and were fitted with gas inlet and outlet ports on opposite walls. By passing gases with different oxygen contents through the boxes, the shake flasks were exposed to different gas atmospheres without excessive stripping of volatile components. The boxes were placed on an orbital shaker operated at 100 rpm. A control with air and two oxygen-enriched atmospheres of 150% the level of oxygen in air (i.e. 31.5% v/v) and 100% v/v oxygen, were tested.

The results for growth of the roots at 25°C over a period of up to 23.5 d are shown in Figure 13.7. The cultures exposed to 150% the level of oxygen in air grew considerably faster than the air control, producing ca. 38% more biomass during the first 11.5 d of culture. In contrast, the atmosphere of pure oxygen was detrimental to growth, with only 34% the mass of roots being produced in oxygen compared with the control over the same period. As oxygen supplementation at the 150% level was effective in enhancing the growth rate of the roots, it must be concluded that oxygen restrictions can exist in shake flask cultures of hairy roots in air. Use of oxygen-enriched air to overcome these limitations must be subject to certain limits, however, as high concentrations apparently induce an inhibitory response.

Effect of Hairy Roots on External Oxygen Transfer Resistance in Shake Flasks

In shake flasks, transport of oxygen from the gas phase to hairy roots submerged in liquid medium involves two major resistances external to the root: the liquid boundary layer at the gas–liquid interface, and the liquid boundary layer surrounding the roots. Under steady-state conditions, rate of oxygen transfer through these boundary layers must be equal to the rate of oxygen uptake by the roots. This situation is represented by the mass balance equation:

$$q_O x = k_L a\, (C_{AL}^* - C_{AL}) = k_S a_S (C_{AL} - C_{AS}) \qquad (2)$$

where q_O is the specific rate of oxygen uptake by the roots, x is the biomass concentration, k_L is the gas–liquid mass transfer coefficient, a is the gas–liquid interfacial area, C_{AL}^* is the equilibrium concentration of oxygen in the liquid, C_{AL} is the concentration of oxygen in the liquid, k_S is the liquid–solid mass transfer coefficient, a_S is the liquid–solid interfacial

area, and C_{AS} is the oxygen concentration in the liquid at the root surface.

Equation (2) is difficult or impossible to apply experimentally to dense root cultures because C_{AL}, the concentration of oxygen in the liquid, is not uniform within or around the root mat. Direct measurements of either $k_L a$ or $k_S a_S$ using standard procedures are therefore virtually meaningless as the results will depend on where in the vessel the measurements are taken. However Equation (2) can be used to estimate the minimum resistance to oxygen transfer in hairy root cultures under conditions of maximum biomass density. The equality of mass transfer rates in the two boundary layers in Equation (2) allows elimination of C_{AL} as follows:

$$C_{AL} = \frac{k_S a_S C_{AS} + k_L a C_{AL}^*}{k_S a_S + k_L a} \qquad (3)$$

As indicated in Figure 13.6b, *A. belladonna* hairy roots were found not to grow when the external oxygen concentration was reduced to 50% air saturation (Yu and Doran, 1994). Therefore, let us assume that growth ceases and the maximum biomass concentration is achieved when the oxygen concentration at the root surface C_{AS} equals $0.5\,C_{AL}^*$. From Equations (2) and (3):

$$x_{max} = \frac{0.5 k_S a_S}{q_O}\left(\frac{k_L a C_{AL}^*}{k_S a_S + k_L a} \right) \qquad (4)$$

where x_{max} is the maximum biomass density supported by the culture system. Rearrangement of Equation (4) yields an equation for $\left(\dfrac{1}{k_L a} + \dfrac{1}{k_S a_S} \right)$, the total resistance to oxygen transfer due to the combination of gas–liquid and liquid–solid boundary layer effects:

$$\frac{1}{k_L a} + \frac{1}{k_S a_S} = \frac{0.5\,C_{AL}^*}{q_O\, x_{max}} \qquad (5)$$

Equation (5) allows estimation of the total mass transfer resistance using known or measurable parameters. C_{AL}^* in liquid medium under 1 atm air and 25°C can be taken as 8×10^{-3} g L^{-1}; x_{max} is the root density measured in culture at the cessation of growth. For *A. belladonna* hairy roots, the maximum value of q_O is shown in Figure 13.6a to be about 9.2×10^{-7} g g^{-1} s^{-1} dry weight. The assumptions used to derive Equation (5) mean that the calculated values for total mass transfer resistance will be rough estimates only. Because hairy roots exhibit a relatively high apparent critical oxygen tension, specific rates of oxygen uptake will be less than the maximum

Table 13.1. Maximum biomass concentrations measured at stationary phase and estimated values of the minimum combined gas–liquid and liquid–solid mass transfer resistance for *Atropa belladonna* hairy root cultures. The flask size was 250 mL; the initial volume of MS medium was 100 mL. The results for mass transfer resistance were calculated using Equation (5). ± indicates standard deviation from triplicate flasks.

Shaker speed (rpm)	Maximum biomass concentration, x_{max} (g L^{-1} dry weight)	Minimum combined mass transfer resistance, $\left(\dfrac{1}{k_L a}+\dfrac{1}{k_S a_S}\right)$ (s)
100	7.17 ± 0.03	610
120	8.16 ± 0.04	530
140	8.68 ± 0.03	500
160	10.11 ± 0.03	430

under most culture conditions. Accordingly, a minimum value of the mass transfer resistance, or best-case scenario, will be obtained by applying in Equation (5) the maximum value for q_O.

Values for x_{max} were measured by growing *A. belladonna* hairy roots in 250-mL shake flasks containing 100 mL MS medium until stationary phase was reached. The flasks were incubated at 25°C on an orbital shaker with maximum linear displacement of 3.5 cm operated at speeds of 100, 120, 140 or 160 rpm. As indicated in Table 13.1, the values for x_{max} were dependent on shaker speed. Because oxygen transfer in flasks is directly affected by shaking speed, this result is consistent with the findings outlined in the previous section that oxygen transfer limitations exist in shake flask cultures of these roots. In addition, the final sugar concentrations at stationary phase ranged between 7.6 and

15.0 g L^{-1} (from an initial level of 30 g L^{-1}), confirming that sugar was not the limiting substrate. The measured values for x_{max} together with the values for the other parameters as mentioned above were applied to estimate the combined mass transfer resistance from Equation (5). The results are listed in Table 13.1; mass transfer resistance in the presence of roots decreased with increasing shaker speed from 610 to 430 s.

The estimated values for $\left(\dfrac{1}{k_L a}+\dfrac{1}{k_S a_S}\right)$ in flasks containing hairy roots can be compared with the mass transfer resistance in shake flasks without hairy roots. This resistance is equal to $^1/_{k_L a}$. $k_L a$ for oxygen was measured at 25°C in 250-mL shake flasks containing MS medium without roots using the dynamic method described by van Suijdam *et al.* (1978) performed in triplicate. The measurements were made on an orbital shaker with 3.5-cm stroke at shaker speeds of 100, 120, 140 and 160 rpm and medium volumes of 50, 100 and 150 mL. The response time of the polarographic dissolved oxygen electrode used in the measurements was compared with the $k_L a$ values according to the method of Philichi and Stenstrom (1989). This analysis showed that any errors in $k_L a$ due to the probe response kinetics were limited to 5–10%. The results for $k_L a$ and $^1/_{k_L a}$ are listed in Table 13.2. As expected, $k_L a$ improved with increasing shaker speed and decreasing liquid volume.

Comparison of the estimated oxygen transfer resistances with (Table 13.1) and without (Table 13.2) hairy roots shows that resistances with roots are 2–3 times higher than in root-free medium. The presence of roots therefore significantly increases oxygen transfer resistance in shake flask cultures. This effect cannot be attributed solely to the development of liquid–solid boundary layers around the biomass; presence of roots in shake flasks can also be expected to reduce the effectiveness of gas–liquid mass transfer.

Table 13.2. Values of $k_L a$ and the gas–liquid mass transfer resistance ($^1/_{k_L a}$) in 250-mL shake flasks containing MS medium without roots.

| Shaker speed (rpm) | Liquid volume (mL) | | | | | |
| | 50 | | 100 | | 150 | |
	$k_L a$ (s^{-1})	$^1/_{k_L a}$ (s)	$k_L a$ (s^{-1})	$^1/_{k_L a}$ (s)	$k_L a$ (s^{-1})	$^1/_{k_L a}$ (s)
100	6.8×10^{-3}	150	3.3×10^{-3}	300	1.1×10^{-3}	930
120	7.7×10^{-3}	130	5.9×10^{-3}	170	2.1×10^{-3}	490
140	9.9×10^{-3}	100	6.1×10^{-3}	160	3.3×10^{-3}	300
160	1.3×10^{-2}	80	7.3×10^{-3}	140	4.8×10^{-3}	210

Conclusions

This chapter presents evidence of oxygen limitations in hairy root cultures, and demonstrates the critical importance of oxygen transfer to hairy roots. Because of the solid-phase nature of roots and the development of oxygen gradients within root tissues, relatively small reductions in dissolved oxygen concentration in the medium can lead to a significant decrease in metabolic activity. Oxygen limitations are also likely to affect shake flask cultures. This is an important observation because most kinetic data for hairy roots have been obtained using shake flasks, and flask cultures continue to be applied as a benchmark for assessment of reactor performance. Although the degree of oxygen limitation will vary depending on the properties of the roots, this work indicates that the adequacy of oxygen transfer in hairy root cultures cannot be generally assumed.

Acknowledgments

This work was supported by the Australian Research Council (ARC). P.M.D. acknowledges additional support from an ARC Queen Elizabeth II Research Fellowship.

References

Armstrong W and PM Beckett (1985). Root aeration in unsaturated soil: a multi-shelled mathematical model of oxygen diffusion and distribution with and without sectoral wet-soil blocking of the diffusion path. *New Phytol* 100: 293–311.

Armstrong W and PM Beckett (1987). Internal aeration and the development of stelar anoxia in submerged roots. *New Phytol* 105: 221–245.

Armstrong W and T Webb (1985). A critical oxygen pressure for root extension in rice. *J Exp Bot* 36: 1573–1582.

Armstrong W, PM Beckett, SHFW Justin and S Lythe (1991). Modelling, and other aspects of root aeration by diffusion. In *Plant Life Under Oxygen Deprivation*, edited by MB Jackson, DD Davies and H Lambers, pp. 267–282. The Hague: SPB Academic.

DiIorio AA, RD Cheetham and PJ Weathers (1992a). Carbon dioxide improves the growth of hairy roots cultured on solid medium and in nutrient mists. *Appl Microbiol Biotechnol* 37: 463–467.

DiIorio AA, RD Cheetham and PJ Weathers (1992b). Growth of transformed roots in a nutrient mist bioreactor: reactor performance and evaluation. *Appl Microbiol Biotechnol* 37: 457–462.

Ermayanti TM, JA McComb and PA O'Brien (1994). Stimulation of synthesis and release of swainsonine from transformed roots of *Swainsona galegifolia*. *Phytochemistry* 36: 313–317.

Flores HE and WR Curtis (1992). Approaches to understanding and manipulating the biosynthetic potential of plant roots. *Ann NY Acad Sci* 665: 188–209.

Gaden EL (1962). Improved shaken flask performance. *Biotechnol Bioeng* 4: 99–103.

Hamill JD, AJ Parr, MJC Rhodes, RJ Robins and NJ Walton (1987). New routes to plant secondary products. *Bio/Technol* 5: 800–804.

Hilton MG and MJC Rhodes (1990). Growth and hyoscyamine production of 'hairy root' cultures of *Datura stramonium* in a modified stirred tank reactor. *Appl Microbiol Biotechnol* 33: 132–138.

Inomata S, M Yokoyama, Y Gozu, T Shimizu and M Yanagi (1993). Growth pattern and ginsenoside production of *Agrobacterium*–transformed *Panax ginseng* roots. *Plant Cell Rep* 12: 681–686.

Kessell RHJ and AH Carr (1972). The effect of dissolved oxygen concentration on growth and differentiation of carrot (*Daucus carota*) tissue. *J Exp Bot* 23: 996–1007.

Kondo O, H Honda, M Taya and T Kobayashi (1989). Comparison of growth properties of carrot hairy root in various bioreactors. *Appl Microbiol Biotechnol* 32: 291–294.

Lee CWT and ML Shuler (1991). Different shake flask closures alter gas phase composition and ajmalicine production in *Catharanthus roseus* cell suspensions. *Biotechnol Techniq* 5: 173–178.

Luxmoore RJ, LH Stolzy and J Letey (1970). Oxygen diffusion in the soil–plant system. II. Respiration rate, permeability, and porosity of consecutive excised segments of maize and rice roots. *Agron J* 62: 322–324.

Mukundan U and MA Hjortso (1991). Growth and thiophene accumulation by hairy root cultures of *Tagetes patula* in media of varying initial pH. *Plant Cell Rep* 9: 627–630.

Payne GF, ML Shuler and P Brodelius (1987). Large scale plant cell culture. In *Large Scale Cell Culture Technology*, edited by BK Lydersen, pp. 193–229. Munich: Hanser.

Philichi TL and MK Stenstrom (1989). Effects of dissolved oxygen probe lag on oxygen transfer parameter estimation. *J Water Poll Contr Fed* 61: 83–86.

Prince CL, V Bringi and ML Shuler (1991). Convective mass transfer in large porous biocatalysts: plant organ cultures. *Biotechnol Prog* 7: 195–199.

Ramakrishnan D and WR Curtis (1995). Elevated meristematic respiration in plant root cultures: implications to reactor design. *J Chem Eng Japan* 28: 491–493.

Repunte VP, M Kino-oka, M Taya and S Tone (1993). Reversible morphology change of horseradish hairy roots cultivated in phytohormone-containing media. *J Ferment Bioeng* 75: 271–275.

Rhodes RP and EL Gaden (1957). Characterization of agitation effects in shaken flasks. *Ind Eng Chem* 49: 1233–1236.

Robins RJ, EG Bent and MJC Rhodes (1991). Studies on the biosynthesis of tropane alkaloids in *Datura stramonium* L. transformed root cultures. 3. The relationship between morphological integrity and alkaloid biosynthesis. *Planta* 185: 385–390.

Taiz L and E Zeiger (1991). *Plant Physiology*. Benjamin/Cummings, Redwood City, USA.

van Suijdam JC, NWF Kossen and AC Joha (1978). Model for oxygen transfer in a shake flask. *Biotechnol Bioeng* 20: 1695–1709.

Williams GRC and PM Doran (1995). The importance of oxygen in hairy root culture. *Australasian Biotechnol* 5: 92–94.

Wilson PDG, MG Hilton, PTH Meehan, CR Waspe and MJC Rhodes (1990). The cultivation of transformed roots from laboratory to pilot plant. In *Progress in Plant Cellular and Molecular Biology*, edited by HJJ Nijkamp, LHW van der Plas and J van Aartrijk, pp. 700–705. Dordrecht: Kluwer.

Yoshikawa T and T Furuya (1987). Saponin production by cultures of *Panax ginseng* transformed with *Agrobacterium rhizogenes*. *Plant Cell Rep* 6: 449–453.

Yu S and PM Doran (1994). Oxygen requirements and mass transfer in hairy-root culture. *Biotechnol Bioeng* 44: 880–887.

Yu S, KH Kwok and PM Doran (1996). Effect of sucrose, exogenous product concentration, and other culture conditions on growth and steroidal alkaloid production by *Solanum aviculare* hairy roots. *Enzyme Microb Technol* 18: 238–243.

Effect of Root Morphology on Reactor Design and Operation for Production of Chemicals

14

Edgard B. Carvalho[1], Sean Holihan[1], Beth Pearsall[1] and Wayne R. Curtis[1,2]

[1]Department of Chemical Engineering, and [2]Biotechnology Institute, The Pennsylvania State University, 519 Wartik Laboratory, University Park, PA 16802-5807, USA; email: wrc2@psuvm.psu.edu

Summary

This chapter starts with a general discussion of different morphological characteristics of transformed root cultures and presents a qualitative discussion of how these characteristics will impact the design and operation of reactors. The second part discusses experimental results examining fluid flow resistance for different root morphologies and preliminary efforts towards characterising root morphology using image analysis.

Introduction

The morphology of transformed roots in culture can be affected by many different factors including the plant species, cultivar, the strain of *Agrobacterium rhizogenes* used in culture induction and, at the molecular level, the location and number of T-DNA integrations. In addition, environmental factors such as nutrients, growth regulators and the general level of stress within the culture can affect root growth patterns. In nature, plant roots have tremendously varied morphology to accomplish various roles, including nutrient and water uptake as well as storage. Given the large interspecies differences, it is not surprising that plant species is a major determinant for root culture morphology. Average root thickness can vary from 0.2 to nearly 1.5 mm, and transformed root cultures of different species display large differences in the extent of root hair development and degree of branching (Hilton *et al.*, 1988; Ramakrishnan and Curtis, 1994). Even for closely related species, there is often a wide range of root morphologies which represent adaptations to different ecosystems (Barlow, 1994). Different cultivars can display very different responses to infection with *Agrobacterium rhizogenes* (Vanhala *et al.*, 1995). In transforming 8 *Psoralea* cultivars, the morphology of root cultures varied from callused, thick and hairy, to profuse lateral branching (Nguyen *et al.*, 1992). Similar variations in transformation success (and resulting root morphology) can be obtained by transforming a single plant cultivar with different strains of *Agrobacterium rhizogenes* (Ciau-Uitz *et al.*, 1994; Drewes and van Staden, 1995). In addition to species and cultivar dependent variations in morphology, multiple transformations using the same species of *A. rhizogenes* can also give rise to many different morphologies and rates of growth (Mano *et al.*, 1989; Jaziri *et al.*, 1994). The basis of these variations are at the level of T-DNA insertion as a

151

result of differences in expression level for different locations of integration on the chromosomes. As will be described in the subsequent sections, there is a tremendous variation in the morphological characteristics of roots which is likely to influence significantly the ability to scale-up root cultures for commercial chemical production.

General Morphology Considerations

Thick Versus Thin

We have observed experimentally that plant species is a dominant factor in determining root thickness. Multiple transformations of radish (*Raphanus sativus*) consistently gave root diameters of about 0.2 mm. Establishment of many lines of *Hyoscyamus muticus* over several years has always yielded cultures with root diameters of 0.7–1 mm. Root cultures of medicinal cucumber established by Hector Flores (The Pennsylvania State University) consistently gave some of the thickest root cultures yet described, some reaching nearly 5 mm in diameter (Savary and Flores, 1994). In establishing root cultures it is only natural to select transformants which have the fastest growth rates. This is justified in part since metabolite production often parallels growth. Jackie Shanks (Rice University) undertook transformation

of four cultivars of *Catharanthus roseus*; out of 150 'clones', five were selected based on the fastest rates of growth. All of the selected lines had "thin, straight, and regular branches with thin tips" (Bhadra *et al.*, 1993). Similarly, Yukimune *et al.* (1994), noted that although initial 'clones' of *Duboisia myoporoides* varied in thickness, after repeated selection for high productivity "the only morphology was fine roots". During the process of adapting roots to liquid culture and subsequent maintenance over long periods of time, it is not unusual to observe significant variations in growth rate. These changes are almost always accompanied by a change in root thickness where the root is thinner during more rapid periods of growth. We have observed this in cultures of *Trichosanthes kirilowii*, *Azidirachta indica* and *Hyoscyamus muticus*. However, a thinner root morphology is not always associated with more rapid growth as we have observed markedly thinner roots under nutrient deprived conditions, consistent with the observation of Yonemitsu *et al.* (1990) that thinner roots of *Lobelia inflata* occurred after growth on suboptimal medium. The observations above are not intended to propose any definitive correlation between growth rate and root thickness, but instead to loosely categorise those factors which might influence root thickness. The importance of root thickness in scale-up of root reactors will recur

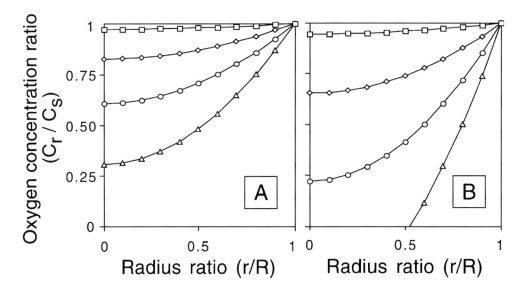

Figure 14.1. Oxygen profiles in root tissue predicted from a model of diffusion in the cylindrical radial direction with zero-order oxygen consumption, yielding the expression: $C_r = C_s - [BOD/(4\, D_{O2})]\, (R^2 - r^2)$, where C_r and C_s are the O_2 concentrations at radius r from the centre of the root and at the root surface, respectively; D_{O2} is the oxygen diffusion coefficient (cm^2 s^{-1}), BOD is the biological oxygen demand (moles O_2 h^{-1} g^{-1} fresh weight), and R is the radius of the root. Respiration rates were: (A) 9×10^{-6} moles O_2 h^{-1} g^{-1} fresh weight, and (B) 1.8×10^{-5} moles O_2 h^{-1} g^{-1} fresh weight, which were calculated based on the stoichiometry of sugar respiration and 'typical' values of 50% carbon yield and 50% carbon content. The diffusion coefficient within the root (D_{O2}) was taken to be 9×10^{-6} cm^2 s^{-1} as measured by Berry and Norris (1949). The curves shown in Figures 14.1A and 14.1B correspond to the following diameter R values: 0.2 mm ($\square$); 0.5 mm ($\diamond$); 0.75 mm ($\bigcirc$); and 1 mm ($\triangle$).

throughout the chapter since this dimension is of fundamental importance in determining the rates of nutrient demand within a root, as well as the area available for nutrient uptake.

Implications for reactor design and operation

The relationship between root thickness and nutrient demand can be of fundamental importance to the growth of plant roots in reactors. For faster growing roots, the thickness of the root may control growth by limiting oxygen availability. Theoretical oxygen profiles in roots of varying diameter are shown in Figure 14.1 for roots with doubling times of 2 and 4 d. In these calculations, oxygen is assumed to move within the root only by diffusion. Such an assumption is logical since there is no leaf transpiration to provide convection, and other source–sink driving forces are eliminated by uniformity of the liquid culture environment. For a root which is growing relatively slowly (subcultured every 3–4 weeks), respiration is sufficiently slow that oxygen should penetrate all but the thickest roots. In contrast, a rapidly respiring root (subcultured every 1–2 weeks) is very likely to have significant oxygen gradients within the root, and possibly complete depletion of oxygen in the innermost tissues. In preliminary studies we have observed a significant reduction in oxygen concentration at the axis of transformed roots using an oxygen sensitive dye (pyrene butyric acid) and laser confocal microscopy (unpublished). If growth is oxygen limited, thinner roots could grow faster due to greater oxygen availability.

The presence of diffusional limitations within the root could result in oxygen limitation even under conditions of high availability. The critical oxygen pressure (COP) for growth is the partial pressure of oxygen below which growth is oxygen limited (for example, a value of 97% of μ_{max}, the maximum specific growth rate, has been used). In liquid systems at equilibrium, the concentration of dissolved oxygen is proportional to the oxygen partial pressure according to Henry's law, allowing the COP to be expressed in terms of dissolved oxygen concentration. For microorganisms, the COP is usually very low, of the order of 0.1–1.6 mg dissolved oxygen per litre (Finn, 1967). This corresponds to less than 19% of the saturation level of oxygen in water under 1 atm pressure. In contrast, Yu and Doran (1994) have reported that *Atropa belladonna* hairy roots do not grow below 50% air saturation, and the COP of excised maize roots has been reported to be greater than air saturation (Saglio *et al.*, 1984). The reason for this difference between microorganisms and root culture is directly related to morphology. For microorganisms, COP is a measure of physiological availability since the organisms are so small that internal transport resistances are minimal. For root cultures, the observed COP is the result of internal mass transfer limitations and is not indicative of the roots' intrinsic affinity for oxygen.

Transport of oxygen within roots is dependent upon the availability of oxygen at the surface. The profiles presented in Figure 14.1 suggest that diffusion may limit oxygen availability even if the medium is fully mixed and oxygenated. In a reactor, this is difficult to achieve. The entwined root matrix impairs liquid mixing, further decreasing the driving force for transport of oxygen to the root. In this regard, a thinner root morphology has a disadvantage since thinner roots will result in a greater resistance to fluid mixing. Roots with larger diameters present lower flow resistance but a greater internal resistance. Qualitatively, the reason for these competing factors is very simple: a thinner morphology provides a greater surface area per mass of tissue. A larger surface area is an advantage for oxygen transport into the root once the oxygen reaches the root surface, but the surface concentration may be lower due to reduced mixing. The principles describing the effect of root diameter on mixing will be developed further in a later section of this chapter. Another important consideration in addition to growth is the potential influence of oxygen availability on secondary metabolite formation. Root thickness could indirectly mediate culture productivity (either positively or negatively) by influencing the aerobic status of the tissue.

Branched Versus Apically Dominant

'Highly branched' is a term which is almost universally used to describe transformed root cultures. The degree of root branching is dependent on the frequency of branch formation as compared to the rate of root elongation. This is the fundamental mathematical characterisation found in structured models of transformed root growth (Taya *et al.*, 1989a; Flint-Wandel and Hjortso, 1993). An equally important aspect which is less easy to capture mathematically is apical dominance which can be seen in the scanned images presented later in this chapter (Figure 14.8). By definition, apical dominance implies that the 'dominant' terminal meristem exerts an influence over the laterals, thus effecting their growth potential. Presumably such an influence would be expressed either as a distance or time dependent function which would require additional kinetic data. For root systems in soil, branching patterns are critically important in determining nutrient and water uptake. As a result, roots display considerable plasticity in order to respond to local environmental variations in soil (Feldman, 1984). Since roots have developed the ability to respond to environmental cues, it is not surprising that the branching of cultured roots can be altered significantly by many different factors.

Implications for reactor design and operation

Essentially all assessments of transformed root growth rate are conducted in gyratory shake flasks at relatively low biomass density. Under these conditions, it is not surprising that high rates of growth tend to correlate with highly branched morphologies. Rapid branching results in faster generation of active growth meristems. However, cultures which display the fastest rates of growth in flask culture may not have the fastest biomass accumulation rates in large-scale reactor systems. It is difficult to achieve the level of mixing on a large scale that is provided in gyratory shake flasks. As convection is reduced with increased biomass, the growth rate becomes governed by mass transfer limitations and not the intrinsic growth rate of the culture. Therefore, an important consideration for maintaining culture productivity during scale-up is to choose the growth habit of roots which delays the onset of mass transfer limitations. A key to minimising mass transfer limitations is to achieve uniform distribution of tissue. This not only avoids problems of local stagnation, but also minimises overall mass transfer resistances throughout the bed. Extremely profuse branching may be detrimental to achieving uniform root tissue distribution in a reactor system as highly branched morphology will result in dense, localised clumps. Subsequent proliferation will be impaired because only the outer regions of the clumps will experience adequate oxygen availability for growth. In contrast, a 'clone' with less frequent branching, i.e. greater apical dominance, will tend to distribute more uniformly as it grows. The entire root mass will experience more uniform nutrient availability and maintain its intrinsic rate of growth for a greater percentage of the culture period.

It is difficult to know *a priori* the extent to which morphology might influence scale-up; however, some preliminary calculations can provide some insight. If growth centres within the reactor are treated as idealised spheres, then the amount of tissue growth will be determined by the rate of respiration and the rate of oxygen transport. A typical biological oxygen demand (BOD) that has been measured for an actively growing root culture is 0.02 mmol h^{-1} g^{-1} fresh weight (Ramakrishnan and Curtis, 1995). The difficulty lies in assessing the rate of oxygen transport. The 'worst case' scenario for oxygen availability within root clumps would result if diffusion were the only transport mechanism. By assuming the oxygen diffusivity in water (D_{O2} = 2.5×10^{-5} cm^2 s^{-1}) as the effective diffusivity of oxygen in the clump (D_{eff}), oxygen profiles within a

spherical clump of radius R_0 would be defined for zero-order kinetics as:

$$\frac{C_r}{C_0} = 1 - \frac{1}{6}\left(\frac{BOD\,\rho_0}{C_0\,D_{eff}}\right)(R_0^2 - r^2) \qquad (1)$$

In this equation, C_r is the oxygen concentration at distance r from the centre of the spherical clump, C_0 is the oxygen concentration at the surface of the clump ($r = R_0$) and ρ_0 is the clump volumetric tissue concentration. Using a conservative estimate of 10% of air saturation for the non-growth oxygen pressure (NOP), any tissue within radial distance $r \leq R_i$, where the dissolved oxygen concentration is less than 10% air saturation, would not grow. Based on oxygen diffusion, the thickness of the root clump outer growing shell R_m (= $R_0 - R_i$) would be exceedingly small. By the time the root ball grew to 3 cm in diameter, the thickness of the outer growing shell would be only 102 µm. This value is far too small to be reasonable, showing that convection (in addition to diffusion) must be contributing to oxygen transport in root clumps. This analysis supports previous experimental demonstrations of the importance of convection in root clumps by Prince *et al.* (1991) and Yu and Doran (1994). If the extent of convection were known, then dispersion (represented by the effective diffusion coefficient) could be used in the preceding analysis. We are presently pursuing methods to conduct this analysis using measurements of root matrix fluid flow resistance (Carvalho and Curtis, 1995), and liquid mixing measurements from tracer analysis (see Ramakrishnan and Curtis, 1994); however, a rigorous approach has not yet been developed. Preliminary studies have demonstrated the severity of liquid mixing problems. Liquid mixing time was measured by monitoring the time dependent dispersion of a dye injection. The elapsed time between injection and 95% of the steady state dye concentration was defined as the mixing time. *S. tuberosum* root cultures were inoculated at 2 g L^{-1} fresh weight in a 15-L bubble column reactor containing a randomly distributed stainless steel matrix for inoculum attachment. Liquid mixing time increased from just over 1 min at inoculation to 40 min when the root tissue reached 180 g L^{-1} fresh weight (Tescione *et al.*, 1995). In addition, specific respiration rates declined by 50% in only 11 d of culture. It is instructive to utilise this information in conjunction with the idealised 'root ball' model to see what value of the effective dispersion coefficient would account for these experimental observations. If the dissolved oxygen concentration at the centre of the root ball is greater than that needed to support growth, then the biomass will grow at its intrinsic growth rate $^{dW}/_{dt} = \mu W$, where W is the root mass (grams) and μ is the specific growth rate (d^{-1}). However, if the outer radius (R_0) exceeds the radius at which the oxygen level has been reduced to below the NOP, then the instantaneous growth rate would be given by:

$$\frac{dW}{dt} = 4\pi\mu\rho_0 R_m^3 \left[\left(\frac{R_0}{R_m}\right)^2 - \left(\frac{R_0}{R_m}\right) + \frac{1}{3}\right] \quad (2)$$

The outer clump radius is defined as:

$$R_0 = \left(\frac{3W}{4\pi\rho_0}\right)^{1/3}$$

and the thickness of the growing outer shell by:

$$R_m = R_0 - \left[R_0^2 - \frac{6(C_0 - NOP)D_{eff}}{BOD\,\rho_0}\right]^{1/2}$$

The early stages of operation of the *S. tuberosum* bubble column reactor gave specific growth rates and BOD values of $0.35\ d^{-1}$ and 1.44×10^{-3} mol O_2 $d^{-1}\ g^{-1}$ fresh weight. For a dissolved oxygen concentration at the clump surface $C_0 = 2.54 \times 10^{-4}$ mol O_2 L^{-1} (air saturation value at atmosphere pressure), NOP = 10% of C_0 and $\rho_0 = 200$ g L^{-1} fresh weight, trial-and-error numerical integration was used to determine the effective dispersion coefficient (D_{eff}) which would meet the observed 50% reduction in respiration. The results are shown graphically in Figure 14.2. The calculations revealed that the effective dispersion coefficient in the system would have to be 50 times the oxygen diffusion coefficient to attenuate the specific respiration by 50% at Day 11 of the culture. This corresponds to an instantaneous specific growth rate at Day 11 of just under $0.2\ d^{-1}$, demonstrating the severity with which mass transfer limitations can attenuate growth. Consistent with the logic presented above, it is common to observe large reductions in growth rate in moving from small- to intermediate-scale reactors. Independent of the assumptions used to analyse these data, the substantial reduction in specific respiration rate clearly shows that a large fraction of the root tissue within the pilot-scale reactor is not contributing to further growth. To this end, the degree of branching can significantly influence the performance of a reactor by affecting time dependent mass transfer limitations.

The conclusions drawn from Figure 14.2 reflect the same branching 'logic' used by plants in soil. Excessive branching results in over-exploitation of a localised area and inefficient use of biomass towards overall biomass accumulation. In contrast, a root morphology which displays some apical dominance will result in a growth pattern which distributes the roots more evenly in the soil. In small-scale experimental systems, it is easy to lose sight of the large-scale objective. It must be kept in mind that improvements achieved at the small scale will only be of significance if they can be translated to an industrial scale.

Profuse Versus Sparse Root Hairs

As the descriptive name suggests, "hairy roots" tend to have very profuse root hairs. On one hand, root hairs provide a greater surface area for nutrient uptake; on the other, these protrusions can create a stagnant boundary layer which can inhibit nutrient mass transfer. Root hairs are very effective for the uptake of nutrients from a soil rhizosphere. In fact, an increase in root hair formation has been observed under nutrient limiting conditions (Foehse and Jungk, 1983; Bates and Lynch, 1996). Although root hairs increase the surface area for nutrient flux, the mechanism of nutrient uptake is not necessarily diffusion controlled. For nutrients such as phosphate or trace metals, these salts are immobilised in the soil and the extension of root hairs represents physical exploration (Tinker, 1976).

In contrast to soil systems, liquid culture provides relatively easy access to inorganic nutrients. In liquid culture, the available flux of inorganics and sugar can be expected to be in excess of metabolic requirements. Some brief considerations follow to support the contention that root hairs have

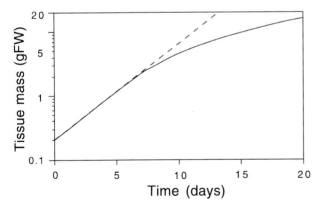

Figure 14.2. A theoretical comparison of growth of a root culture in which all the root tissue remains available for growth ($\mu = 0.35\ d^{-1}$) (dashed line), and a culture which develops a tight spherical root mass in which only an outer shell (where the dissolved oxygen is higher than the non-growing oxygen pressure, NOP) is actively growing (continuous line). The initial root mass was taken as 0.2 g fresh weight. On the 11th day after inoculation, the specific respiration rate is 50% of the initial value, as observed experimentally in a 15-L reactor (see text).

no constructive role in liquid culture. Since the medium surrounding the roots is quite rich in most nutrients and a substantial amount of water is accumulated within the biomass, passive uptake of medium could supply a substantial amount of many nutrients. Quick calculations reveal that a typical root segment (0.5 mm diameter, μ_{av} = 0.28 d^{-1}) can take up as much as 5×10^{-4} mL water per day per cm root length. Using phosphate as an example, we have shown that root cultures of *H. muticus* are not phosphate limited if the intracellular level stays above 50 μmol g^{-1} dry weight, a value which appears to be relatively constant for different plants (Dunlop and Curtis, 1991). Using the phosphate level in Gamborg's B5 medium of 1.11 mM, the level of phosphate that would accumulate passively in the tissue would give a concentration of 17.8 μmol g^{-1} dry weight or roughly 35% of the total required! Plant cultures in fact have a relatively high affinity for phosphate and other trace minerals because they must be extracted from very dilute concentrations in natural systems. Therefore, with only a minimal amount of mixing, plant root cultures extract the additional nutrients to complement passive uptake. The observation that cultured plant cells including roots rapidly accumulate inorganics in great excess of metabolic requirements provides experimental evidence that there is little concern for mass transfer limitations of inorganics (Curtis, 1988).

Correlations for mass transfer coefficients under defined hydrodynamic conditions provide an alternative means of assessing nutrient availability. Using phosphate once again as a basis for mass transfer analysis, evidence to support our contention about the role of root hairs can be found by estimating the liquid velocities required to provide mass transfer rates that meet the root biological phosphate demand. A well defined hydrodynamic environment is forced convective liquid flow past a submerged root. The mass transfer process in such an environment can be described by the following equation:

$$N_{phosphate} = k_L (C_{bulk} - C_{surf})_{mean} \qquad (3)$$

where $N_{phosphate}$ is the phosphate flux (μmol cm^{-2} s^{-1}); k_L is the mass transfer coefficient (cm s^{-1}); and $(C_{bulk} - C_{surf})_{mean}$ is the mean driving force (μmol cm^{-3}) in terms of bulk and surface phosphate concentrations. The mass transfer coefficient can be related to liquid velocity using an appropriate correlation representing fluid flow through a packed bed of cylinders (Ramakrishnan and Curtis, 1995). Based on experimentally measured tissue phosphate levels, growth rate and geometry of root tip images, a maximum phosphate flux of 2.5×10^{-6} μmol cm^{-2} s^{-1} was calculated. Using Equation (3), a mass transfer coefficient (k_L) of 3×10^{-6} cm s^{-1} is then

obtained for a typical root bed (void fraction = 0.8), assuming a driving force of 75% of the bulk concentration. This mass transfer coefficient corresponds to a superficial velocity of 1×10^{-11} cm s^{-1}, which is a ridiculously low rate of convection. In fact, this velocity is much lower than the linear extension rate of the root, which for *H. muticus* is about 7×10^{-7} cm s^{-1} ($\approx$ 1 mm d^{-1}). The purpose of these calculations is to provide a direct contrast to subsequent calculations regarding oxygen mass transfer. As will be discussed in the next section, the convection required to overcome oxygen mass transfer limitations is quite high, showing that for all but nutrient depleted medium, oxygen availability is limiting for root culture growth. If dissolved inorganics are limiting, this can be remedied by adding nutrients to the medium without concern that they will not be available to the root due to mass transfer limitations at the root surface. Therefore unlike roots in soil, there is no need to increase interfacial area for dissolved nutrient uptake. In the next section we will take this observation one step further and suggest that the profuse root hairs of so-called 'hairy roots' do not contribute to enhanced growth rates of transformed root cultures.

Implications for reactor design and operation

In addition to the likelihood that root hairs are not contributing to nutrient uptake, the profuse root hairs observed on many transformed root cultures may be inhibitory towards scale-up. While most nutrients can be supplied at the beginning of the batch culture period, the dissolved oxygen in a 200 g L^{-1} fresh weight culture will be depleted in less than 4 min unless replenished from the gas phase. As a result, mass transfer of oxygen becomes the most significant concern with regard to scale-up and reactor design. Using the approach described above for phosphate, it is possible to show that superficial velocities of the order of 1 cm s^{-1} are needed to provide oxygen fluxes to meet the localised root tip biological oxygen demand (0.266 mmol h^{-1} g^{-1} fresh weight) of a typical root bed (void fraction = 0.8; root diameter = 0.7 mm) (Ramakrishnan and Curtis, 1995). At this flow rate, the boundary layer thickness for a smooth cylinder is of the order of 10^{-2} mm. Since the root hairs have a length dimension (0.5–0.7 mm) much larger than the boundary layer thickness, their presence will increase stagnation and tend to impede oxygen transport to the root.

The effects of root hairs are not limited to submerged culture. In submerged culture, root hairs are distended; however, if the medium is removed, the delicate hairs mat along the surface analogous to removing a furry animal from water. We refer to this pictorially as the "wet cat effect" as shown

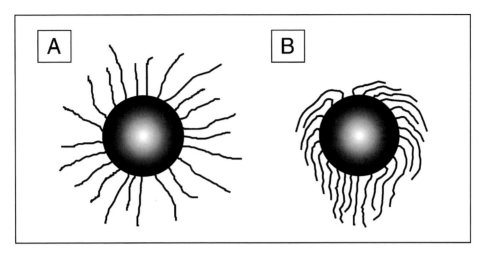

Figure 14.3. Schematic showing a comparison of root hairs in (A) a gas-dispersed (submerged), and (B) a liquid-dispersed environment. Without the buoyancy of submersion, root hairs can mat at the surface of the root to produce the 'wet cat effect'.

graphically in Figure 14.3. The root hairs and the medium they retain present an additional boundary layer which must be overcome for nutrient delivery to the internal root tissue.

It should be recognised that root hairs are not uniformly produced along the root axis. In non-transformed roots, root hairs develop in the maturing region of the elongation zone, and are believed to have a major role in water uptake from soil (Cailloux, 1974). In older portions of the root, the root hairs appear to be sloughed off. Patterns of root hair development can be similar in transformed root culture. In the case of *H. muticus*, root hairs are absent within 2 mm from the apical meristem; root hairs are also lost on older segments of root tissue even though the roots do not undergo secondary development. Despite the fact that root hairs may not be present on a majority of tissue in a root culture, they may still have a substantial influence due to their presence near the meristematic regions which undergo the most rapid growth.

Some transformed root cultures do not have prolific root hairs. An absence of root hairs was noted in root cultures of licorice (*Glycyrrhiza glabra*) (Toivonen and Rosenqvist, 1995), and we maintain several transformed root cultures which have insignificant root hair formation: medicinal Chinese cucumber (*Trichosanthes kirilowii*) and *Gaillardia pulchella*. It is likely that the extent of root hair formation can also be affected by environmental factors such as accumulation of gases (CO_2, ethylene) and nutrient status. Our laboratory and others have observed root hair loss under sparged conditions for different species, possibly due to shear (Weathers *et al.*, 1989; McKelvey *et al.*, 1993). Since root hairs

may hinder mass transfer, it could be possible to improve reactor performance by altering the culture medium to minimise root hair proliferation. Elimination of root hairs might influence scale-up in other ways. Damage to the fragile root hairs could initiate signalling to the main root axis that would greatly impair overall root growth. A phenomenon such as this has been suggested to be the basis of poor growth performance of *Catharanthus roseus* roots in stirred tanks where the root tissue was isolated from direct mechanical damage, but appeared to experience indirect damage due to the flow field produced by the radial flow impellers (Nuutila *et al.*, 1994). As culture systems are scaled up, it becomes difficult to distribute the power for oxygen transfer. This is true for bubble columns as well as for stirred tank systems. For example, in a 100-fold scale-up of a bubble column with the same aspect ratio, there is only a 22-fold increase in cross-sectional area. As a result, introduction of gas at the same volumetric gassing rate (vvm = volume of gas per volume of medium per minute) takes place in a proportionately smaller area, so that the gas superficial velocity increases with size, increasing the potential for damage. The dilute nature of biologically derived secondary metabolites such as therapeutics requires production on a scale of thousands of litres. The presence of root hairs appears to be detrimental towards this goal both in terms of mixing and sensitivity to shear.

Flexible Versus Rigid

In addition to differences in physical appearance, root cultures also have very different mechanical properties. A mechanical property which varies

Table 14.1. Measurement of flexibility for a broad range of transformed root cultures, as indicated by the force required to cause a 0.5-cm displacement of a 2-cm root segment. The corresponding bending modulus is calculated using the root diameter.

Root culture*	Diameter (mm)	Bending force (dyn)	Bending modulus ($\times 10^{-7}$ dyn cm^{-2})
L. esculentum	0.25	1.5 ± 1.1	42 ± 31
L. erythrorhizon	0.40	6.4 ± 13	28 ± 54
A. indica	0.60	84 ± 41	71 ± 34
T. kirilowii (var. *Japonicum*)	0.70	76 ± 59	47 ± 65
H. muticus (line: A4C17)	0.91	37 ± 32	6.3 ± 5.6
(line: HM90T)	0.70	16 ± 21	7.2 ± 9.3
G. pulchella (dark)	0.80	27 ± 13	7.1 ± 3.4
(light)	0.68	20 ± 10	11.7 ± 6.7
R. sativus	0.17	0.92 ± 1.40	82 ± 97
N. tabacum	0.25	1.7 ± 1.2	47 ± 33
S. tuberosum	0.55	10.6 ± 5.8	12.6 ± 6.9
B. vulgaris	0.33	2.9 ± 1.4	27 ± 13

* All cultures were maintained in Gamborg's B5 medium in 125-mL gyratory shake flasks (120 rpm, 1 inch stroke) with subculture every 2 weeks. All cultures were grown in media with sucrose as carbon source, except *B. vulgaris* was grown on fructose.

significantly for different roots is resistance to bending. Some roots easily support their own weight, while others are very limp and flaccid. To quantify these observed differences, a simple measurement was compiled for several different root cultures. By pressing a 2-cm root segment cantilevered horizontally against the edge of a beaker placed on a balance, the force (weight) produced by a 0.5-cm displacement was tabulated (Table 14.1). These results show that the force required to bend different roots can vary by a factor of 90. The large variances obtained for 6–12 replicates indicated that there is considerable heterogeneity within a culture. Despite the simplicity of this approach, these data can provide a first approximation for further engineering analysis. Analogous to the bending of cantilever beams, (Nash, 1972), these measurements can be used to determine the root modulus of elasticity (Table 14.1) as shown below:

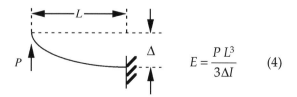

$$E = \frac{P L^3}{3 \Delta I} \qquad (4)$$

where E is modulus of elasticity (dyn cm^{-2}); P is the net force applied to the free end of the root segment (dyn); L is the length of the root segment/cantilever (cm); Δ is the maximum deflection of the root segment in the direction of the applied force (cm); and I is the moment of inertia of the cylindrical beam ($I = \frac{\pi}{64} D_r^4$ where D_r is the root diameter). As a basis for reference, a polymeric material such as polyethylene would have a modulus of 620×10^7 dyn cm^{-2}, showing that hairy roots are quite flexible. As shown in Table 14.1, there is an order of magnitude range in both the force required for bending, as well as the elastic modulus. However, the two are not necessarily correlated so that some tissues clearly have different intrinsic flexibilities. Roots derived from the neem tree (*Azidirachta indica*) and radish (*Raphanus sativus*) displayed the greatest resistance to bending (highest elastic modulus). In the case of *A. indica*, the measurements could be conducted on root tip segments; for radish the high modulus is partially an artifact since this root culture is so thin and highly branched that 2-cm segments could only be obtained by excising lateral roots along the older, most developed segments. A more valid comparison would be for roots of similar mean diameter such as *A. indica* and *H. muticus*. Despite the similar diameters, the elastic moduli for

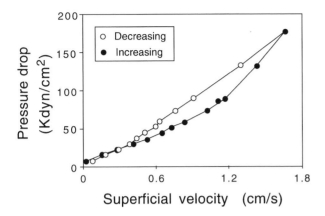

Figure 14.4. Fluid velocities achieved at different hydrostatic heads for *B. vulgaris* roots at a tissue density of 292 g L^{-1} fresh weight ($\varepsilon = 0.708$). Measurements were taken during an increase in pressure drop followed by a steady decrease to identify hysteresis effects (adapted from Holihan, 1993). Hydrostatic head was produced using a continuous overflow device described elsewhere (Ramakrishnan and Curtis, 1994). Transformed roots were grown in gyratory shake flasks and placed in a 30-cm-long × 2.5-cm-ID tube. The tissue bed length was 7.6 cm.

these two species are nearly an order of magnitude different. The diameter of the root is an extremely important factor affecting bending since its functional dependence is introduced to the fourth power from the moment of inertia. For this reason, a root such as *H. muticus* which displays nearly the lowest intrinsic resistance to bending (low elastic modulus) requires a moderate force for bending because it has a relatively large diameter. In terms of application to reactor design, it is the extent of bending that is of interest, so inaccuracies introduced due to diameter measurements are minimised.

Implications for reactor design and operation

Rigid roots will not tend to deform. If a root is flexible, this can influence reactor performance by permitting localised compactions. These localised restrictions will then dominate the resistance to flow and hinder mixing to a greater extent than would otherwise be experienced if the tissue were rigid. An example of localised compaction is demonstrated in Figure 14.4, where hysteresis is observed for flow through a packed bed of *Beta vulgaris* roots. Volumetric flow rate was measured at a series of increasing pressure drops, followed by a steady decrease in pressure drop. Volumetric flow during the decreasing phase was considerably lower than that observed during the initial increasing phase. This demonstrates that the resistance to flow had increased in the bed due to deformations caused

by the higher flow rates. The flow rates in this experiment reached a maximum superficial velocity of only 1.67 cm s^{-1}. This corresponds to a interstitial velocity of 2.35 cm s^{-1}, which is comparable to the velocities required to overcome boundary layer mass transfer limitations (Ramakrishnan and Curtis, 1995). This demonstrates experimentally that localised compaction effects can be observed under meaningful operational conditions in a reactor. Using the information provided in Table 14.1, the analysis of bending can be further expanded to provide some quantification of the extent to which deformation will be encountered. Since superficial velocities of about 1 cm s^{-1} are required to provide oxygen, this gives a basis for calculation of the fluid force that would be exerted on a root placed perpendicular to the flow field. Under these conditions, the drag force on a cylindrical root segment is given by:

$$F_D = C_D A_r \left(\frac{\rho_r u^2}{2} \right) \tag{5}$$

where F_D is the drag force per unit length of root segment (dyn cm^{-1}); C_D is the drag coefficient; A_r is the projected area per unit length of the root normal to the direction of fluid flow ($= D_r$) (cm^2 cm^{-1}); ρ_r is the density of root tissue (g mL^{-1}); and u is the superficial fluid flow velocity (cm s^{-1}). To compare the deflections on a common basis, calculations were made using a 2-cm root segment fixed at one end. This corresponds to a cantilever with uniformly distributed load. The maximum deflection experienced by this single root segment can be calculated by the following expression (Nash, 1972):

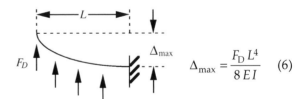

$$\Delta_{max} = \frac{F_D L^4}{8 E I} \tag{6}$$

where Δ_{max} is the deflection of the free end of the root segment (cm); F_D is the drag force per unit length of root segment (dyn cm^{-1}); L is the length ($= 2$ cm); E is the modulus of elasticity (dyn cm^{-2}); and I is the moment of inertia of the cylindrical beam ($\pi/_{64} D_r^4$). The results are tabulated in Table 14.2, showing the bending that would be experienced by a 2-cm root due to fluid flow of 1 cm s^{-1}. Although the deflections are quite small, for radish (*Raphanus sativus*) such a displacement would be significantly greater than the root diameter. For other root cultures such as tobacco (*Nicotiana tabacum*), tomato (*Lycopersicum esculentum*) and beet (*Beta*

Table 14.2. Assessment of the deflection that could be encountered due to fluid flow in submerged root culture. Deflection was calculated for a 2-cm cantilevered root segment subject to a superficial liquid velocity of 1 cm s^{-1}.

Root culture	Deflection (mm)
L. esculentum	0.17
L. erythrorhizon	0.05
A. indica	0.00
T. kirilowii (var. Japonicum)	0.00
H. muticus (line: A4C17)	0.02
(line: HM90T)	0.03
G. pulchella (dark)	0.02
(light)	0.02
R. sativus	0.36
N. tabacum	0.15
S. tuberosum	0.03
B. vulgaris	0.10

Table 14.3. Assessment of the deflection that could be encountered due to entrainment of medium in a liquid-dispersed reactor such as a trickle bed. Deflection is calculated for a 2-cm cantilevered root segment covered with a liquid film. The thickness of the liquid film was calculated based on experimental measurements of medium entrainment for clumps of roots in which the medium was permitted to drain by gravity to constant weight (Ramakrishnan and Curtis, 1994).

Root culture	Liquid retention (mL g^{-1} fresh weight)	Deflection (mm)
H. muticus (line: A4C17)	1.25	1.39
S. tuberosum	1.04	1.74
B. vulgaris	6.06	7.97

vulgaris), the calculated deflection is comparable to the root diameter. Such a deflection would be significant in a packed bed of roots where tissue can occupy 50% of the reactor volume (500 g L^{-1} fresh weight). This effect can be compounded since localised compactions can have profound effects on flow resistance.

Flexibility could also play a role in liquid-dispersed reactors. In a reactor such as a trickle-bed where medium is sprayed over the top of the root bed, the combination of lack of buoyancy and the weight of the entrained liquid can result in compaction of the growing root centres. A simple quantitative assessment of root flexibility in trickle bed reactors can be made by comparing the extents to which different roots would deflect due to entrained liquid and tissue weight for a specified length (2 cm) of root tissue supported as a cantilever (similar to the calculations above). Table 14.3 shows the extent of deflection that would result for several root cultures based on previously published data of liquid retention by these root types (Ramakrishnan and Curtis, 1994).

As in the case of flow-induced compaction of submerged roots, deformation resulting from gravity will decrease the uniformity of tissue distribution and increase the likelihood of localised nutrient depletion. Specifically in the case of a trickle-flow reactor, root bending will result in a greater number of contact points between adjacent roots. On one hand, the meniscus at these points of contact will tend to retain medium; on the other, the contact points can provide a contiguous flow path for more rapid medium drainage. As a result of these competing effects, the impact of bending on draining characteristics could be very complex. We have observed altered liquid retention characteristics during the switch of a reactor from bubble column to an ebb-and-flow operation mode (Cuello, 1994). In these experiments, the rate of liquid draining was measured by pumping the drained liquid to a balance. The set-up was arranged to reduce the lag time involved in transferring liquid from the bottom of the reactor to the balance, as well as to avoid altering the drainage profile by the pumping action. Draining was measured repeatedly for two reactors: one operated as a bubble column, and another operated using a cycle consisting of 1 min filling, 2 min operation under submerged conditions, 1 min draining, and 2 min operation under dispersed liquid conditions. Both reactors were continuously sparged with air at 0.12 vvm. Root biomasses were 283 g L^{-1} fresh weight and 267 g L^{-1} fresh weight for the ebb-and-flow and bubble column reactors, respectively, at the time the measurements were made. The results are shown in Figure 14.5 for six consecutive measurements. For the ebb-and-flow reactor, the drainage characteristics were the same each time. The amount of drained liquid from the bubble column increased as the number of draining cycles increased, indicating adaptation of the bed to a less flow-resistant configuration. The enhancement in draining could result from a tendency of the bending roots to take on a more vertical orientation once the effects of buoyancy are removed from the submerged culture. If plant tissues are subject to stresses, they respond by becoming stiffer through a phenomenon termed thigmomorphogenesis (Jaffe

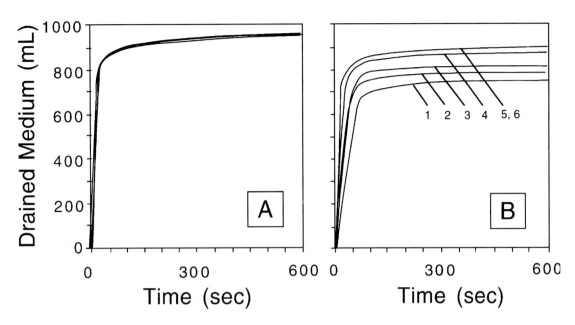

Figure 14.5. Repeated liquid draining rates measured from 2-L reactors grown under (A) ebb-and-flow conditions, and (B) bubble column conditions (unpublished, Cuello, 1994, obtained with the assistance of Jason van Wees). Numbers in (B) indicate the sequence of draining experiments performed.

and Forbes, 1993). As a result, the history of the growth conditions within a reactor will influence the physical mechanical properties of bending and the ease with which the tissue can be broken.

Brittle Versus Fibrous

Another mechanical property of root cultures which varies tremendously is the ease with which the roots can be broken. For those who have experience in maintaining many different root cultures, this becomes obvious from the ease/difficulty of removing root segments for serial subculture. To quantify this qualitative observation, a simplified tensile test was conducted by measuring the weight required to break individual root segments. These measurements were carried out by holding the root segments in a Hoffman screw clamp covered with rubber tubing and weighted with large metal washers. Measurements were repeated 5–8 times with the clamp pressure adjusted to hold the roots but avoid breakage at the point of clamping. The root was then slowly pulled upwards with the reduction in weight noted at the time of root failure. Yield stress was calculated based on the visually estimated root diameter. Although this experiment is rather simplistic (ignoring complexities of elastic strain and plastic deformation), it produces a reasonable indicator of root strength. The yield force is an absolute measure of the force required to break the root, whereas the yield stress (or pressure) is an intrinsic property taking into account the cross-sectional area of the

Table 14.4. Measurement of the brittleness of a broad range of transformed root cultures, as indicated by the force required to break a root segment pulled from both ends. The corresponding yield stress under tension is calculated from the root cross-sectional area: $\sigma = F/\text{area}$. If the roots are considered isotropic, the maximum stress (τ_{max}) to break a root under shear can be calculated as half the yield stress: $\tau_{max} = \sigma/2$ (Timoshenko and Gere, 1972).

Root culture	Field force F ($\times 10^{-7}$ dyn)	Yield stress σ ($\times 10^{-5}$ dyn cm^{-2})
L. esculentum	3.4 ± 3.3	69 ± 68
L. erythrorhizon	0.84 ± 0.52	6.7 ± 4.2
A. indica	4.4 ± 1.2	15 ± 4.1
T. kirilowii (var. Japonicum)	14 ± 5.8	36 ± 15
H. muticus (line: A4C17) (line: HM90T)	4.1 ± 2.4 3.5 ± 1.1	6.3 ± 3.7 9.0 ± 2.9
G. pulchella (dark) (light)	14 ± 11 12.7 ± 4.1	28 ± 21 35 ± 11
R. sativus	0.18 ± 0.08	7.8 ± 3.6
N. tabacum	5.6 ± 2.8	114 ± 57
S. tuberosum	23.4 ± 4.0	98 ± 17
B. vulgaris	2.6 ± 1.5	30 ± 17

root segment being broken. The force required to break the roots varied over two orders of magnitude, while the yield stress varied over just one decade (Table 14.4). The most difficult root to break was potato, which required nearly twice the force of medicinal cucumber (*T. kirilowii*) or *G. pulchella*, despite being much thinner than either of these roots. In contrast, *A. indica* is rather easy to break, being both rigid (high elastic modulus) and inflexible. Brittleness is very species dependent: *H. muticus* and tomato require nearly the same force to break despite having nearly an 8-fold difference in cross-sectional area. At the other extreme, *G. pulchella* and *H. muticus* (HM90T) have nearly the same cross-sectional area, but *G. pulchella* requires nearly four times more force to break.

Implications for reactor design and operation

For most reactor systems, the forces acting within the system are not so large that they would disrupt the tissue. Since it is difficult to generate flow rates which are sufficiently high to overcome the boundary layer mass transfer resistances in root beds, for most reactor systems, the forces exerted by flow in the system would be insufficient to cause damage to the roots. However, in the case of a stirred tank reactor, the presence of stresses sufficient to disrupt the roots is much more likely. The shear rates produced by an impeller are proportional to the impeller tip speed or rate of rotation (van't Riet and Smith, 1975; Croughan et al., 1987). Utilising the relationships developed in these references, the conditions that our group has used to scale-up plant cell suspension cultures in stirred tanks results in maximum shear rates of the order of 100 to 300 s^{-1}. For an isotropic material that does not deform substantially before it breaks, the maximum shear stress at failure is equal to half the tensile yield stress (Timoshenko and Gere, 1972). Referring to Table 14.4, one can see that the shear stresses required to break the 10 species of root listed would be of the order of 10^6 dyn cm^{-2}. This is several orders of magnitude higher that the shear stresses produced by impellers under typical culture conditions. The root tissue can be disrupted if the agitation conditions become comparable to what would be used for homogenisation. In fact, we have shown that homogenisation can be used to prepare large amounts of root tissue for inoculation of pilot-scale reactor systems (Ramakrishnan et al., 1994). In utilising this technique, we have observed that *H. muticus* tissue was much easier to disrupt than *S. tuberosum*. This qualitative observation is consistent with the large difference in yield stresses observed for these two root cultures.

Despite the use of relatively low impeller speeds, disruption of transformed roots has been observed

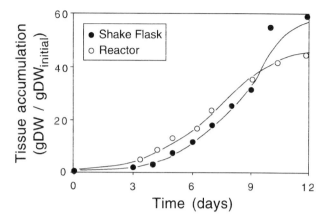

Figure 14.6. Growth of *H. muticus* root cultures in a 5-L stirred tank reactor (New Brunswick, BioFlo III) fitted with a 45° pitch blade axial flow impeller. Conditions for the culture were: Gamborg's B5 medium with 3% sucrose; impeller diameter 10 cm and height 7.5 cm; rotation rate 50 rpm; gassing rate 0.12 vvm. Initial and final weights were measured gravimetrically. Tissue accumulation rates from the bioreactor were determined by correlation from measurements of medium electrical conductivity. The controls were 125-mL shaker flasks containing 50 mL of Gamborg's B5 medium. Control points represent the average of three replicates. Standard deviations not shown.

in many cases for several different root cultures (Jung and Tepfer, 1987; Taya et al., 1989b; Wilson et al., 1990). We have cultivated *Hyoscyamus muticus* roots for 12 d in a stirred tank reactor without a protective mesh. At the end of the run, the root tissue consisted of root fragments 1–2 cm in length. Starting from an inoculum of 4 g L^{-1} fresh weight, the final tissue concentration was 245 g L^{-1} fresh weight, giving an effective specific growth rate for this culture over the 12 d period of 0.274 d^{-1}. This growth rate is nearly the same as would be experienced in a gyratory shake flask (Figure 14.6) despite the fact that the cultures were clearly stressed. Since the shear rates were not sufficient to directly disrupt the roots, the tissue break-up observed was most likely facilitated indirectly. Under stressful conditions, plant root cultures have a tendency to become more callus-like, presumably with a much lower yield stress. Such a physiologically stressed condition would likely result from damage to root hairs (as noted in the preceding section). Independent of the mechanism of break-up, the ability to proliferate fragmented root tissue could be used either in production systems (if the stress is not excessive) or in seed vessels which provide inoculum for large-scale production systems such as a trickle-bed.

Characterisation of Fluid Flow Resistance

Many of the morphological characteristics which have a detrimental impact on reactor design and scale-up result in increased mass transfer resistances. The root matrix severely impairs bulk liquid mixing, limiting the delivery of nutrients to core regions of tissue clumps. The observed correlation between increasing mixing time and declining specific respiration rate strongly implicates bulk convection as the dominant mass transfer mechanism in root cultures. Characterisation of flow resistance in submerged cultures can be accomplished by analysing the pressure drop due to liquid flow through a root packed bed. Due to the complexity of flow in packed beds, the mathematical description of this phenomenon requires a semi-empirical approach. Different studies vary in the assumptions used to describe the pathways and the cross-sectional areas available for fluid flow, from D'Arcy's phenomenological approach (1856) to the semi-empirical Ergun's equation (Ergun, 1952). This latter equation considers the pressure drop due to liquid flow through a packed bed to be the result of the contributions of two terms, viscous and inertial, allowing application to a wide range of flow conditions. For this reason we have applied this equation in the analysis for root culture (Holihan, 1993; Carvalho and Curtis, 1995). For a packed bed of long cylinders, the equation can be written as:

$$\frac{\Delta P}{\Delta x} = K_1 \frac{16\mu(1-\varepsilon)^2}{d_p^2 \varepsilon^3} v + K_2 \frac{4\rho(1-\varepsilon)}{d_p \varepsilon^3} v^2 \quad (7)$$

where μ, ρ and v are the medium viscosity, density and velocity, respectively; ε is the bed void fraction; d_p is the cylinder diameter, K_1 and K_2 are constants; and $\Delta P / \Delta x$ is the pressure drop per bed length. The effect of packing diameter on pressure drop can be verified in Equation (7). Considering that velocities of about 1 cm s^{-1} are required to minimise mass transfer limitations in submerged root clumps (Prince et al., 1991), the major contribution to pressure drop is expected to come from the first term of the right hand side of Equation (7), which accounts for the viscous contributions to the total pressure drop. We have shown that the use of the root diameter as a characteristic dimension underpredicts the flow resistance in root packed beds (Carvalho, 1995). The reason is most likely due to the complex morphology of the roots. The Ergun equation has primarily been applied to spherical materials. Roots have hairs, a distribution of diameters, interlocked structure and the potential for localised compaction. For these reasons, in applying Equation (7) to root cultures, the diameter should be viewed as an *effective diameter*, which is indicative of the *effective resistance to*

Table 14.5. Summary of flow resistance parameters regressed for three different root species. The effective diameter was obtained by fitting experimental data to the Ergun equation; hydraulic permeability was obtained by fitting experimental data to D'Arcy's equation. The effect of apparatus intrinsic resistance was taken into account. The root cultures selected represent different morphologies: *H. muticus*, moderately branched; *S. tuberosum*, apically dominant with minimal root hairs; *B. vulgaris*, highly branched and profuse root hairs.

Root culture	Diameter (mm)	Effective diameter (mm)	Hydraulic permeability* (cm^2)
H. muticus (line: A4C17)	0.99	0.092 ± 0.030	1.95 × 10^{-5}
S. tuberosum	0.27	0.056 ± 0.032	1.37 × 10^{-5}
B. vulgaris	0.18	0.025 ± 0.010	1.02 × 10^{-6}

* Based on a bed porosity of 60% ($\varepsilon = 0.6$).

flow. The smaller the effective diameter, the greater the resistance presented by the roots.

Holihan (1993) studied the resistance to flow of three transformed root cultures: *Beta vulgaris*, *Solanum tuberosum* and *Hyoscyamus muticus*. These species were chosen to represent different morphologies. *Beta vulgaris* is highly branched and hairy, whereas *Solanum tuberosum* has few branches and hairs. *Hyoscyamus muticus* has an intermediate morphology in respect to the two former species. The apparatus used in the experiments has already been described by Ramakrishnan and Curtis (1994). Roots were placed in a 30-cm-long glass cylinder of diameter 2.5 cm, and the liquid flow rate through the bed was measured at different hydrostatic levels. Effective diameters were regressed using the Ergun equation:

$$\frac{\Delta P}{\Delta x} - \frac{Q\alpha\mu}{\Delta x} = K_1 \frac{16\mu(1-\varepsilon)^2}{d_p^2 \varepsilon^3} v + K_2 \frac{4\rho(1-\varepsilon)}{d_p \varepsilon^3} v^2 \quad (8)$$

where α accounts for the intrinsic resistance to flow of the apparatus itself, and Q is the fluid flow rate. The results from this study are presented in Table 14.5.

Regressed diameters for the three species were 5–10 times lower than the microscopically measured diameters. This confirms that the resistance to flow presented by the roots is significantly higher than would be predicted by an estimate based on the physical diameter. The contribution of root morphology to the flow resistance can be evaluated by comparing the pressure drop for transformed cul-

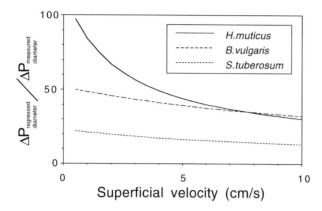

Figure 14.7. Comparison of the pressure drops for three transformed root species calculated from Equation (8) based on a regressed effective diameter ($\Delta P_{\text{regressed diameter}}$) with the pressure drop calculated from Equation (8) using the microscopically measured root diameter ($\Delta P_{\text{measured diameter}}$).

tures of *Beta vulgaris* and *Solanum tuberosum*, which have very different morphologies. *B. vulgaris* has a complex geometry involving a high degree of branching and profuse root hairs; *S. tuberosum* has few branches and sparse root hairs. As a result of the more complex geometry, the hydraulic permeability for *B. vulgaris* is an order of magnitude lower than that for *S. tuberosum*. The fluid flow resistance is much greater for the more complex root geometry. In addition to root hairs and branching, *B. vulgaris* is also one of the easier roots to bend (see Table 14.1); therefore, localised compaction could also be contributing to increased flow resistance. *H. muticus* roots would be expected to have a substantially lower

flow resistance because of their larger diameter. However, the hydraulic permeability of *H. muticus* is similar to that of *S. tuberosum*, showing that root complexities such as intense branching and profuse hairs increase flow resistance. The influence of root geometry on fluid flow resistance is shown more clearly in Figure 14.7. This figure presents the ratio of the pressure drop that would be predicted by Equation (8) based on the effective diameter regressed from experimental pressure drop–fluid velocity data, to the pressure drop calculated from Equation (8) based on the microscopically measured root diameter. This ratio is greater than one for all the roots tested, showing that the actual flow resistance is higher than would be predicted based on the measured root diameter. Roots which have the more complex geometries (*H. muticus* and *B. vulgaris*) have much higher ratios, showing that complexity of root morphology contributes additional resistances to fluid flow. At higher flow rates, the ratio of actual to predicted pressure drop declines because the inertial term of the Ergun equation (second term on the right hand side of Equation 8) has a weaker dependency on diameter. As discussed previously, a flow velocity of roughly 1 cm s^{-1} is needed to overcome the oxygen boundary layer mass transfer resistances. These results clearly show that microscopically measured root diameters are insufficient for characterisation of the geometry of roots in relation to their contribution to fluid flow resistance.

In an effort to provide some insight into potential methods of geometric characterisation, we undertook image analysis studies. *S. tuberosum* (potato) and *B. vulgaris* (beet) cultures were chosen since they displayed the greatest difference in root morphology. Figure 14.8 shows the scanned images of

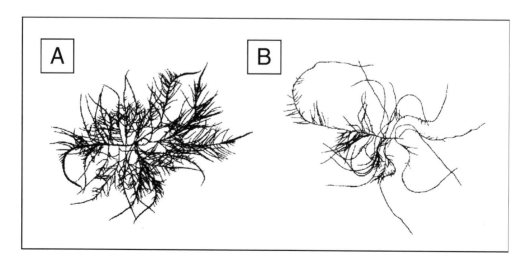

Figure 14.8. Images of (A) *B. vulgaris*, and (B) *S. tuberosum* root cultures scanned at 200 dpi after staining with 0.16 g L^{-1} neutral red for 30+ min. (From Pearsall, 1994). Apical dominance degrees vary from intense (*S. tuberosum*) to weak (*B. vulgaris*).

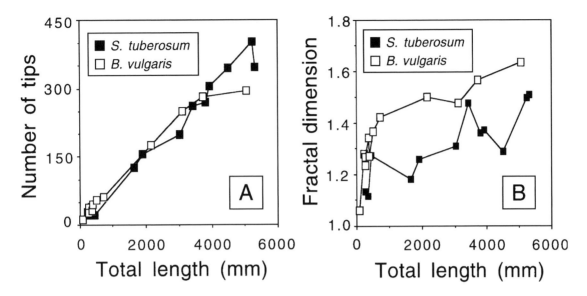

Figure 14.9. (A) Number of tips, and (B) fractal dimension of *B. vulgaris* and *S. tuberosum* roots, as a function of total root length. (From Pearsall, 1994).

root structures that resulted from the growth of 1-cm root segments. These images were analysed using Metamorph™ (Universal Imaging Corp., USA) and Delta-T Scan (Delta-T Scan Corp., USA) software to examine such parameters as total root length, number of root tips, and fractal dimension. Fractal analysis is used to describe irregularly shaped objects like sand, powder or porous particles. The analysis is based on the use of fractal dimensions, which are structures formed by an infinite number of parts, each one upon magnification resembling the whole structure (Sandami, 1991). By inoculating a group of 12 flasks with individual root tips of approximately 1 cm in length and periodically harvesting the cultures for analysis, the progression of root tip formation and fractal dimension could be followed as the cultures grew. The increase in number of root tips and fractal dimension is shown in Figure 14.9 as a function of total root length in the flasks. Surprisingly, despite the significant difference in appearance of the root morphologies, the results revealed that both cultures had nearly the same number of tips per unit length (1.2 cm per tip) and this remained constant as the cultures grew over the 16 d period of the analysis. The difference in root geometry is reflected in the fractal dimension. The beet cultures showed a consistently higher fractal dimension indicating a more complex geometry. The fractal dimension for both cultures initially increased with time and tended to level off at longer root lengths. These results show a 10–20% difference in fractal dimension between the beet and potato cul-

tures. Although such a difference in fractal dimension is significant, it is not clear how useful such a parameter might be for characterising root geometry for reactor design purposes. Since the geometries of beet and potato represent very different degrees of apical dominance, it was anticipated that these roots would perform very differently in reactors due to differences in distribution and clump formation within the vessel. These predictions have been verified visually by observing growth of these respective cultures in a 15-L bubble column; however, insufficient information has been compiled to make quantitative comparisons. To be useful in characterising such differences, fractal dimension would have to provide a reproducibly discernible difference. This remains to be determined. For some aspects of reactor design, fractal dimension would clearly be insufficient. For example, it appears that root hairs play an extremely important role in fluid flow resistance. The contribution of root hairs would not be captured by this level of fractal analysis since the resolution does not extend to that scale.

The preceding analyses demonstrate not only the complexity of roots as catalysts, but also how little is known towards characterising these systems for the purpose of reactor scale-up for commercial chemical production. Nonetheless, sufficient qualitative observations have been made to anticipate the major problems of scale-up, and the principles are in place to provide a reasonable 'first order' design of a large-scale system if the economics warrant industrial implementation.

Acknowledgments

Supported by National Science Foundation (Grant no. BCS 9110288 and NSF Young Investigator Award no. BCS 9358452), National Institute of Health, Biotechnology Training Grant program (Grant no. GM08358-02), Research Experience for Undergraduate program (S.H., B.L.P.), and Partners of the Americas (E.B.C.). Growth of root cultures in the stirred tank reactor was conducted by Derek Luyk, Divakar Ramakrishnan and Todd Nester. Hysteresis draining effects were conducted by Joel Cuello (PSU College of Agriculture Inter-College Research Grant, Project 3211) facilitated with the help of Jason Van Wees as part of a NSF research experience for undergraduate (Grant no. EEC-9300374). Computer templates and classifiers for fractal analysis in Metamorph™ were devised by Dr Siglinda Snapp in the laboratory of Dr Jonathan Lynch (PSU Department of Horticulture). Measurements of phosphorus diffusion and oxygen gradients within the root were initiated by Michelle Brincat as part of a NSF-DOE-USDA Collaborative Research in Plant Biology Program (BIR-9220330) with the assistance of Dr Simon Gilroy (PSU Department of Biology). The breadth of information from numerous root types was made possible due to the large transformed root culture collection of Dr Hector Flores (PSU Department of Plant Pathology). We would like to acknowledge the help of our colleagues in preparing various aspects of the manuscript: Patrick Asplund, oxygen profiles and mechanical property measurements; Lia Tescione, mass transfer mediated growth attenuation; Divakar Ramakrishnan, nutrient flux calculations and mechanical property analysis; Julie Bordonaro and Colleen Merritt, mechanical property measurements.

References

Barlow PW (1994). The origin, diversity and biology of shoot-borne roots. In *Biology of Adventitious Root Formation*, edited by TD Davis and BE Haissig, pp. 1–23. New York: Plenum.

Bates TR and JP Lynch (1996). Stimulation of root hair elongation in *Arabidopsis thaliana* by low phosphorous availability. *Plant Cell Environ* 19: 529–538.

Berry LJ and WE Norris (1949). Studies of onion respiration. II. The effect of temperature on the apparent diffusion coefficient in different segments of the root tip. *Biochim Biophys Acta* 3: 607–614.

Bhadra R, S Vani and JV Shanks (1993). Production of indole alkaloids by selected hairy root lines of *Catharanthus roseus*. *Biotechnol Bioeng* 41: 581–592.

Cailloux M (1974). Metabolism and absorption of water by root hairs. In *Structure and Function of Primary Root Tissues*, edited by J Kolek, pp. 315–322. Czech Republic: Veda (Slovak Academy of Sciences), Bratislava.

Carvalho EB (1995). *Measurement of Fluid Flow Resistance in a Submerged Root Reactor*. MS thesis, The Pennsylvania State University, USA.

Carvalho EB and WR Curtis (1995). Characterization of fluid-flow resistance in root cultures with a 'plug flow' bioreactor. Submitted to *Biotechnol Bioeng*.

Ciau-Uitz R, ML Miranda-Ham, J Coello-Coello, B Chi, LM Pacheco and VM Loyola-Vargas (1994). Indole alkaloid production by transformed and non transformed cultures of *Catharanthus roseus*. *In Vitro Cell Dev Biol* 30: 84–88.

Croughan MS, JF Hamel and DIC Wang (1987). Hydrodynamic effects on animal cells grown in microcarrier cultures. *Biotechnol Bioeng* 29: 130–141.

Cuello JL (1994). *Design and Scale-up of Ebb-and-Flow Bioreactor (EFBR) For "Hairy-Root" Cultures*. PhD thesis, The Pennsylvania State University, USA.

Curtis WR (1988). *Kinetics of Phosphate Limited Growth of Poppy Plant Suspension Cultures*. PhD thesis, Purdue University, USA.

D'Arcy HPG (1856). *Les Fontaines Publiques de la Ville de Dijon*. Victor Dalmont, Paris.

Drewes FE and J van Staden (1995). Initiation of solasodine production in hairy root cultures of *Solanum mauritianum* Scop. *Plant Growth Reg* 17: 27–31.

Dunlop DS and WR Curtis (1991). Synergistic response of plant hairy-root cultures to phosphate limitation and fungal elicitation. *Biotechnol Prog* 7: 434–438.

Ergun S (1952). Fluid flow through packed columns. *Chem Eng Prog* 48(2), 89–94.

Feldman LJ (1984). Regulation of root development. *Ann Rev Plant Physiol* 35: 223–242.

Finn, RK (1967). Agitation and aeration. In *Biochemical and Biological Engineering Science*, edited by N Blakebrough, pp. 69–100, vol 1. New York: Academic Press.

Flint-Wandel J and M Hjortso (1993). A flow cell reactor for the study of growth kinetics of single hairy roots. *Biotechnol Techniq* 7: 447–452.

Foehse D and A Jungk (1983). Influence of phosphate and nitrate supply on root hair formation of rape, spinach and tomato plant. *Plant Soil* 74: 359–368.

Hilton MG, PDG Wilson, RJ Robins and MJC Rhodes (1988). Transformed root cultures: fermentation aspects. In *Manipulating Secondary Metabolism in Culture*, edited by RJ Robins and MJC Rhodes, pp. 239–245. Cambridge: Cambridge University Press.

Holihan S (1993). *Hydraulic Permeability for Packed Bed of Roots*. BS Honours thesis, The Pennsylvania State University, USA.

Jaffe MJ and S Forbes (1993). Thigmomorphogenesis: the effect of mechanical perturbation on plants. *Plant Growth Reg* 12: 313–324.

Jaziri M, J Homes and K Shimomura (1994). An unusual root tip formation in "hairy root" culture of *Hyoscyamus muticus*. *Plant Cell Rep* 13: 349–352.

Jung G and D Tepfer (1987). Use of genetic transformation by the Ri T-DNA of *Agrobacterium rhizogenes* to stimulate biomass and tropane alkaloid production in *Atropa belladonna* and *Calystegia sepium* roots grown *in vitro*. *Plant Sci* 50: 145–151.

Mano Y, H Okawa and Y Yamada (1989). Production of tropane alkaloids by hairy root cultures of *Duboisia leichhardtii* transformed by *Agrobacterium rhizogenes*. *Plant Sci* 59: 191–201.

McKelvey SA, JA Gehrig, KA Hollar and WR Curtis (1993). Growth of plant root cultures in liquid- and gas-dispersed reactor environments. *Biotechnol Prog* 9: 317–322.

Nash WA (1972). *Strength of Materials*, 2nd ed. McGraw-Hill, New York.

Nguyen, C, F. Bourgaud, P. Forlot and A. Guckert (1992). Establishment of hairy root cultures of *Psoralea* species. *Plant Cell Rep* 11: 424–427.

Nuutila AM, L Toivonen and V Kauppinen (1994). Bioreactor studies on hairy root cultures of *Catharanthus roseus*: comparison of three bioreactor types. *Biotechnol Techniq* 8(1), 61–66.

Pearsall BL (1994). *Image and Fractal Analysis of Bioreactor Root Cultures*. BS Honours thesis, The Pennsylvania State University, USA.

Prince CL, V Bringi and ML Shuler (1991). Convective mass transfer in large porous biocatalysts: plant organ cultures. *Biotechnol Prog* 7: 195–199.

Ramakrishnan D and WR Curtis (1994). Fluid dynamic studies on plant root cultures for application to bioreactor design. In *Advances in Plant Biotechnology*, edited by DDY Ryu and S Furusaki, pp. 281–305. Amsterdam: Elsevier.

Ramakrishnan D and WR Curtis (1995). Elevated meristematic respiration in plant root cultures: implications to reactor design. *J Chem Eng Japan* 28: 491–493.

Ramakrishnan D, J Salim and WR Curtis (1994). Inoculation and tissue distribution in pilot-scale plant root bioreactors. *Biotechnol Techniq* 8: 639–644.

Saglio PH, M Rancillac, F Bruzan and A Pradet (1984). Critical oxygen pressure for growth and respiration of excised and intact roots. *Plant Physiol* 76: 151–154.

Sandami G (1991). The simple rules of complexity. *Chem Eng* 98(7), 30–35.

Savary BJ and HE Flores (1994). Biosynthesis of defense-related proteins in transformed root cultures of *Trichosanthes kirilowii* Maxim. var *japonicum* (Kitam.) *Plant Physiol* 106: 1195–1204.

Taya M, M Kino-oka, S Tone and T Kobayashi (1989a). A kinetic model of branching growth of plant hairy root. *J Chem Eng Japan* 22(6), 698–700.

Taya M, A Yoyama, O Kondo, T Kobayashi and C Matsui (1989b). Growth characteristics of plant hairy roots and their cultures in bioreactors. *J Chem Eng Japan* 22(1), 84–89.

Tescione L, EB Carvalho, D Ramakrishnan and WR Curtis (1995). Oxygen transfer limitations in pilot-scale reactors for plant roots. *Abstr Kanto Chapter Soc Chem Engineers Meeting*. Kanto, Japan, #C214: 266–267.

Timoshenko SP and JM Gere (1972). *Mechanics of Materials*, 1st ed. Van Nostrand Reinhold, New York.

Tinker PB (1976). Roots and water: transport of water to plant roots in soil. *Phil Trans Royal Soc London* B 273: 445–461.

Toivonen L and H Rosenqvist (1995). Establishment and growth characteristics of *Glycyrrhiza glabra* hairy root cultures. *Plant Cell Tiss Organ Cult* 41: 249–258.

Vanhala L, R Hiltunen and K-M Oksman-Caldentey (1995). Virulence of different *Agrobacterium* strains on hairy root formation of *Hyoscyamus muticus*. *Plant Cell Rep* 14: 236–240.

van't Riet K and JM Smith (1975). The trailing vortex system produced by Rushton turbine agitators. *Chem Eng Sci* 30: 1093–1105.

Weathers P, A DiIorio and RD Cheetam (1989). A bioreactor for differentiated plant tissues. *Proc Biotech USA Conference*. Conference Management Corporation, San Francisco, pp. 247–256.

Wilson PDG, ME Hilton, PTH Meehan, CR Waspe and M Rhodes (1990). The cultivation of transformed roots from laboratory to pilot plant. In *Progress in Plant Cellular and Molecular Biology* edited by HJJ Nijkamp, LHW van der Plas and J van Aartrijk, pp. 700–705. Dordrecht: Kluwer Academic.

Yonemitsu H, K Shimomura, M Satake, S Mochida, M Tanaka, T Endo and A Kaji (1990). Lobeline production by hairy root culture of *Lobelia inflata* L. *Plant Cell Rep* 9: 307–310.

Yu S and PM Doran (1994). Oxygen requirements and mass transfer in hairy-root culture. *Biotechnol Bioeng* 44: 880–887.

Yukimune Y, YH Hara and Y Yamada (1994). Tropane alkaloid production in root cultures of *Duboisia myoporoides* obtained by repeated selection. *Biosci Biotechnol Biochem* 58: 1443–1446.

Mathematical Modelling of Hairy Root Growth

15

Martin A. Hjortso

Department of Chemical Engineering, Louisiana State University, Baton Rouge, LA 70803, USA;
email: hjortso@che.lsu.edu

Introduction

A huge number of mathematical models have been developed for various types of cultures and bioreactors. The majority of these models were developed for microbial or cell cultures and not for root cultures. Since root growth differs from microbial growth in several important aspects, most of the bioreactor models in the literature are not suitable for describing growth of hairy roots in these reactors.

Whereas microbial cultures contain mostly identical cells which all pass repeatedly through identical cell division cycles, roots contain a wide range of different types of cells. From a purely kinetic standpoint, roots contain two kinds of cells: tip cells or cells in the apical meristem which continuously pass through a cell division cycle, and the remaining cells which do not divide. The non-dividing cells originally arise from tip cells, but elongate and differentiate into the various tissues of the root, a process referred to as terminal differentiation. The bulk of the root thus consists of cells that do not divide. Meristematic cells outside of the tip can start dividing to form a new tip, which results in formation of a lateral branch as cells within it start the terminal

differentiation process. It is the branching process: the formation of new tips and lateral branches, that determines the overall growth rate of the root biomass. Models of the kinetics of branch formation are therefore an essential part of any model of root growth.

The most commonly used models in biokinetics rely on the assumption that the biophase is well mixed. This assumption is often reasonable for microbial cultures, but for roots where there are large differences between cells in different parts of the root, it is not reasonable. Models that do account for the differences between cells in a culture are know as segregated models. The term 'corpuscular' has recently been advocated as a more appropriate term than 'segregated' (Ramkrishna, 1994). For roots and other tissues in which the cells remain attached to each other rather than form corpuscles, this terminology does not seem appropriate. A general approach to writing segregated models of microbial cultures already exists (Fredrickson *et al.*, 1967) and this approach can be adapted to root growth with only minor modifications. The key variable that one seeks to determine with these models is the distribution of states. This is a frequency function, possibly multi-dimensional, which gives the fraction of all the cells in the culture which are in a given cell state.

In this chapter we will first discuss a simple approach to modelling the branching process before introducing some simple segregated models of hairy root growth. Finally, a stochastic model of root branching in the presence of tip death will be described.

Modelling the Branching Process

The branching process or formation of new tips in roots is kinetically very analogous to the cell division process in microorganisms. Both are the source of new reproducing units, tips or cells, and both can be modelled using similar concepts. In order to model either process, one must have a verbal description, based on experimental observations, of when cell division or root branching occurs. For instance, for a binary fission organism, one can state that cell division occurs some fixed time, the cell cycle time Δt, after a cell is born. If one initially has N_0 cells in the culture, it is self evident that the total number of cells as a function of time can be described by the difference equation

$$N_i = 2 \cdot N_{i-1}$$

where N_i equals the number of cells at time $i \cdot \Delta t$. It is also easy to show that the specific growth rate of cell number is

$$\mu = \frac{1}{\Delta t} \cdot \ln 2$$

A very similar modelling approach can be used to model the total number of tips in a root culture. A simple model would be to state that roots branch by bifurcating at the tip, and the number of tips therefore doubles over some characteristic time (Taya *et al.*, 1989). This results in a model similar to that for binary fission cells but is not an accurate description of the branching process. Branching occurs, not at the tip, but in the region where terminally differentiating cells cease to elongate and, once commenced, branching continues to occur in this region. There is thus a delay between the first appearance of a branch and the time at which this branch itself starts to form laterals, a delay between the appearance of the first lateral and the second lateral and so forth. Description of the dynamics of root branching therefore requires multiple characteristic times, as opposed to the cell cycle of binary fission organisms which needs only a single characteristic time, the cell cycle time.

A verbal description of branching process can be summarised succinctly by a set of branching rules. Such rules specify, directly or indirectly, all the delays or characteristic times of the branching process. For instance, branching rules could be statements of the form "The first lateral branch initially appears on the parent branch a period Δt_1 after the tip of the parent branch appeared", "The second lateral branch appears a period Δt_2 after the appearance of the first lateral" and so on. The characteristic times specified this way will generally change with the growth conditions and the physiological state of the root. They will also be subject to random fluctuations even under fixed growth conditions. Both of these complicating factors will be ignored in the initial analysis. Detailed experimental observations are required to identify branching rules, but even though lateral branch formation has been studied since 1936 (Thimann, 1936), all of these studies have been on untransformed roots. One should be careful about extending conclusions from these observations to transformed roots, which are characterised by a higher frequency of branching. Branching rules for transformed roots must therefore at this point in time necessarily be somewhat speculative.

Using branching rules, one can formulate a difference equation for the number of tips similar to the equation given above for the number of cells in a culture of a binary fission organism. For the binary fission organism, the time step of the difference equation is necessarily equal to the single characteristic cell cycle time Δt. However, for a branching process which is modelled by several characteristic times, it is not obvious what time step to base the difference equation on. The difference equation must establish the number of tips at a time t as a function of the number of tips at previous times where all the previous times equal the time t minus integer multiples of the time step. The time step of the difference equation is essentially the largest interval which is still so small that all the characteristic times of the branching process are integer multiples of this time interval. To illustrate this fact, consider the following simple branching rules with only two characteristic times.

(1)   The delay between the first appearance of a tip and the subsequent appearance of the first lateral on the branch formed by this tip is Δt_1.
(2)   The delay between the appearance of subsequent lateral branches on the parent branch is Δt_2.

For illustrative purposes, let $\Delta t_1 = 3$ and $\Delta t_2 = 2$ in some appropriate units, and consider an unbranched root on which the first lateral appears at time zero. By keeping track of all tips and all past tips appear-

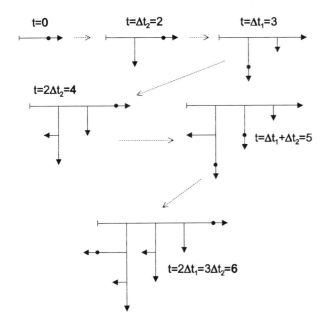

Figure 15.1. Evolution of a root for which the first lateral branch on a parent branch forms a period Δt_1 after the tip of the parent branch first appeared and all subsequent laterals appear a period Δt_2 after the previous lateral first appeared. For the root shown, $\Delta t_1 = 3$, $\Delta t_2 = 2$ and the initial condition is an unbranched root on which the first lateral is just appearing. Tips are indicated by arrows and points at which tips appear by filled circles. A schematic of the root is shown at all the times for which a new tip appears, up to time $t = 6$. At time 5, two new tips appear. These tips result from two different and unrelated branching events but appear simultaneously because both form from the initial root after a delay of length $\Delta t_1 + \Delta t_2$. Similarly, the two next tips which appear on the root appear simultaneously at time 6. One (the fourth lateral on the original branch) appears after a delay of length $3\Delta t_2$, the other (the first lateral on the first lateral on the first lateral of the original branch) appears after a delay of length $2\Delta t_1$. With the choice of parameter values, these two delays are identical. The key observation is that new tips can only appear after some number of delays of length Δt_1 plus some number of delays of length Δt_2, so new tips appear at times that are the sum of integer multiples of the two characteristic times.

ance events, the evolution of this root under the given branching rules can be sketched. This is done in Figure 15.1 where tips are indicated by arrows and filled circles indicate points at which new tips appear. The times at which new tips appear are all integer multiples of 1, the largest time interval which is an integer fraction of both Δt_1 and Δt_2. This must necessarily be the case because all tips appear only after a multiple of the delay Δt_1 plus another multiple of the delay Δt_2 have passed since

the initial time. The difference equation for this root must therefore be based on a time step of size 1.

This argument can easily be extended to branching rules with an arbitrary number of characteristic times. One can assume without loss of generality that the zero of the time axis is defined in such a way that a tip appears at time zero. All subsequent tips will then necessarily appear at times that are sums of integer multiples of the characteristic times specified in the branching rules, viz.

$$t_{\text{tip appearance}} = \sum N_i \cdot \Delta t_i$$

where the N_i are integers. Let Δt equal the largest number which is an integer fraction of all the characteristic times, i.e. $\Delta t = \Delta t_i / M_i$ where M_i are integers. Then

$$t_{\text{tip appearance}} = \sum N_i \cdot M_i \cdot \Delta t = \Delta t \cdot \sum N_i \cdot M_i = \Delta t \cdot N$$

where N is some integer. This result shows that tips only appear at times that are integer multiples of Δt, which is therefore the obvious choice for the time step in the difference equation for the total number of tips. The parameter Δt is a purely mathematical help variable which will be used in writing a mathematical model of the branching kinetics and it should not be mistaken for a characteristic time of the overall root growth dynamics.

The method for writing the difference equation for the number of tips is described in more detail elsewhere (Kim *et al.*, 1995). As an illustration of the method, we will here consider the same branching rules as used above. The difference equation is now obtained by writing balance equations on the number of tips, one equation for each rule. Let the number of laterals which appear at time $i \cdot \Delta t$ as the first laterals on the parent branch be $F(i \cdot \Delta t) = F_i$, and let the number of laterals which form at the same time as siblings to an older lateral be $S(i \cdot \Delta t) = S_i$. From the first branching rule it then follows that

$$F(i \cdot \Delta t) = F(i \cdot \Delta t - \Delta t_1) + S(i \cdot \Delta t - \Delta t_1)$$

or using that $\Delta t_1 = M_1 \cdot \Delta t$

$$F_i = F_{i - M_1} + S_{i - M_1}$$

while, similarly, it follows from the second branching rule that

$$S_i = F_{i - M_2} + S_{i - M_2}$$

The total number of tips which appear at time $i \cdot \Delta t$, T_i, is the sum of F_i and S_i and is found by adding the two number balances just given. After simplifications, this becomes

$$T_i = T_{i-M_1} + T_{i-M_2}$$

The total number of tips at time $i \cdot \Delta t$ is the sum of the tips which appear at this time plus the tips which appeared at all previous times, i.e. the sum of all the T_j's up to and including T_i. This sum also satisfies the difference equation above as can easily be verified by substitution into the equation. The difference equation above can therefore be used to calculate both the number of new tips which appear at a given time, and the total number of tips at any time.

Given an initial condition, this equation can be solved either analytically using standard methods for difference equations (Henrici, 1962), or solutions up to large values of i can be found on a computer by implementation of the equation itself. The solution only specifies the number of tips at times that are integer multiples of Δt. However, since tips only appear at these times, the number of tips at any other time equals the number of tips at the closest, previous time equal to an integer multiple of Δt.

The Age Distribution

The simplest possible segregated model is the model for the age distribution. The age distribution in a batch reactor is governed by the partial differential equation (Gurtin and MacCamy, 1974)

$$\frac{\partial f}{\partial t} + \frac{\partial f}{\partial a} = 0$$

where $f(t, a)$ is the age distribution at time t, and t and a are time and age, respectively. To fully specify the problem, one must also specify an initial condition of the form

$$f(0, a) = f_0(a)$$

and a boundary condition at age zero, an equation often referred to as the renewal equation. This equation specifies the rate at which cells are being born. Considering only the cells that are undergoing terminal differentiation, these cells are born when tip cells divide and one of the newborn cells leaves the tip and starts differentiation. Thus, the rate at which these cells are born must be proportional to the total number of tips in the system. Furthermore, at constant growth conditions as assumed here, the proportionality constant does not change with time and the renewal equation becomes

$$f(t, 0) = C \cdot T(t)$$

where the proportionality constant C is the rate at which cells leave a single tip. The solution for the age distribution can now be found (Kim et al., 1995)

$$f(t, a) = \begin{cases} f_0(a - t) , & a > t \\ C \cdot T(t - a) , & a < t \end{cases}$$

The value of the age distribution at any age less than the current time equals the constant C times the number of tips that were present when cells of the specified age were born. Examples of age distributions are shown in Figure 15.2 for several different choices of model parameters. All the distributions in the figure are normalised for comparison purposes, are shown at the same time, and were calculated using as initial condition a root in the state at which the first lateral (or second tip) appears. The branching parameters and time are given without units, as dimensionless parameters and dimensionless time. The curves shown thus apply to any root with the same ratio of branching parameters, Δt_1 and Δt_2, as those used in a given plot.

Age distributions given by the formula above are piecewise constant functions. Each interval over which the distribution remains constant corresponds to cells that were born while the tip number remained constant. The discontinuities between these intervals result from the appearance of new tips. The relative position of the discontinuities and the value of the function over the constant intervals do not change. The shape of the age distribution therefore does not change with time and its transient behaviour is quite simple. The distribution translates with the velocity 1, (the rate of change of age with respect to time), towards greater ages while new parts of the distribution appear at age zero. The constant shape is completely specified by the branching rules or the associated difference equation and, once this shape is determined, the problem of determining the age distribution at a given time is reduced to determining the location along the age axis of this shape.

Two of the distributions shown in Figure 15.2 are for almost identical values of the model parameters, Δt_1 equal to 5 for both and Δt_2 equal to 1 and 1.05, respectively. Using a result given later, it can be shown that the dimensionless specific growth rates are also almost equal: 0.281 and 0.276, respectively. The shapes of the two distributions are, not surprisingly, similar except for the fact that the constant intervals of the distribution with $\Delta t_2 = 1.05$ are smaller than for the distribution with $\Delta t_2 = 1$. This difference occurs because the value of the help variable Δt differs greatly between the models: 1 and 0.05, respectively. This illustrates the fact mentioned above that Δt is purely a mathematical help variable and

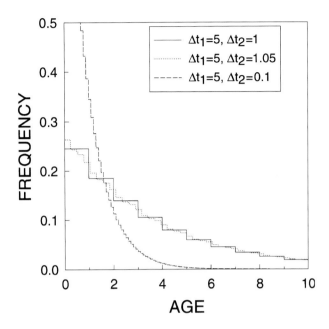

Figure 15.2. Normalised age distributions for various dimensionless values of the model parameters. All distributions are shown at a dimensionless time of 50 and the initial condition in each case is a root with a single tip in which the age of the oldest cell equals Δt_1. In other words, the initial condition is a root on which the second tip is just appearing.

not a characteristic time for the growth dynamics of the root.

The last of the three distributions shown in Figure 15.2 differs from the other two by its smaller value of Δt_2, i.e. by a much higher frequency of tip formation. As expected, the dimensionless specific growth rate for this distribution, 0.576, is higher at approximately twice the value of the two other distributions shown. The higher growth rate is also evidenced by the fact that this age distribution is shifted towards younger ages as compared to the two other distributions.

Exponential Specific Growth Rate

The specific growth rate in cell number is formally obtained by integrating the age distribution over all ages to obtain the total number of cells versus time, differentiating this result with respect to time to obtain the absolute growth rate, and dividing the absolute growth rate by the total number of cells to obtain the specific growth rate. It is evident that the discontinuities of the age distribution will cause the specific growth rate to change discon-

tinuously each time tips appear on the root. This is an artifact of the simplifying assumption that the characteristic branching times are not subject to variation, and tip appearances are therefore strictly synchronised. In actuality, tip appearance is not controlled so strictly and a better measure of the observed specific growth rate is probably found by averaging the specific growth rate, calculated as described above, over a time period of length Δt. At large times, growth becomes exponential and it can be shown that the specific exponential growth rate can be obtained from the difference equation for the tip number as follows (Kim *et al.*, 1995). Make the substitution $T_i = \lambda^i$ in the difference equation and multiply or divide through by powers of λ to obtain an expression where the lowest power of λ is 0. The result is the so-called characteristic polynomial of the difference equation. Find the unique, positive root of this polynomial and let this root be λ_0. The specific growth rate μ is then given by

$$\mu = \frac{1}{\Delta t} \cdot \ln \lambda_0$$

This specific growth rate is only observed in the limit as the number of tips become large and before changes in the limiting substrate concentrations start to alter the values of the characteristic times of the branching dynamics. It is the specific growth rate observed during the exponential growth phase, after the lag phase but before transition to the stationary phase. If the inoculum is large compared to the amount of medium, the lag phase and the transition phase may even overlap so that a specific exponential growth rate will not be observable.

Lag phase refers to the initial transient before exponential growth is attained. The term comes from microbial kinetics and will be used here although it can be a misnomer when it comes to roots because it implies a specific growth rate lower than the specific exponential rate during this period. However, for roots, the model indicates that the opposite can in fact be the case. The depressed growth rate that is observed in microbial cultures during the lag phase is attributed to the fact that cells in the culture must adapt their metabolism to the new environment before they can reproduce at their maximum rate. The same metabolic transient can be expected in root cultures but in addition, the branching dynamics will exhibit a transient which will affect the lag phase.

A root inoculum typically consists of tip segments that have been cut from old roots so that the segments do not have lateral branches or, if they do, they have only a very low number of generations of laterals. When no laterals are present, a root grows by elongation. At a constant rate of elongation, this results in a linear increase in the number of cells with time: in other words, zero order growth kinetics. After first generation laterals start to form, growth changes to some other type of kinetics, the precise type depending on the pattern of lateral branch formation. Only in the limit, as the number of generations of laterals becomes very large, does the growth become exponential. This transient in the branching dynamics is explained nicely through mathematical analysis of the difference equation for the number of tips. The complete solution to the difference equation can be written as a sum of linearly independent basis solutions or functions and, as time increases, some of these solutions become small relative to others until eventually only one of the basis solutions contributes significantly to the complete solution. It is this process which gives rise to the transient in the branching dynamics and exponential growth is attained when a single basis solution has become the only significant part of the complete solution.

This transient in the branching dynamics does in fact give rise to an initial specific growth rate which is greater than the final, exponential specific growth rate. To see this, note that an inoculum which is made of tip segments is equivalent to an old root from which all the parts that no longer form new branches have been removed. Removing these parts of the root does not alter the absolute growth rate which is determined by the total number of tips. It does, however, decrease the total number of cells, and the inoculum specific growth rate (the absolute growth rate divided by the number of cells) must therefore be greater than the specific growth rate of the root from which the inoculum was derived. Conversely, as the inoculum grows, it will add non-branching sections to the root. These sections represent an increase in cell number without a concomitant increase in absolute growth rate, i.e. a decrease in the specific growth rate. Therefore, when roots grow under constant growth conditions in which the characteristic times of the branching process do not change, the branching dynamics give rise to a specific growth rate which is a continuously decreasing function of time.

Effects of Tip Death

The branching rules used in the example above are quite simple and therefore fail to reflect some of the features of hairy root branching dynamics which have been observed. Studies of a single hairy root of *Tagetes erecta* in a flow cell reactor (Flint-Wändel and Hjortso, 1993) suggest that as tips age, the rate of elongation of the root branch decreases until the tip dies. Tip death is then followed by simultaneous

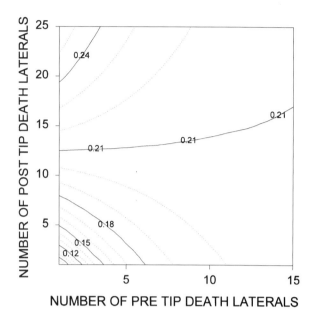

Figure 15.3. Exponential specific growth rate of a root in which the tip of a branch dies after a fixed number of laterals have formed on the branch and, a fixed time period thereafter, additional laterals simultaneously appear. The model from which the plot was calculated employs three characteristic times given by the branching rules: (1) The time between appearance of the parent tip and appearance of the first lateral on the parent branch is Δt_1, (2) The time between the appearance of the last pre-death lateral and the simultaneous appearance of the post-death laterals is Δt_2, and (3) The time between the appearance of pre-death latreals is Δt_1. The values used for the plot are $\Delta t_1 = 8$, $\Delta t_2 = 5$ and $\Delta t_3 = 1$.

formation of several lateral branches, presumably due to loss of apical dominance resulting from death of the parent tip. These dynamics can be summarised by branching rules, a difference equation for the total tip number can then be derived, and the exponential specific growth rate of the root calculated (Kim *et al.*, 1995). The result of such a calculation is shown in Figure 15.3 in the form of a contour plot. Each contour or curve represents a constant specific growth rate, plotted versus the number of laterals formed before the death of the parent tip and the number of laterals formed after. The dimensionless value of the specific growth rate is indicated on several of the curves. Perhaps surprisingly, it turns out that the exponential specific growth rate with tip death can exceed the rate without tip death. For the model parameters used in calculation of the figure, the dimensionless exponential specific growth rate without tip death equals 0.2087. This is less than the rate found when the number of post-death laterals is high. Thus the simultaneous formation of several laterals shortly after tip death can more than compensate for the loss of laterals that would have formed at later times if the parent tip had remained viable.

Sources of Variation

Because roots are so much larger than microbial cells, root cultures contain far fewer individual roots than microbial cultures contain cells. Therefore, the effects of variations in growth among different roots in a culture are far more evident than the effects of similar variations in microbial cultures where the law of large numbers prevails. The variation among different roots is particularly unmistakable in comparisons of single roots grown in different shake flasks under identical conditions. In our lab, we have often seen substantial differences in the time it takes for roots in different flasks to grow to fill the flasks. We have also observed significant variations among flasks in specific growth rates and have found the amount of variation to depend on the growth conditions. Both of these observations have simple explanations in terms of mathematical models of the branching dynamics.

The characteristic times defined by branching rules are no more true constants than the cell cycle time is for a microorganism. Variations are bound to occur between different roots and different branching events. This immediately explains why roots in some shake flasks grow quickly while others grow more slowly. With an inoculum in each flask consisting of just a few branches with few or no laterals, random variations between the branching times will delay the onset of exponential growth in some flasks and advance it in others. This effect will become more pronounced when the number of tips in the inoculum is decreased. Similarly, as the number of tips becomes large, the random fluctuations in the characteristic branching times will become unnoticeable as one starts to observe an average effect of a large number of branchings.

A mathematical model which takes into account stochastic variations in the characteristic branching times will be statistical in nature and will invariably be more complex than the simple deterministic model based on fixed characteristic branching times described above. Such statistical models are obviously required to model roots with only few tips. However, they are unnecessarily complex when the number of tips is large enough to conceal variations in branching times. In order to stay with the simple model for such cases, one can try and use the average values of the branching times in lieu of the true stochastic branching times. It can be shown by counter example (Kim *et al.*, 1995) that this approach does not yield the correct values for the exponential

specific growth rate. However, simulations carried out using various probability functions to model stochastic variations in branching times exhibit specific growth rates that are very close to those that are found using the average branching times in a deterministic model (Kim, 1993). Thus, for most practical purposes, the average characteristic branching times give a sufficiently accurate description of root growth in cultures where the number of tips is large.

Partially random branching also affects the age distribution of a root. Without variations in branching times, new tip formation is perfectly synchronised and only occurs at times that are integer multiples of the variable Δt. However, just as variations in cell cycle times cause a loss of synchrony in microbial cultures, variations in branching times result in the loss of synchronised tip formation in roots. Thus, the discontinuities of the age distribution which reflect tip appearances will, as the root grows, distribute themselves more evenly over all ages. The effect of this is to make the age distribution 'smoother' in appearance. An expression for this smooth form is obtained by demanding that it gives rise to the same specific exponential growth rate μ as that determined from the branching rules. This is only possible if the age distribution has the form

$$f(t,\ a) = e^{\mu \cdot t} \cdot g(a)$$

where $g(a)$ is the smooth approximation of the age distribution. Substituting this expression into the partial differential equation for $f(t,\ a)$ gives

$$g(a) \approx e^{-\mu \cdot a}$$

and thus

$$f(t,\ a) \approx e^{\mu \cdot (t-a)}$$

Both this expression, and the previously presented exact model solution of the age distribution, should be regarded as approximations to the true age distribution, valid at different stages of growth. The exact solution is an approximation to the real age distribution in the sense that the solution is based on tip numbers calculated from branching rules which do not reflect the random component of the branching times. The error produced by this becomes more severe as the number of tips increases and the synchrony between tip appearances vanishes, something which the solution completely fails to reflect. The exact model solution is therefore best used at early times, for roots with few tips. In fact, for roots that are not yet very branched, it is possible to visually observe the number of tips as well as their time of appearance. The actual number of tips

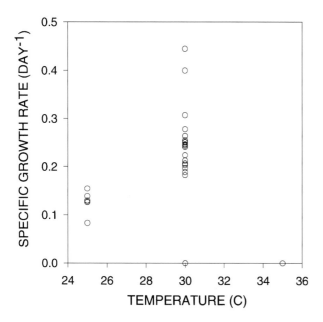

Figure 15.4. Observed specific growth rate versus temperature for *Tagetes erecta* hairy roots grown in shake flasks from an inoculum consisting of a single tip. The medium used was MS (Murashige and Skoog, 1962) with Gamborg's vitamins (Gamborg *et al.*, 1968) and cultures were incubated in the dark on a shaker operated at 120 rpm.

can then be used instead of the calculated number of tips to provide a correct expression for the age distribution for that root. By the same token, the approximate exponential function solution given above is the best approximation to the true age distribution at later times when the number of tips is large and synchrony between tip appearances has vanished.

The environmental effect on the extent of variation in growth kinetics is illustrated by the data in Figure 15.4. This figure shows the effect of temperature on the specific growth rate of *Tagetes erecta* hairy roots grown in shake flasks from an inoculum consisting of a single tip. Details of the culture conditions and data analysis are given elsewhere (Ludwig Hopper, 1995). At 25°C the specific growth rate varied between 0.08 d^{-1} and 0.15 d^{-1}; at 30°C it varied between 0 d^{-1}, indicating that the root died, and 0.44 d^{-1}; at 35°C all roots died. What appears on the plot to be a single data point at 35°C is in fact several points plotted on top of each other. Qualitatively, these observations can be explained by assuming that increasing temperature decreases the characteristic branching times but increases the probability that a tip will die. If all tips on a root die, the root itself is dead. At low temperatures, the probability of tip death is low and the characteristic

branching times are large resulting in a low specific growth rate. The low probability of tip death means that the likelihood of a whole root dying is negligible. The rare death of a tip will occasionally decrease the observed growth rate of a root, giving rise to a small growth rate variation among the roots. At higher temperatures, the probability of tip death is so high that for a given root the rate of new branch formation may be closely matched by the rate of tip death. Such a root can teeter between dying and living until it either dies or acquires so many tips that the probability of all tips dying become insignificant. Roots grown at such temperatures will therefore split into two subpopulations, those that ultimately die and those that live. The roots that do live can emerge from the teetering phase at widely separated times resulting in large variations in observed growth rate among the roots. Also, the roots that do live exhibit a higher growth rate than roots grown at lower temperatures because the characteristic branching times are shorter at higher temperatures. Finally, at the highest temperatures, the probability of tip death is so large that it dominates the growth dynamics and all roots die.

A quantitative mathematical model of this behaviour is not hard to derive. However, to keep the mathematics as simple as possible, it will be assumed that a single characteristic branching time, the time required for a doubling of the tip number, is sufficient to describe the growth. This is true only if branching occurs by bifurcation at the tip, a crude approximation of reality, but the resulting model does successfully describe the observed behaviour and it is far simpler and easier to use than a model which incorporates several characteristic branching times.

Let the characteristic time over which the number of tips doubles in number, assuming none of the tips dies, be Δt, and let the probability that a given tip will double over this time period instead of die be p. Starting with a large number of identical roots, each with a single tip, we seek the fraction of roots with N live tips after i periods of length Δt. Let this fraction be

$F(i, N)$ = Fraction of roots with N live tips at time $i \cdot \Delta t$.

This fraction can be calculated from the fraction one time step previous as

$$F(i,N) = \sum_{M=N/2}^{2^{i-1}} F(i-1,M) \cdot P(M,N,p)$$

where $P(M, N, p)$ is the probability that a root with M tips will give rise to a root with N tips after one time step of length Δt. Since the number of tips can at most double in a time step, the lower limit on the sum must be $N/2$, and since the number of tips after $i-1$ time steps can at most be 2^{i-1}, this must be the upper limit on the sum. To find $P(M, N, p)$, note that each tip can either double with probability p, or die with probability $1-p$, and that the outcome is independent of the fate of the other tips. It is a well known result from probability calculus that the number of tips which double must then be binomially distributed. Thus

$$P(M,N,p)=\frac{M!}{(N/2)!\cdot(M-N/2)!}\cdot p^{N/2}\cdot(1-p)^{M-N/2}$$

where the expression on the right hand side is the binomial distribution of the probability that out of M tips, $N/2$ will double when the probability of doubling equals p. The fraction of roots with a given number of tips is thus given by the expression

$$F(i,N)=\sum_{M=N/2}^{2^{i-1}}F(i-1,M)\cdot$$

$$\frac{M!}{(N/2)!\cdot(M-N/2)!}\cdot p^{N/2}\cdot(1-p)^{M-N/2}$$

This result does not hold for the fraction of dead roots, which must be calculated by summing over the dead roots produced after each time step

$$F(i,0)=\sum_{j=1}^{i}\left\{\sum_{M=1}^{2^{i-1}}F(j-1,M)\cdot(1-p)^M\right\}$$

Examples of two model predictions are plotted in Figure 15.5. The model predicts that tips either double or die over each time step, so the number of tips predicted by the model must necessarily be an even number. The two graphs in Figure 15.5 therefore show only the distribution of even numbers of tips. Both graphs show the tip number frequency after 8 periods of length Δt. Notice the different vertical scales on the graphs. For the top graph, the probability that a tip will survive and double over the period Δt is relatively low, 0.9. A large fraction of the roots, 0.11, have died and the roots which are still alive have tip numbers that range over a wide interval of values. For the bottom graph, the probability of survival of individual tips is relatively high and a large majority of the roots have close to the maximum number of tips, 256. Some roots have died, but these constitute a small fraction, 0.01, of the population.

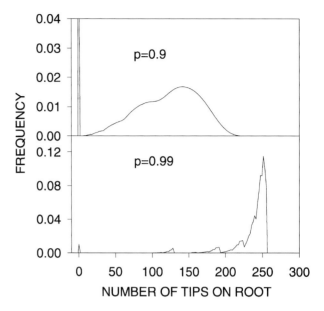

Figure 15.5. Frequency of tip numbers on roots grown under conditions with two different probabilities of tip death. The probability that a tip over a fixed time period, Δt, will double and form two tips is p. Tips that do not double over this period die. Frequency functions are shown after 8 time periods for roots that initially have a single tip.

The relative magnitude of the specific growth rates associated with each graph in Figure 15.5 cannot be inferred from the graphs themselves. The graphs are plotted for cultures which have passed through the same number of potential doubling times, Δt, and only when the relative magnitude of the Δt's are known can the magnitude of the specific growth rates be inferred. For instance, if the Δt for the two graphs differed by a factor of 2, with the smaller value associated with the smaller probability p of tip doubling, then the graph for $p = 0.9$ after 8 periods should be compared to the graph for $p = 0.99$ after 4 periods (not shown). However, after 4 periods, the maximum number of tips equals $2^4 = 16$, which evidently is much less than the average number of tips for the top graph in Figure 15.5. Calculations show that the average tip numbers are 110 for $p = 0.9$ after 8 periods and 15.4 for $p = 0.99$ after 4 periods. The specific growth rates based on tip numbers therefore differ by a factor of $\ln(110)/\ln(15.3) \approx 1.7$.

A consequence of the kind of growth dynamics described by this model is that optimal growth conditions are not well defined for individual roots. Conditions which are optimal in the sense that they minimise death may result in very low growth rates, while conditions that give high growth rates carry with them an elevated risk that the root will die. The optimal growth conditions are well defined for a large number of roots, and can be defined as the conditions which give the highest average growth rates. A better way of thinking about optimal conditions than focusing on a single best condition, is to think in terms of an optimal growth strategy. For the type of growth kinetics modelled above, an optimal strategy could possibly be to initially grow the roots at conditions which minimise death while favouring only a low growth rate. However, once the roots have attained a sufficiently large number of tips, the likelihood of all tips, and therefore the whole root, dying becomes insignificant and conditions can be switched to those that give a higher growth rate without running the risk of killing a substantial fraction of the roots in the culture.

Acknowledgments

This research was supported by the National Science Foundation under grant No. BCS-9024412.

References

Flint-Wändel J and MA Hjortso (1993). A flow cell reactor for the study of growth kinetics of single hairy roots. *Biotechnol Techniq* 7: 447–452.

Fredrickson AG, D Ramkrishna and HM Tsuchiya (1967). Statistics and dynamics of procaryotic cell populations. *Math Biosci* 1: 327–374.

Gamborg OL, RA Miller and K Ojima (1968). Nutrient requirements of suspension cultures of soybean root cells. *Exp Cell Res* 50: 151–158.

Gurtin ME and RC MacCamy (1974). Non-linear age-dependent population dynamics. *Arch Ration Mech Anal* 54: 281–300.

Henrici P (1962). *Discrete Variable Methods in Ordinary Differential Equations*. New York: John Wiley.

Kim S (1993). *Mathematical Models of Hairy Root Growth*. PhD dissertation, Chemical Engineering, Louisiana State University, Baton Rouge, USA.

Kim S, E Hopper and M Hjortso (1995). Hairy root growth models: effect of different branching patterns. *Biotechnol Prog* 11: 178–186.

Ludwig Hopper E (1995). *Growth Dynamics of Transformed Roots of Tagetes erecta Under Various Conditions* MS thesis, Chemical Engineering, Louisiana State University, Baton Rouge, USA.

Murashige T and F Skoog (1962). A revised medium for rapid growth and bio assays with tobacco tissue cultures. *Physiol Plant* 15: 473–497.

Ramkrishna D (1994). Towards a self-similar theory of microbial populations. *Biotechnol Bioeng* 43: 138–148.

Taya M, M Kino-oka, S Tone and T Kobayashi (1989). A kinetic model of branching growth of plant hairy root. *J Chem Eng Japan* 22: 698–700.

Thimann KV (1936). Auxins and the growth of roots. *Am J Bot* 23: 561–567.

The Pilot-Scale Cultivation of Transformed Roots

16

Peter D.G. Wilson

Institute of Food Research, Norwich Laboratory, Norwich Research Park, Colney, Norwich NR47UA, UK; email: PeterDG.Wilson@bbsrc.ac.uk

Background, History and Scope

The Plant Kingdom is a rich source of biochemicals used by the human race for nutrition, healing and recreation. As our knowledge of the natural world has developed, so our understanding of the specific roles of these compounds has increased. Although the traditional use of herbal remedies has been replaced, at least in the Western World, by 'scientific' medications, many of these are still based upon plant-derived compounds (Thengane and Mascarenhas, 1987). This move to 'scientific' medicine has led to an industrialisation of healing, with pharmaceutical preparations being produced by centralised organisations. In order to continue to reap the curative benefits of plant-derived compounds, pharmaceutical producers have had to collect plants from the wild, cultivate them in plantations, or chemically synthesise their active components. The advent of biotechnology provides a further alternative — direct *in vitro* cultivation of plant cells. Developments in the field of molecular biology provide additional opportunities for manipulating the machinery of the cell to improve production yields and even to modify the structure of the desired compound. The technology of transformed root culture provides the twin advantage of a stable biochemical platform on which to perform these manipulations, and a built-in genetic vector for introducing foreign DNA. These specific issues are discussed elsewhere in this monograph.

The Institute of Food Research became involved in transformed root research at an early stage, resulting from its interest in plant cells as sources of food ingredients, although it had always been clear that, for many reasons, the first commercial exploitation of the technology is likely to be in pharmaceutical production. My involvement as a biochemical engineer began in 1986 when I joined a multidisciplinary team looking at bioreactor issues in plant cell culture. Our initial work on cell suspension and immobilised cultures very quickly began to focus on transformed root cultures and led to the group's first publication (Rhodes *et al.*, 1986) on the growth of hairy roots of *Nicotiana* in a one-litre bioreactor. Our understanding of the important issues in the cultivation of this novel material developed over the succeeding years, and took us through a number of reactor configurations including packed columns (Wilson *et al.*, 1987), modified stirred tank reactors, (Hilton *et al.*, 1988; Hilton and Rhodes, 1990), droplet phase reactors (Wilson *et al.*, 1990), and even growing roots in sealed plastic bags (unpublished). Our research culminated with the development of a pilot-scale bioreactor system, which we believed was capable of addressing all the issues relevant to cultivation of transformed roots on an industrial scale. The purpose of this chapter is to describe the development, construction and operation of this bioreactor system, and present some of the art, as well as the science of the pilot-scale cultivation of transformed roots. A more general discussion of bioreactors for plant cells may be found elsewhere (Wilson and Hilton, 1995; Wilson and Rhodes, 1992), and other design concepts for transformed root bioreactors are presented within this monograph. The development of our design approach was the product of a multidisciplinary team, which brought together scientific expertise in biochemistry, molecular biology, plant physiology, microbiology and biochemical engineering, with the technical expertise of IFR's Engineering Development Group and the skills of our engineering workshop. The source material for this chapter is thus the product of the endeavours of many people, whose input is more specifically recognised below.

Development of Concepts

Before describing the design and operational details of the pilot-scale bioreactor, it will be useful to outline some of the underlying concepts and design principles which led us to develop the system in the way we did.

A key design concept arose from the observation by Hector Flores (Flores *et al.*, 1987) that the biosynthetic potential of transformed roots is intimately linked to the structural integrity of the plant system. Expanding on Flores' observations, it became clear that, in simplistic terms, the biosynthetic capabilities of transformed root cultures are not greatly affected by their growth environment as long as the organised nature of the root morphology is maintained. This is not to say that growth conditions are not important to productivity, but rather that the descent into a disorganised state usually disrupts the cellular control mechanisms to such an extent that productivity is completely destroyed. Thus, a necessary condition and one of our key measures of achievement for any new bioreactor system was its ability to maintain the 'rooty' morphology. One factor capable of destroying this organisation is mechanical damage, which leads to the typical wound-response of production of undifferentiated callus-type tissue. Our experience with stirred reactors (Wilson *et al.*, 1987) showed that interactions between root material and reactor impellers can induce root disorganisation and so lead to loss of biosynthetic capacity. The difficulties of

both quantifying and predicting shear fields in three-phase systems led us to design the pilot-scale reactor without an impeller, although other work in our laboratory (Hilton and Rhodes, 1990) has since led to modifications of the stirred tank reactor which are able to mitigate the effects of shear.

From our earliest experiments, it became clear that not only are hairy roots capable of growing to high densities, but that the morphology of the tissue is such that the biomass reaches a stage where it becomes effectively self-immobilising. This is easily seen even in flask cultures grown on orbital shakers where, at later stages of growth, the root mat stops rotating with the liquid flow. However, the main function of the bioreactor is to produce biomass, and so in the early stages there is unlikely to be a sufficient biomass density to self-immobilise. Thus, without some form of immobilisation support from the outset, the eventual spatial distribution of the self-immobilised roots becomes difficult to control. This concept is perhaps not evident when experiments are carried out in small vessels, because a single inoculation point of root material is often capable of filling a volume approaching the size of the vessel. However, our experience with growing roots in larger vessels, say more than 10 litres, is that without a support matrix, the initially free-floating roots tend to collect at one or two points in the vessel. The final biomass produced in such a system is consequently highly localised at these points, and often tends to have a core of material which has lost its rooty morphology. The rest of the liquid volume is devoid of root material and so effectively constitutes an unused region of the reactor. From these observations, we chose to include an immobilisation matrix in the design for the pilot-scale reactor. The spacing of immobilisation points was determined by a consideration of the "effective volume" that could be filled by roots from a single inoculation point. It is our experience that this volume will be dependent on the species of root under cultivation. Our transformed root lines of *Datura stramonium*, *Solanum tuberosum*, *Nicotiana tabacum*, and *Beta vulgaris* were able to fill a volume of approximately $10 \times 10 \times 10$ cm without significant callus formation at the inoculation point. We thus constructed an immobilisation support to provide one inoculation point per litre of the vessel working volume. For other species such as *Cinchona pubescens* and *Panax ginseng*, this effective volume appears to be much smaller, thus potentially requiring a higher density of inoculation points. Furthermore, it was clear that many species of root did not easily release their products to the growth medium, and so a production process would require some downstream processing of the biomass to extract the required product. It was thus important to be able to release the root mass from the immobilisation matrix for subsequent comminution and extraction of the products. The design we chose to achieve this is described in more detail below.

Another design issue which relates to harvesting is the task of removing the biomass from the vessel. In industrial-scale microbial and cell suspension cultures this is easily achieved by means of a pump, and in laboratory-scale root cultivation by means of forceps or a gloved hand. However, industrial-scale root cultivation would need a more appropriate technology. To solve the problem, we designed the immobilisation support to be capable of supporting the weight of the biomass so that it could be used to remove the roots from the vessel. In our 10-litre scale vessels, we were able to achieve biomass densities in excess of 50% (v/v). Under these conditions, the roots are so tightly packed that it becomes difficult to harvest the biomass due to the friction with the vessel walls. In order to overcome these problems, we designed the growth chamber of the reactor to have tapering sides, to allow the root mass to be withdrawn easily from the top of the vessel.

Our experience with small-scale vessels led us to believe that root cultures exhibit good growth characteristics when grown in a mist or droplet phase reactor. (Mist phase reactors have been studied extensively by Pam Weathers and her colleagues, and are discussed in more detail elsewhere in this monograph. The droplet size we employed was much larger than those studied by Weathers, and would be described by the person in the street more as a 'drizzle' than a mist.) The very large area of gas–liquid interface in a droplet or mist reactor appears to overcome oxygen transfer limitations which are often evident in microbial reactors. Indeed, we have attempted to measure the oxygen mass transfer coefficient (k_La) in our pilot-scale reactor by the standard dynamic technique (Taguchi and Humphrey, 1966), but were unable to do so because our polarographic oxygen probes could not respond fast enough (Paul Meehan, personal communication).

One feature of organ culture is that it is often difficult to measure the progress of growth. In our laboratory we have often used a 'drain and weigh' procedure for submerged root cultures, where the growth medium is periodically removed aseptically from the fermenter, the vessel and its contents are weighed, and the growth medium returned to the reactor. In this way it is possible to follow the growth of the roots by knowledge of the weight of the empty vessel. Other workers working with organ cultures and cell suspension cultures have used conductivity measurements, which are based on the stoichiometry of removal of electrolytes (usually nitrate) from the growth medium. A consequence of using a reactor with a dispersed liquid phase (i.e. as droplets or a mist) is that if the bulk of the liquid inventory is kept in a separate vessel, it is possible to estimate the increase in root mass directly by measuring the weight of the growth chamber.

Thus, the observations of good growth characteristics, high oxygen transfer, and the ability to follow the progress of growth non-invasively led us to decide to use droplet phase culture in the pilot-scale reactor. This decision has inevitable consequences for the operation of the system. In order to fill the reactor with roots, it is necessary to provide a certain amount of growth nutrients. If these are prepared as a batch to the same concentration as used in small-scale experiments, then the liquid inventory requires a second storage vessel at least as large as the reactor itself. This did not seem to be an economically viable option for industrial use, let alone for a research vessel. It is possible to increase the concentration of the nutrients to a certain extent, although flask experiments have demonstrated that elevation of sucrose concentrations can lead to the accumulation of starch in certain species of roots (Hilton and Rhodes, 1994). It was consequently decided to operate the reactor as a fed-batch system, i.e. by adding nutrients to a smaller liquid inventory during the course of root growth. (Incidentally, the choice of fed-batch operations makes it difficult to use stoichiometric measures of growth such as conductivity due to the confounding effects of the added nutrients.)

To run a fed-batch reactor successfully, it is necessary to either use a direct feedback control system where the composition of the growth medium is measured and supplements added to maintain the desired composition, or alternatively to employ an indirect control system where the composition of the growth medium is inferred from the progress of the growth. Although the former strategy is more secure in control terms, samples need to be removed frequently from the vessel and analysed immediately if the control strategy is to be effective. The expected time course for changes in the nutrient composition can be estimated from the ratio of the size of the growth chamber to the volume of the liquid inventory. For a 10:1 ratio, the rate of change of liquid composition would be expected to be 10 times higher than that observed in batch culture. So for a batch cultivation which might typically last for 20 d, sampling and supplementing the medium on a daily basis could still lead to daily swings in nutrient composition equivalent to half the mean value, especially when the biomass loading is high. Without automatic on-line analysis, such a strategy would therefore lead to unacceptably high demands on staff time, and a high risk of contamination from the frequent sampling regime. It was therefore decided to operate a feeding strategy based upon the

Table 16.1. Media compositions.

Component	Immobilisation medium (mg L^{-1})	'Datura' medium (mg L^{-1})
CaCl$_2$.6H$_2$O	110	772
MgSO$_4$.7H$_2$O	125	3224
NaH$_2$PO$_4$.2H$_2$O	85	1700
(NH$_4$)$_2$SO$_4$	67	1140
KNO$_3$	1500	19,765
Nicotinic acid	0.5	5.0
Thiamine HCl	5.0	50.0
Pyridoxine HCl	0.5	5.0
Inositol	50	250
CoCl$_2$.6H$_2$O	0.025	0.13
CuSO$_4$.5H$_2$O	0.025	0.13
H$_3$BO$_3$	3.0	15.0
KI	0.75	3.75
ZnSO$_4$.7H$_2$O	2.0	10.0
NaMoO$_4$.2H$_2$O	0.25	1.25
MnSO$_4$.4H$_2$O	13.2	66.0
Sucrose	20 g L^{-1}	150 g L^{-1}

increase in biomass, as estimated by the increasing weight of the growth vessel and its contents.

To achieve this fed-batch operation it was necessary to devise a stoichiometrically-balanced feeding medium for the species under cultivation. Small-scale cultures were used, and the stoichiometric yield coefficients determined for each of the inorganic nutrients as well as for sucrose (Hilton and Wilson, 1995). Furthermore, it was necessary to determine a yield coefficient for water, an often neglected nutrient, but one which is significantly consumed in high-density culture. The raw stoichiometric coefficients obtained from this exercise were used to devise the feeding medium. The medium was not based strictly on the stoichiometric yields obtained from the batch experiments, as it was evident from the time course of batch growth and nutrient depletion that some of the inorganic nutrients (most notably phosphate) were being rapidly removed from the medium and stored in intracellular pools, resulting in an overestimation of the levels required in a balanced medium. Instead, a judgement was made and trial feeding media were tested on a smaller scale (Meehan *et al.*, in preparation). Details of the composition of the '*Datura*' medium used are given in Table 16.1.

Finally, it was apparent that solving the problem of inoculation of transformed root bioreactors would be a major obstacle in the development of a commercial process. In the laboratory scale, we inoculated our vessels by removing the headplate in a laminar flow cabinet, and aseptically transferring the roots from the inoculum flasks with sterile forceps. In this way, the inoculum could be positioned, if required, directly on the immobilisation support. Whilst extremely effective on a laboratory scale, this was clearly impractical for any commercial operation. To overcome this problem we developed a seed vessel and transfer system to enable aseptic inoculation of the pilot reactor, as well as an operational procedure to ensure even distribution of the biomass on the immobilisation matrix.

The construction of the pilot bioreactor, immobilisation support, seed vessel and transfer system is described below, together with a typical operational protocol and results for the cultivation of transformed roots of *Datura stramonium*.

Construction of the Bioreactor

The bioreactor has a volume of 500 litres, and is essentially a tapered stainless steel vessel with an aspect ratio of 2:1. The bottom of the vessel is sealed by a hinged domed cap, which houses the sparge ring, a number of probe ports, and a centrally-located drain port for recycle of growth medium. The sparge assembly consists of two semi-circular tubes fitted with a number of nozzles, each tube fed by its own air supply. Manipulation of the air flows thus allows the liquid flow pattern in the vessel to be controlled to provide either bubble-column type aeration, or effectively an airlift configuration (without a baffle) capable of generating a liquid circulation pattern in either of two directions. The use of the sparge system to control inoculum distribution is described further below.

The tapered sides of the reactor are jacketed for *in situ* steam sterilisation and temperature control, and are fitted with three vertical sightglasses enabling visual inspection of the entire height of the vessel contents. This was considered important in a research vessel, to visualise the immobilisation of the root mass. The provision of full-height sightglasses does however complicate the engineering design leading to increased capital costs, and could be omitted for a commercial system. The vessel walls are also equipped with a number of ports to accommodate probes and pipework connections, and on the interior are fitted with supports to locate the immobilisation matrix.

The top plate of the vessel is also removable, and is fitted with a number of sample, addition and probe ports, a pressure relief valve, a stainless steel

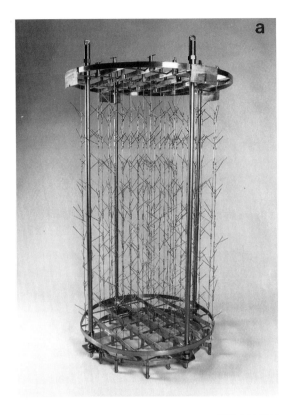

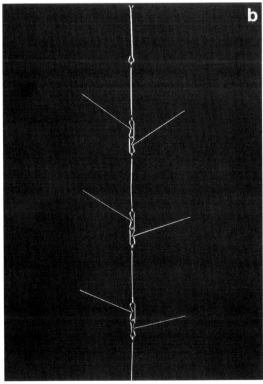

Figure 16.1. (a) Immobilisation assembly (70 cm diameter, 120 cm height). (b) Immobilisation barbs (barb spacing 10 cm).

off-gas condenser and filter, and a central 50-mm port for inoculation. The inside of the top plate also houses two spray assemblies. Each assembly consists of two concentric rings of spray nozzles fed independently by a high pressure (10 barg) diaphragm pump. The system was designed to allow each spray assembly to produce different droplet sizes, and also to provide a backup system in case of blockage of spray nozzles. In practice, only one spray assembly was ever required in use.

Located within the reactor is an immobilisation assembly (Figure 16.1a). The assembly consists of a number of barbs (Figure 16.1b) constructed from stainless steel wire and forming a series of chains. During the inoculation and growth phases, the chains are held in tension by two stainless steel supports at the top and bottom of the immobilisation matrix (see Figure 16.1a), causing the barbs to extend and form immobilisation points where the root inoculum can lodge and become entangled. After growth, the complete immobilisation assembly containing the biomass is removed from the vessel, thus facilitating harvest. The chains are then removed from the top support by means of a quick-release mechanism, and the lower support is detached from the four vertical stay-rods. The chain and barb assembly can then be removed from the root mass as a single unit. The design of the barbed chain allows the barbs to fold flat during this process, whilst the roots are held between the top support and a third grid located immediately above the lower chain support (Rhodes *et al.*, 1992). In this way, the whole biomass can be manipulated for further processing.

Construction of the Seed Vessel and Transfer Tube

A seed vessel was designed to both grow the inoculum and to render it into a pumpable form. Of approx. 10 litres capacity, the vessel (Figure 16.2a) comprised a cylindrical glass tube (ca. 22 cm diameter, 30 cm height) mounted upon a stainless steel conical section and surmounted by a stainless steel headplate. To enable inoculation, the whole seed vessel is transferred into a laminar flow cabinet, and root material introduced via a specially designed port in the top plate (Figure 16.2b). The conical base houses a conical stainless steel helical screw with a blade attached to the leading edge (Figures 16.2c and 16.2d). The screw can be turned by means of a shaft running vertically through the vessel. The helical screw was designed so that as its diameter decreased towards the base of the vessel, so its pitch increased, maintaining an approximately constant cross-sectional area for transport of the comminuted root material. This feature was included to prevent compressional damage to the root mass

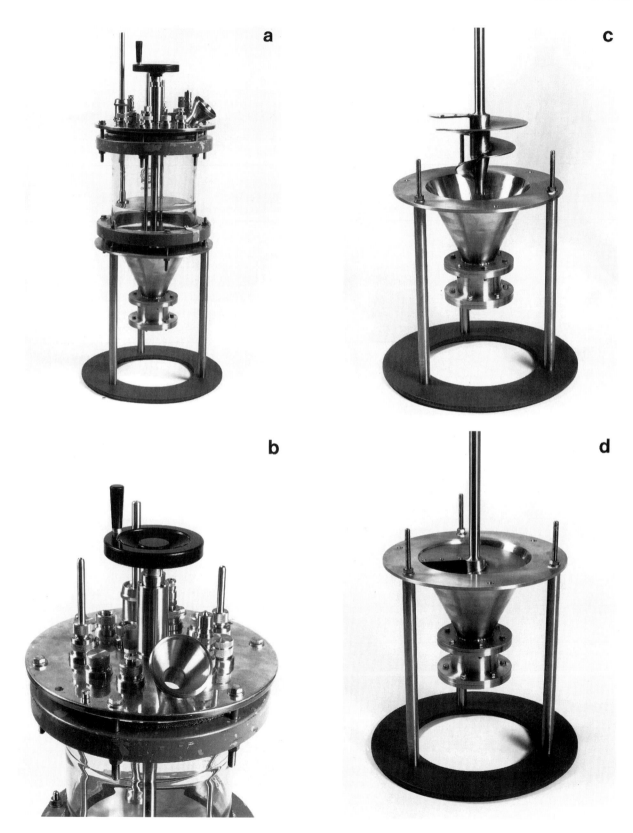

Figure 16.2. (a) Seed vessel. (b) Headplate, showing inoculation port. (c) 'Exploded' view of helical screw and cutter. (d) Helical screw in normal operating position.

during transfer to the pilot vessel. A circular perforated stainless steel plate of a diameter slightly less than the internal diameter of the glass cylinder was located within the seed vessel. This was connected to a vertical shaft passing through the headplate, and can be seen in Figure 16.2a. To maintain sterility, the whole shaft assembly is contained within a flexible, autoclavable plastic sheath. The plate is used to push the root mass onto the cutting blade of the screw conveyor during inoculation.

The base of the conical section of the seed vessel is terminated with a 20-mm ball valve. During inoculation, this is connected to the top of the pilot reactor by means of a transfer tube. The transfer tube comprises a section of 50-mm diameter glass pipe with PTFE expansion bellows at each end. After installing the pipework, the transfer tube is steam-sterilised *in situ*. The PTFE bellows allow for expansion during sterilisation, and can also compensate for slight misalignment of the seed and pilot vessels. Early experience of inoculum transfer had demonstrated that, despite the seed vessel being vented to atmosphere via a sterilising air filter, a partial vacuum was created during inoculum transfer, leading to a greatly increased risk of contamination. To prevent this problem, the headplates of both vessels were joined by a flexible tube (stainless steel braided PTFE, sterilised *in situ*) to equalise the pressure in the vessels. A feed of sterile growth medium to the top of the seed vessel from the pilot reactor allows comminuted root material to be washed through into the growth vessel, ensuring complete transfer of the inoculum.

Operational Protocol

The following is an account of a batch cultivation of transformed roots of *Datura stramonium*. Approximately two weeks before the planned inoculation, the pilot vessel was filled with 500 litres of immobilisation medium (Table 16.1) and sterilised *in situ*. This medium was held at growth temperature and aerated for the 15 d leading up to inoculation to confirm successful sterilisation. On the following day, the seed vessel was sterilised and charged with 10 litres of sterile Gamborg's B5 medium containing 3% (w/v) sucrose. Using a laminar flow cabinet to maintain sterility, the vessel was inoculated with three flasks of a 9-d culture of *D. stramonium* roots. The vessel was sparged with sterile air and held at 25°C in a temperature-controlled laboratory for 14 d to provide the inoculum for the pilot vessel. The relatively low biomass density (i.e. pre-stationary phase) produced during this time not only provided a rapidly growing inoculum for the pilot vessel, but allowed the seed vessel to be operated without an immobilisation support.

In order to perform the inoculation, the seed vessel was moved to a position above the pilot scale reactor, immediately above an inoculation port, and supported on a removable gantry. The transfer tube with its PTFE expansion bellows was used to connect the base of the seed vessel with the inoculation port on the pilot reactor headplate (Figure 16.3a). The transfer tube and associated valves were steam-sterilised *in situ*. Once the tube had been sterilised and allowed to cool, inoculation could take place as described above.

The comminuted root material, consisting of pieces of root up to ca. 4 cm in length together with smaller fragments, was circulated in the immobilisation medium (Table 16.1) by passing air through one side of the sparge ring assembly (Figure 16.3b). This circulation pattern caused the root inoculum to become lodged on the immobilisation barbs within a few minutes. The seed vessel assembly was removed from the pilot plant, and the circulation pattern allowed to continue for 3 d. During this time, sufficient root growth occurred to firmly attach the root inoculum to the half of the immobilisation matrix in the downward flow path. The air flow was then manipulated to reverse the direction of the liquid circulation pattern (Figure 16.3c) to allow root material to lodge on the other side of the immobilisation matrix. The immobilised root material is shown in Figure 16.4a. The system was then left for 21 d to allow growth to proceed in submerged culture. At this time, the vessel was drained of the immobilisation medium (Figure 16.3d) and the reactor operated in the droplet mode (Figure 16.3e).

Results and Conclusions

For the following 40 d (and nights) the vessel was operated in the droplet mode, with growth medium drawn from a 75-litre holding vessel, sprayed over the root mass, and finally recirculated to the holding vessel (Figure 16.5). During this time, the weight of the growth vessel was monitored continuously by means of load cells. This measurement was fed to a computer control system, which added a medium concentrate to the holding vessel in an attempt to replenish those nutrients used for growth. For each 100 g rise in the weight of the growth vessel due to root growth, 100 g of *Datura* medium (Table 16.1) was added to the holding vessel. Measurements of pH, temperature and dissolved oxygen were also logged. Figure 16.6 shows the progress of root growth during the 40 d of droplet operation, together with the associated changes in pH. It became evident by the eighth day of droplet operation that the growth rate of the culture was declining. It was thought that this was due to ineffective replenishment of the nutrients used by the roots. Due to the increased

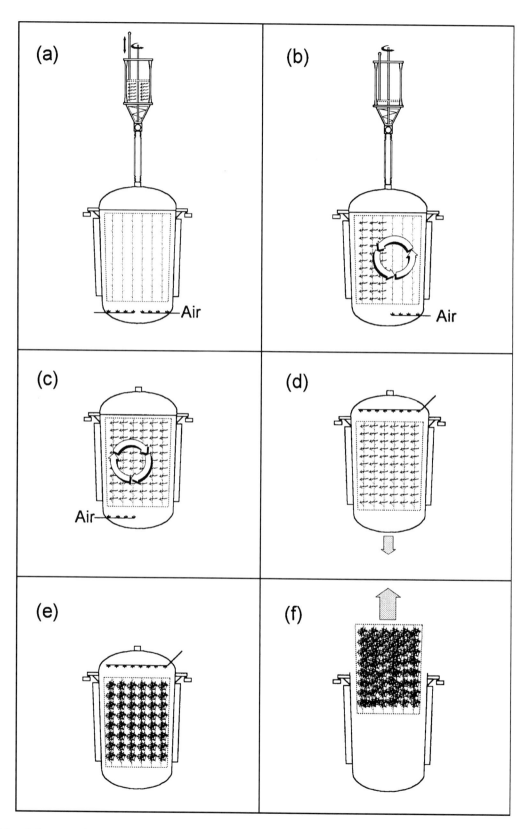

Figure 16.3. Schematic diagram illustrating stages of (a) inoculation, (b, c) immobilisation, (d, e) growth, and (f) harvest.

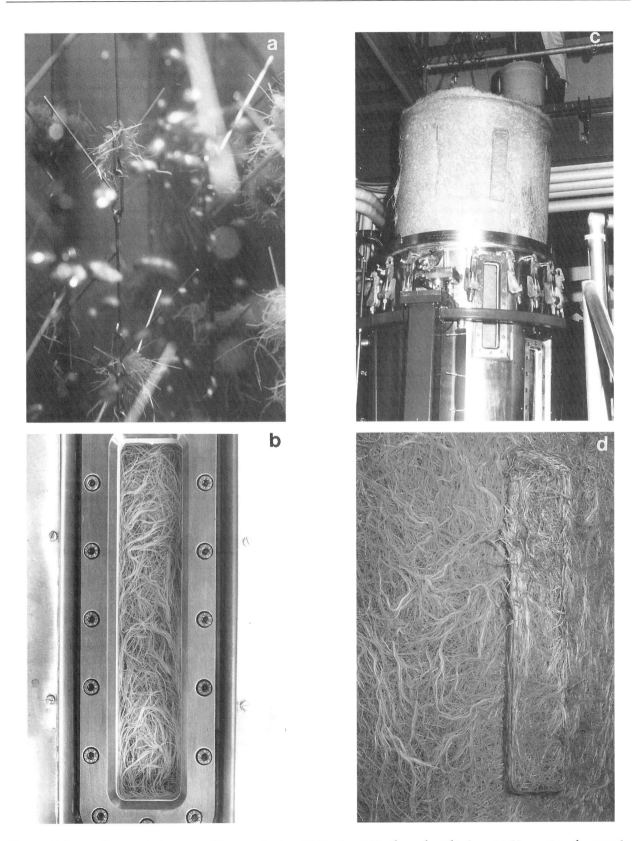

Figure 16.4.   (a) Root material immobilised on barbs. (b) Roots visible through sightglass. (c) Harvesting the vessel. (d) Root mass after harvesting showing colonisation of sightglass indentation.

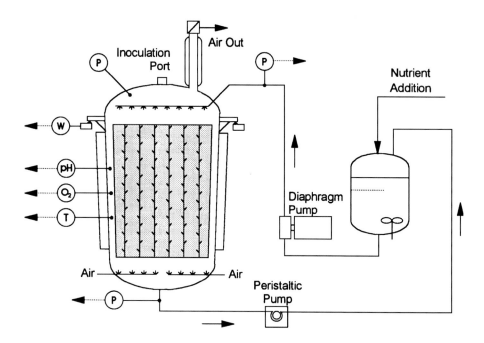

Figure 16.5. Schematic diagram of the bioreactor system.

risk of contamination involved in sampling the fermenter, "blind" interventions were made to the medium composition on Day 8 and Day 12, firstly by adding *Datura* medium in case the roots had used the available nutrients, and secondly by add-

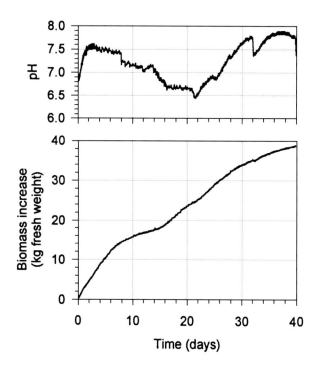

Figure 16.6. Time course of pH and growth during the droplet phase.

ing water in case the medium concentration had risen and inhibited growth. As neither of these interventions restored the initial growth rate, it was decided to take a sample of growth medium from the vessel and analyse it for sucrose, glucose and level of inorganic nutrients. The analysis showed that the sucrose concentration had fallen from an initial level of 3% (w/v) to 0.03%, with a corresponding rise in glucose. The concentration of nitrate had however been successfully maintained by the feeding strategy. Previous flask-scale studies had shown this inversion of sucrose, and very low growth rates obtained in the absence of sucrose. This evidence led to the decision to supplement the medium with sucrose at this stage. This action effectively restored the initial growth rate. During the rest of the 40 d growth in droplet mode, a number of samples were removed from the vessel for analysis, and nutrient interventions made. These analyses and interventions are documented in Table 16.2.

At the end of 40 d, substantial growth had occurred which could be observed through the sightglasses (Figure 16.4b). The immobilisation frame containing the root material was removed from the vessel (Figure 16.4c), and the roots removed from the frame. A total of 39.8 kg (fresh weight) of roots were harvested. The reactor configuration had been successful in maintaining the organised morphology of the roots. Figure 16.4d shows a close-up view of the harvested roots from an area adjacent to one of the sightglasses, and shows how the roots had colonised the vessel volume to take up the form of the sightglass indentation. The growth yield was

Table 16.2. Growth of *Datura stramonium* — interventions and analyses.

Day	Observation	Analysis			Action	Result
		Sucrose (%)	Glucose (%)	Inorganics		
8	Growth rate slowing	–	–	–	Added 2 litres *Datura* medium[a]	None
12	Growth rate slowing	–	–	–	Added 15 litres sterile water	None
14	Growth rate slowing	0.028	1.29	Normal	Added 2.5 kg sucrose[b]	Growth rate increased
21	Small reduction in growth rate	0.020	2.50	Normal	Removed 20 litres medium Added 2.5 kg sucrose[b]	Growth rate increased
25	Growth rate maintained	0.55	3.46	Nitrate 50% Phosphate 25%	Added 2.5 kg sucrose[b]	None
27	–	2.01	3.72	–	–	
28	–	1.55	3.84	–	Added 2.5 kg sucrose[b] Liquid volume measured as 77 litres.	None
32	–	1.8	4.2	Low	Added 15 litres *Datura* medium[a]	
33	–	4.2	4.2	–		

[a] See Table 16.1.
[b] Added as 5 litres of 50% (w/v) sterile, aqueous solution.

however significantly lower than we expected to achieve based on our experience at the laboratory scale, and growth had proceeded at a slower rate. We believe this to be the result of a non-optimal feeding strategy, resulting in a medium composition incapable of supporting rapid growth. It can be seen from the glucose analyses in Table 16.2 that a considerable amount of the sucrose provided had been inverted to glucose. Although a fructose analysis was not carried out on these samples, laboratory studies had previously shown that both fructose and glucose accumulate during the growth of *D. stramonium*, and that it was likely therefore that a considerable amount of fructose was also present. If a fed-batch strategy is to be successful for the growth of *Datura* roots, a more thorough study of factors affecting the utilisation and conversion of the carbon source would be of great importance. The strategy for replenishing nutrients in this experiment, i.e. the addition on a weight-for-weight basis of a medium designed to be "stoichiometrically equivalent" to the root mass, proved reasonably success-

ful at maintaining the concentrations of inorganic nutrients. However, it should be noted that any error in determining the stoichiometric equivalents of the nutrients required will be cumulative during the growth of the roots, and may lead to instabilities in the nutrient composition. A better strategy could be developed by using regular, or even on-line samples of the growth medium for analysis of nutrient composition to allow direct feedback control of the medium composition, providing that sampling could be guaranteed to be aseptic.

Despite the lower rate and extent of growth than expected, a number of other conclusions may be drawn from the experience of this experiment.

Transfer of inoculum: We have demonstrated that it is possible to grow an inoculum in a seed vessel, render it into a pumpable form, and transfer it aseptically to a pilot-scale vessel whilst maintaining the viability of the roots. Our observation of the roots soon after inoculation revealed little evidence of callus formation in the transferred root material, indicating that the cutting and screw-conveying carried out in our seed vessel caused little mechanical damage to the plant material.

Distribution and immobilisation of roots: We have shown that the use of an immobilisation matrix can successfully lead to the even distribution of roots throughout the volume of a pilot-scale growth vessel. Whilst we chose to use a growth medium to allow the positioning of the roots and a limited amount of growth over a number of days, it may be conceivable and perhaps advantageous to use a cheaper liquid (perhaps an osmotically-balanced saline) to perform the immobilisation more rapidly, thus allowing an earlier move to droplet or mist operation.

Growth in droplet phase: We have shown that it is possible to grow transformed roots on a pilot scale by the use of a droplet phase reactor. Not only does such a dispersed liquid phase have advantages of oxygen transfer and opportunities for product release into a small liquid inventory, but it allows the growth of the root biomass to be determined by direct measurement of the weight of the growth vessel.

Harvesting: We have demonstrated the use of an immobilisation matrix to facilitate the harvest of root biomass from a pilot-scale fermenter. This technique allows easy transfer of roots from the growth vessel to a post-processing stage of product extraction. The material is easily and safely handled, with little or no damage to the root material which could otherwise lead to product loss or the creation of hazardous aerosols.

The Future

We hope that our experiences with large-scale cultivation of transformed roots will encourage the many workers around the world with whom we share a vision of commercial production of phytochemicals through transformed root technology. Like many other advances in technology, successful commercial operation will only be possible by a coordinated multidisciplinary approach, bringing together skills from biochemistry, molecular biology, biochemical engineering and mechanical design.

Acknowledgments

The developments presented in this chapter are the result of the efforts of a large number of people, in particular: Martin Hilton, Christopher Waspe and Paul Meehan, with whom I worked on the bioreactor development; IFR's engineering staff of Ian Rudgley, David Steer, David Pennington and Terry Hurn; Engineering design consultant Reginald Thirkettle; and the Plant Biotechnology Group, led by Mike Rhodes and including Richard Robins, John Hamill, Nick Walton, Adrian Parr, Judy Furze and Abigael Peerless. The work was funded by the UK Office of Science and Technology.

References

Flores HE, MW Hoy and JJ Pickard (1987). Secondary metabolites from root cultures. *Trends in Biotechnol* 5: 64–69.

Hilton MG and MJC Rhodes (1990). Growth and hyoscyamine production of 'hairy root' cultures of *Datura stramonium* in a modified stirred tank reactor. *Appl Microbiol Biotechnol* 33: 132–138.

Hilton MG and MJC Rhodes (1994). The effect of varying levels of Gamborg's B5 salts and temperature on the accumulation of starch and hyoscyamine in batch cultures of transformed roots of *Datura stramonium*. *Plant Cell Tiss Organ Cult* 38: 45–51.

Hilton MG and PDG Wilson (1995). Growth and uptake of sucrose and mineral ions by transformed root cultures of *Datura stramonium, Datura candida × aurea, Datura wrightii, Hyoscyamus muticus* and *Atropa belladonna. Planta Med*. 61: 345–350.

Hilton MG, PDG Wilson, RJ Robins and MJC Rhodes (1988). Transformed root cultures — fermentation aspects. In *Manipulating Secondary Metabolism in Culture*, edited by RJ Robins and MJC Rhodes, pp. 239–245. Cambridge: Cambridge University Press.

Rhodes MJC, MG Hilton, AJ Parr, JD Hamill and RJ Robins (1986). Nicotine production by 'hairy root' culture; fermentation and product recovery. *Biotechnol Lett* 8: 415–420.

Rhodes MJC, RJ Robins, CR Waspe, DC Steer, MG Hilton and PDG Wilson (1992). Fermenter. UK Patent GB2238551B.

Taguchi H and AE Humphrey (1966). Dynamic measurement of the volumetric oxygen transfer coefficient in fermentation systems. *J Ferm Tech* 44: 881–889.

Thengane S and AF Mascarenhas (1987). Plant cell culture in the pharmaceutical industry: a biotechnological appraisal. *Indian Drugs* 24(10): 460–470.

Wilson PDG and MJC Rhodes (1992). Plant cell bioreactors. *Adv Plant Cell Biochem Biotechnol* 1: 105–150.

Wilson PDG and MG Hilton (1995). Plant cell bioreactors. In *Bioreactor System Design*, edited by JA Asenjo and JC Merchuk, pp. 413–439. New York: Marcel Dekker.

Wilson PDG, MG Hilton, RJ Robins and MJC Rhodes (1987). Fermentation studies of transformed root cultures. In *Bioreactors and Biotransformations*, edited by GW Moody and PB Baker, pp. 38–51. London: Elsevier.

Wilson PDG, MG Hilton, PTH Meehan, CR Waspe and MJC Rhodes (1990). The cultivation of transformed roots from laboratory to pilot plant. In *Progress in Plant Cellular and Molecular Biology*, edited by HJJ Nijkamp, LHW van der Plas and J van Aartrijk, pp. 700–705. Dordrecht: Kluwer.

Laboratory-Scale Studies of Nutrient Mist Reactors for Culturing Hairy Roots

17

Pamela J. Weathers[1], Barbara E. Wyslouzil[2] and Michelle Whipple[2]

[1]Department of Biology and Biotechnology, and [2]Department of Chemical Engineering,
Worcester Polytechnic Institute, Worcester, MA 01609, USA; email: weathers@wpi.edu

Problems of Hairy Root Cultures and Advantages in Using Mists

Plant organs are poorly suited to the conventional, submerged stirred tank reactors typically used for growing yeast or bacteria. This environment is characterised by high shear which damages the roots, poor control of critical gas concentrations, insufficient nourishment of the biomass at high concentrations due to poor liquid circulation, and, when the reactor is densely packed, chemical gradients in the medium. Because many products are produced at low levels, the high ratio of medium to metabolite volume complicates separation of the products. Reactors in which liquid is the dispersed phase appear to offer ideal conditions for growth and productivity of root cultures because of the high gas content available to growing roots (McKelvey et al., 1993; DiIorio et al., 1992a, 1992b; Kondo et al., 1989; Whitney, 1992). For example, aeroponics minimises many of the problems associated with a continuous liquid phase by using gas as the continuous phase.

There are four primary environmental factors which affect aeroponic root culture: moisture, temperature, mineral nutrition, and the gas phase composition (mainly carbon dioxide, ethylene and oxygen) (see review by Weathers and Zobel, 1992). Other factors may be involved, but are less affected by the mode of culture. Moisture is required by all plant tissues; optimum timing (duration and cycle) and intensity of moisture applications vary with species, temperature and configuration of the spray apparatus (Weathers and Zobel, 1992). Of particular importance is the thickness of the liquid film deposited on the surface of the roots. When this water film is thick enough it can limit nutrient and gas transfer to the tissue. The use of on \ off moisture application cycles as well as the use of mists as opposed to sprays (nominal droplet diameter 0.01–10 µm and 10 µm–1 mm, respectively) decrease the development of thick water films on roots. Because the droplet size is usually quite small (average of 7–10 µm), mists may require little or no off cycle. Mists can be produced by using either ultrasonics, or nozzles and compressed air. In addition to avoiding damage from the fluid shear associated with sprays, mists produced by ultrasonics are more water efficient, thus eliminating the need for extensive recirculation equipment.

One of the major advantages of aeroponics is the complete control of gases in the culture environment. This is important for the maintenance of

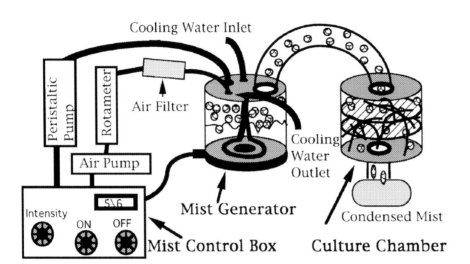

Figure 17.1. Nutrient mist bioreactor set up for continuous operation.

healthy root cultures because gases play a major role in root metabolism. Oxygen is essential for respiration needed for growth. Hairy roots may be oxygen limited unless cultures are aerated with oxygen-enriched air (Yu and Doran, 1994). Oxygen depletion triggers anaerobiosis and the production of ethylene. The latter is involved in lateral root but not basal root initiation and stimulates adventitious rooting. Since ethylene is relatively insoluble in water, the formation of thin films at the surface of roots can cause a localised build-up to concentrations far in excess of normal. Carbon dioxide has been shown to stimulate root growth of many species (Weathers and Zobel, 1992; DiIorio et al., 1992a). Optimal concentrations and duration of application vary with species, although concentrations of 0.1–0.5% are often stimulatory (Weathers and Zobel, 1992). Overall, adequate moisture cycling and use of droplet sizes small enough to avoid excessive film build-up are critical for proper control of gases in root culture.

Basic Components of a Nutrient Mist Bioreactor

The mist bioreactor used in our experiments can be configured to provide a wide variety of growth environments for hairy roots (DiIorio et al., 1992b; DiIorio, 1991). The parameters that can be adjusted to change the culture conditions include gas flow rate and composition, tubing geometry and direction of flow through the culture chamber, mist cycle times, and mist intensity. A liquid level control allows continuous operation of the reactor with very little supervision.

Figure 17.1 (Whipple, 1995) illustrates a typical set-up for hairy root growth within the nutrient mist bioreactor (Mistifier™, Manostat Corp., New York).

The key components are the mist generator, culture chamber, control box, medium reservoir, peristaltic pump, air pump and rotameter. Mist is generated by an ultrasonic transducer (1.5 MHz) located in the bottom plate of the 7.6 cm I.D. × 10.2 cm high mist generator. The top plate of the mist generator contains holes for the cooling water inlet and outlet, medium inlet, filtered gas inlet and mist outlet.

Two pumps are used to produce gas flow rates from 1–6 L min^{-1}, and a rotameter is used to regulate the gas flow rate. The gas stream is usually air which can be supplemented with other gases (e.g. CO_2). Filtered gas flows through the mist generator and carries the mist through tubing to the culture chamber where the hairy roots grow. The culture chamber usually consists of a 12.7 cm I.D. × 9.5 cm high (1.2 L) polycarbonate cylinder containing a simple trellis support for the growing roots. The dimensions of the culture chamber have not yet been optimised for hairy root growth. Mist can be fed either from the top or bottom of the chamber. Condensed mist is collected at the outlet of the chamber and returned to the medium reservoir if the reactor is operating in batch mode. If the reactor is operating in continuous mode, this liquid is discarded and fresh medium is added to the mist generator to make up the loss. The exit gas stream travels through a mist coalescer to remove any remaining atomised medium and is then vented to the atmosphere.

A simple control box regulates misting cycles, the liquid level within the mist generator, and the mist intensity. A timing circuit operates the bioreactor on\off misting cycles by controlling the power to the mist generator and air pump. The liquid level within the mist generator is monitored by measuring the conductance between two rods. When the liquid level drops below the set point (~2 cm), the peristaltic pump turns on and transfers medium

Figure 17.2. *Beta vulgaris* hairy roots cultured in a nutrient mist bioreactor for 14 d at 25°C in Gamborg's B5 medium with 3% sucrose. The culture was supplemented with 1% CO_2.

from a reservoir into the mist generator until the liquid reaches a set maximum depth (~4 cm). The mist intensity can be adjusted by varying the power delivered to the transducer. On a scale of one to ten, the lightest mist density corresponds to a setting of one.

Biological Responses of Roots Grown in a Nutrient Mist Reactor

Growth and Biomass Changes

Figure 17.2 shows *Beta vulgaris* hairy roots grown for 14 d in a nutrient mist bioreactor. Compared with roots grown in liquid (shake flasks) or on plates, roots grown in mists show significant changes in root hair morphology. Cultures on semi-solid medium (plates) have long, profuse root hairs. Roots grown in liquid show little or no root hair development until the culture becomes so dense that aerial

Table 17.1. Growth responses in some mist/spray reactors.

Culture conditions	Mist cycle	Doubling time (d)	Biomass increase (X_t/X_0)	% Dry weight in the biomass	Biomass productivity (g L^{-1} d^{-1} dry weight)
Beta vulgaris (DiIorio et al., 1992b; DiIorio, 1991)					
Nutrient mist bioreactor (without CO_2)	0.19 mL min^{-1}				
7 d	5\6 min	3	3.5	17.0	0.33
14 d	5\6 min	4	8.0	26.2	0.56
Shake flasks					
7 d	n.a.	2	3.5	8.0	0.29
14 d	n.a.	2	8.4	8.0	0.63
Nutrient mist bioreactor (with 1% CO_2)	0.19 mL min^{-1}				
7 d	5\6 min	3	4.0	n.d.	n.d.
Plates	n.a.	2	n.a.	5.7	n.d.
Hyoscyamus muticus (McKelvey et al., 1993)					
Spray reactor	1.14 mL min^{-1}				
18 d	continuous[a]	n.d.	53.25	n.d.	0.19
Nicotiana tabacum (Whitney, 1992)					
Spray reactor	250 mL min^{-1}				
28 d	40\120 s	2.4	3403	n.d.	0.73[b]
Shake flask					
28 d	n.a.	2.8	1084	n.d.	0.27

n.a., not applicable; n.d., not determined. X_t/X_0 is the ratio of final to initial fresh weight. Mist cycle is denoted as time on\time off. Productivity is based on reactor volume.
[a] Since no spray cycle was stated, it is assumed the application of the mist was continuous.
[b] Dry weight assumed to be 8% of fresh weight.

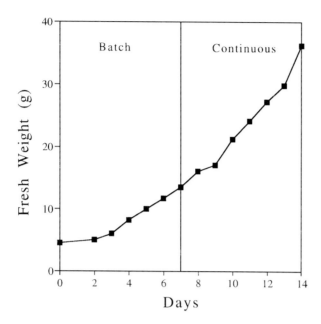

Figure 17.3. Continuous culture of *Beta vulgaris* hairy roots in mists. Hairy root tissue (inoculum: 4.5 g fresh weight) was cultured in batch for 7 d in Gamborg's B5 medium with 30 g L^{-1} sucrose prior to shifting to continuous culture with the same medium. Fresh weight of the culture was determined by weighing the entire growth chamber on a top-loading balance. No mass reading was taken on Day 1 because the culture chamber had not yet equilibrated; coalescence of incoming mist was still altering the tare weight. The misting cycle was 5 min on\6 min off, with an air flow rate of 1500 mL min^{-1}. Air flow ceased during the mist-off time period.

roots protrude above the surface of the liquid. Significant hair development similar to that on plates is then observed. Root hairs found in mist-grown cultures are somewhat stubby and significantly less profuse than roots grown on plates, but more profuse than those found in liquid cultures. Since root hairs provide more surface area for transport of nutrients and waste products, culturing roots in mists may facilitate mass transport, although excessive root hair growth, especially in densely packed beds, can lead to capillary liquid hold-up which may impair reactor scaleability (Ramakrishnan and Curtis, 1994).

Biomass increases, measured in terms of fresh weight for *Beta vulgaris* hairy roots grown in batch (7 d + 1% CO_2) then continuous mode (8–14 d, no CO_2) in nutrient mists, were comparable to increases in shake flasks up to 14 d (Table 17.1) (DiIorio *et al.*, 1992b; DiIorio, 1991). However, this was probably due to an increase in the dry weight to fresh weight ratio (% dry weight in biomass) in the mist cultures, as doubling times were lower in mists than in flasks (Table 17.1). In fact the doubling time increased by 33% during the second week of growth (Table 17.1, Figure 17.3) which resulted in a biomass productivity for mist cultures slightly greater than in shake flasks at 7 d but lower at 14 d. Since the medium feed rate was about 0.19 mL min^{-1} for the entire mist on\off cycle, nutrients probably limited growth especially in the second week. Both mist and liquid cultures had a lag phase of about 2 d (Figure 17.3). Roots grown in a mist reactor typically need a minimum inoculum density of 6 g L^{-1} fresh weight

Table 17.2. Effect of misting cycle and culture mode on hairy roots grown in a nutrient mist bioreactor for 7 d. (From DiIorio, 1991; DiIorio *et al.*, 1992b).

Culture mode	Inoculum (g fresh weight)	Mist cycle (min)	Biomass increase (x_t/x_0)	Number of doublings
Carthamus tinctorius				
Continuous	9.0	5\2	2.5	1.3
	9.0	5\5	2.9	1.6
	9.0	5\10	2.2	1.2
	9.0	5\20	1.2	0.2
	4.5	5\5	2.2	1.2
Batch	9.0	5\5	3.2	1.8
	4.5	5\5	3.1	1.7
Beta vulgaris				
Continuous	4.5	5\6	2.8	1.5
	4.5	5\6	3.5	1.8
Batch	4.5	5\2	3.4	1.8
	9.0	5\6	3.3	1.7

x_t/x_0, ratio of final to initial fresh weight. Doublings are based on dry weight. Batch culture mode is 100% medium recycle; continuous culture is 100% fresh medium fed to root bed.

grown in batch mode for the first week (Figure 17.3). Subsequent culture can be in continuous mode (Table 17.1, Figure 17.3). Other plant species grown in either our mist reactor or other mist reactors also showed equivalent or better growth than in shake flasks (Table 17.1). For tobacco, doubling times in mist decreased compared with shake flasks; biomass productivity was very high ($0.73 \, g \, L^{-1} \, d^{-1}$ dry weight) most probably due to a high feed rate of nutrients to the culture ($250 \, mL \, min^{-1}$) (Whitney, 1992).

Mist Cycle Responses

Provision of intermittent mists may improve growth for some species probably because liquid layer build-up on the roots is minimised resulting in more optimised nutrient transport. Studies with *Carthamus tinctorius* hairy roots in our nutrient mist bioreactor showed optimum growth over 7 d with a mist cycle of 5 min on, 5 min off; nearly continuous misting (5 min on, 2 min off) inhibited growth by 14% (Table 17.2; DiIorio, 1991). This sensitivity to nearly continuous misting was less obvious for *Beta vulgaris* (Table 17.2). Whitney (1992) also employed an intermittent mist for *Nicotiana*, but no data were provided to indicate if this was an optimised factor. McKelvey *et al.* (1993) apparently used a continuous spray for *Hyoscyamus* resulting in growth comparable to shake flasks. For optimising growth of root cultures in mists, mist cycle responses should be measured.

Responses to Changes in Gas Composition

Although there are three gases of major importance to plant roots, only responses to CO_2 in mists have been studied systematically. CO_2 stimulated growth of *Beta vulgaris* and *Carthamus tinctorius* hairy roots (DiIorio, 1991; DiIorio *et al.*, 1992a). This response was not due to an increase in growth rate but rather to a decrease in lag time of the cultures (Figure 17.4). Loss of the lag phase resulted in an apparent increase in biomass when cultures were harvested at the same time point (Figure 17.4). Different species have different optimal concentrations of CO_2 at which they exhibit maximum growth. For example, the optimum for *B. vulgaris* is about 1.4%; for *C. tinctorius* it is about 1% (DiIorio *et al.*, 1992a). Consequently, addition of CO_2 to mist grown cultures is probably only necessary for the first 3–4 d of culture to eliminate the lag phase.

Recent experiments with *Artemisia annua* grown on plates suggest that CO_2 only slightly increases biomass yields (unpublished data). However, in the presence of gibberellic acid, the generation time was significantly decreased in shake flasks. Gibberellic acid has been shown to be a substitute for CO_2 enrichment of the root zone resulting in increases of root growth (Arteca and Dong, 1981; Zobel, 1989).

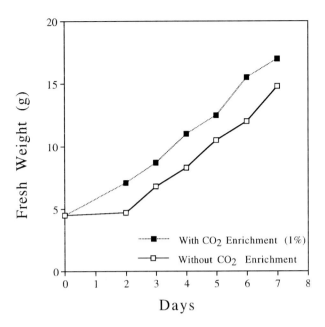

Figure 17.4. The effect of CO_2 on the growth kinetics of *Beta vulgaris* hairy roots cultured in batch in nutrient mists. Biomass measurements were as described in Figure 17.3. The inoculum was 4.5 g fresh weight in Gamborg's B5 medium with $30 \, g \, L^{-1}$ sucrose.

Growth of *A. annua* in mists supplemented with CO_2 has not yet been investigated. Studies are needed to measure the growth responses of hairy roots in aeroponics with altered concentrations of the other major metabolic gases: oxygen and ethylene.

Changes in Secondary Product Levels

Culturing plant tissues in mists instead of in a liquid can alter the profile of secondary metabolites significantly (Kilby *et al.*, 1992; Weathers and Giles, 1988). Stem nodal explants of *Cinchona ledgeriana* grown in mists showed a 400% increase in cinchonidine but a 75% decrease in cinchonine compared with liquid cultures, whereas the quinine and quinidine alkaloids were about the same (Kilby *et al.*, 1992). Yet for hairy roots of *B. vulgaris*, there was no significant difference in betanin content for cultures grown in mists ($2.8 \, mg \, g^{-1}$ dry weight) or in shake flasks ($2.9 \, mg \, g^{-1}$ dry weight) (DiIorio, 1991). These data indicate that the mode of culture can alter alkaloid profiles and should be considered when choosing a culture system.

Engineering Aspects

Mist Production and Transport to the Culture Chamber

In the nutrient mist bioreactor, an ultrasonic transducer produces the micron-size, dense mists used to feed the roots. Despite the widespread use of these transducers, formation of the aerosol from the acoustic energy transmitted by the transducer to the fluid is still not well understood. The two physical mechanisms thought to be most important are cavitation and geyser formation. A simple correlation developed by Lang (1962) relates the number mean diameter, D_n, to the liquid properties and transducer frequency, v, by:

$$D_n = 0.34 \ (8 \ \pi \ \sigma / \rho \ v^2)^{1/3} \tag{1}$$

where σ is the surface tension, and ρ is the liquid density. Unfortunately this equation does not predict the width of the size distribution or the number density of the mist, nor can it incorporate the effects of transducer power on mist production. Equation (1) appears to be valid (Lang, 1962) up to 800 kHz. Above this frequency D_n appears to depend upon the power density at the liquid surface (Tarr et al., 1991), and most transducers used for mist production usually operate near 1.5 MHz. Finally, any prediction of mist production is complicated by the fact that the amount of liquid reaching the culture chamber is really a combination of the rate at which mist is generated by the ultrasonic transducer, and the efficiency with which mist is transported in the carrier gas stream to the growth chamber. In light of these uncertainties, mist production must be characterised experimentally.

There are two complementary ways to characterise mist production. The easiest is to measure the total mass or volume of medium leaving the mist generator over some reasonable time period. We refer to this as an 'integrated' mist production method. The second way is to measure the particle size distributions directly using specialised aerosol measuring equipment. As discussed later, particle size distributions are important for understanding and predicting particle transport and deposition in dense root beds.

We conducted both integrated mist production experiments and particle size distribution measurements under a wide variety of conditions. We varied the air flow rate, Q_A, exit tube diameter, D_{tube}, exit tube length, L_{tube}, tube geometry, and transducer intensity, I, and measured the liquid flow rate, Q_L. Figure 17.5 (DiIorio, 1991) summarises results from integrated mist production experiments when D_{tube} and Q_A were varied at constant mist intensity. For the conditions investigated in Figure 17.5, Q_L increases linearly with Q_A but is not strictly proportional to Q_A, i.e. increasing Q_A by a factor of 3 does not increase Q_L by the same amount. In general, flow arrangements that increase the impaction efficiency of particles as they try to leave the generator

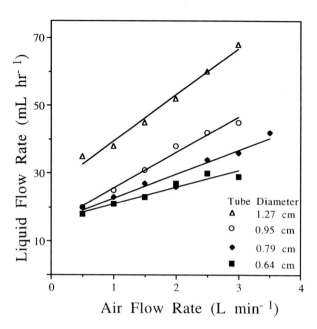

Figure 17.5. Integrated measurement of liquid flow rate as a function of air flow rate for several mist outlet tube diameters.

and enter the tube will decrease the liquid flow rate. For example, particles impact onto surfaces more readily at high velocities. For a fixed D_{tube}, an increase in Q_A leads to an increase in velocity and that, in turn, increases the impaction efficiency of the largest particles. The lack of strict proportionality between Q_L and Q_A is due, at least in part, to this phenomenon. It also explains why Q_L decreases with D_{tube} at a fixed Q_A. There is, however, also an upper limit to Q_L as a function of D_{tube}. Increasing D_{tube} must eventually decrease Q_L because the velocity of the carrier gas will drop below the settling velocity of the largest particles and they will no longer be swept out of the generator.

Geometric effects in the transport tube that will decrease the total amount of liquid entering the culture chamber include bends and coils. Particles that cannot follow the streamlines around sharp corners will impact at bends. Coils will induce secondary flows that move droplets from the centre of the stream closer to the edges where they can deposit on the wall more readily.

Aerosol size distribution measurements were made using both a Phase Doppler Particle Analyzer (PDPA, Aerometrics, Sunnyvale, CA USA) and an APS33 Aerodynamic Particle Sizer (TSI, St. Paul, USA). Figure 17.6 shows typical number and mass particle size distributions measured using the APS33. The mist number densities range from 10^4 to 10^6 cm^{-3} with $4 < D_n < 6 \ \mu$m and mass mean diameters, D_m, in the range 4–9 μm. The mists with the highest D_n

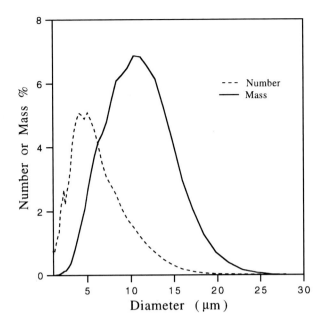

Figure 17.6. Number and mass particle size distributions measured using the TSI APS33. The mist generator was operated in a downflow configuration with a 1" I.D. tube, air flow rate at 2 L min^{-1}, mist intensity of 1, and using tap water.

and D_m, number densities and Q_L were produced at the highest power inputs to the transducer. In contrast to the results of integrated experiments, dramatic changes in Q_L with Q_A were not observed in the PDPA measurements except for the largest tube diameter ($1^5/8$"). Furthermore, Q_L derived from size resolved data using either measurement method are typically an order of magnitude lower than the integrated liquid flow rates. We believe this large difference is due to a relatively small number of very large particles that dominate the total liquid flow rates but are out of the measurement range of either instrument.

Maximising the liquid flow rate reaching the culture chamber from the mist generator is not necessarily the best strategy for successful root growth. If the roots are suddenly exposed to a high density of large particles, dense root beds may clog and both liquid and gas exchange will be hindered. Understanding how to adjust the operating parameters to meet the nutrient demand of hairy roots is far more important.

Mist Deposition in Dense Root Beds

The deposition of particles (mist) onto fibres (roots) in a flowing stream is a classic problem in filtration (Fuchs, 1964). The three physical mechanisms most responsible for deposition under typical mist bioreactor conditions are impaction, interception and Brownian diffusion. Particle settling may also contribute to deposition in some situations such as at the end of a mist-on cycle. The dominant deposition mechanism is a function of particle size, particle velocity, and the ratio of particle to fibre diameter.

Particles larger than 1 μm will deposit by a combination of interception and impaction. Interception is strictly a geometric effect where droplets are assumed to move with the gas stream lines around the root. If the droplet radius is greater than the distance from the streamline to the root, the droplet will touch the root and adhere to its surface. Impaction is a consequence of particle inertia by which droplets with sufficient momentum leave the streamlines and therefore impact onto the root. Particle diffusion to the root surface due to Brownian motion is only an effective mechanism for particles smaller than 1 μm.

In order to develop analytical expressions for deposition efficiency, the flow field around the roots must be determined. In these slow-moving mists, creeping flow near the root surface is appropriate. If the roots are far enough apart so that flow near the surface of each root is independent, a single fibre model such as the Kuwabara model can be used (Kuwabara, 1959; Bruschke and Advani, 1993; Kirsch and Fuchs, 1967) to calculate the single fibre deposition efficiency, η. The overall bed efficiency, η_{bed}, can then be derived by a simple particle balance over the length of the filter bed:

$$\eta_{bed} = 1 - N_{out}/N_{in} = 1 - \exp\left[-4\alpha\eta L/(\pi(1-\alpha)D_f)\right] \tag{2}$$

where N_{in} and N_{out} are the number of particles passing into and out of the root bed respectively, α is the packing density, L is the bed length, and D_f is the fibre diameter. Expressions for calculating the individual deposition efficiencies are readily available in the literature and will not be repeated here (see for example, Flagan and Seinfeld, 1988).

Because of the sharp transition between particle sizes that deposit and those that do not, filtration devices are most often characterised by a single parameter, D_{50}, the diameter at which 50% of the particles deposit. From particle size measurements made with and without a root bed, we are able to calculate the size dependent deposition efficiencies across different root bed depths and calculate D_{50}. As shown in Table 17.3, our measured values of D_{50} compare quite well with the analytical results based on a simple aerosol deposition model that considers diffusion, impaction and interception (Whipple, 1995).

Once particles deposit they will form a thin layer of liquid on the roots and eventually this liquid will begin to drain. Large droplets deposit most effi-

Table 17.3. Comparison of measured D_{50} values with analytical results from a simple aerosol deposition model with 50% packing density of hairy roots of *Artemisia annua*. Measurements were conducted using 5 min on\6 min off misting cycles.

Bed depth (cm)	D_{50} Analytical	Measured
1	7.3	n.d.
2	5.6	7.5
4	4.7	5.2
6	4.3	4.9
8	4.0	3.6
10	3.8	3.5

n.d., not determined. Roots were grown in shake flasks and manually packed into the tube for the deposition measurements.

ciently onto the first few centimetres of the root bed and under continuous misting conditions this can lead to blockage or plugging of the root bed. Typically, for a root bed of *Artemisia annua* at 50% packing density we observe complete blockage after about 1 hour of continuous misting. For growing roots, we avoid this problem by running an intermittent misting cycle.

Potential Pitfalls

There are a number of potential problems which must be solved before nutrient mist (or spray) reactors will be useful on a commercial scale. Initially a uniform distribution of roots must be inoculated into the reactor to maximise biomass production in the growth chamber. Uneven inoculation will lead to localised centres of growth and premature localised nutrient deprivation (Ramakrishnan and Curtis, 1994). Good distribution of inoculum will probably have to be initiated by pumping a liquid suspension of root 'pieces' into the growth chamber, allowing for attachment to some support matrix or trellis. After attachment, the medium can be drained and the mist culture environment initiated.

As the roots proceed to pack the growth chamber with biomass, concern about penetration of nutrient mists (or sprays) into dense mats of roots arises. As discussed earlier, measurements of mist (or spray) penetration into root beds exceeding 50% packing density (fresh weight) are in progress. At this time, little is known about how mist droplet size influences root morphology.

The mode of mist production and transport becomes important especially as one approaches scale-up. Current production is via ultrasonics (for mists) or nozzles (for sprays). Although the clogging and shear problems frequently encountered with nozzles are obviated with ultrasonic mists, the fluid dynamics of both the droplet phase and the gas phase must be carefully studied and controlled to ensure adequate delivery of nutrients to the root bed, especially as the root packing density increases to commercially useful levels.

Finally, until basic studies are completed, the economics for ultrasonic mist-fed reactors are unclear. It is possible that despite basic problems inherent in the operation of compressed air generated aerosols, they may be more cost effective energetically.

Acknowledgments

We would like to thank Professor Douglas P. Hart of the Massachusetts Institute of Technology for the use of the PDPA for the initial droplet deposition studies. We would also like to thank Professor Douglas B. Walcerz and Chinmay Chatterjee for their help. This work was supported in part by grants from the National Science Foundation BES-9414858 and the National Institutes of Health 1R15AI34131-01.

References

Arteca RN and CN Dong (1981). Increased photosynthetic rates following gibberellic acid treatments to the roots of tomato plants. *Photosyn Res* 2: 243–249.

Bruschke MV and SG Advani (1993). Flow of generalized Newtonian fluids across a periodic array of cylinders. *J Rheol* 37: 479–498.

DiIorio AA (1991). *Betacyanin Production and Efflux From Transformed Roots of Beta vulgaris in a Nutrient Mist Bioreactor*. PhD thesis, Worcester Polytechnic Institute, USA.

DiIorio AA, RD Cheetham and PJ Weathers (1992a). Carbon dioxide improves the growth of hairy roots cultured on solid medium and in nutrient mists. *Appl Microbiol Biotechnol* 37: 463–467.

DiIorio AA, RD Cheetham and PJ Weathers (1992b). Growth of transformed roots in a nutrient mist bioreactor: reactor performance and evaluation. *Appl Microbiol Biotechnol* 37: 457–462.

Flagan RC and JH Seinfeld (1988). *Fundamentals of Air Pollution Engineering*. Prentice Hall, New Jersey; Chapter 7.

Fuchs NA (1964). *Mechanics of Aerosols*. New York: Pergamon.

Kilby NJ, DL Griggs and SF Berry (1992). Assessment of inter- and intra-population variation in quinoline alkaloid profile of juvenile shoot cultures of *Cinchona ledgeriana*: effect of method of nutrient delivery on alkaloid profile. *Plant Cell Tiss Organ Cult* 28: 275–280.

Kirsch AA and NA Fuchs (1967). Studies on fibrous aerosol filters. *Ann Occupat Hygiene* 10: 23–30.

Kondo O, H Honda, T Taya and T Kobayashi (1989). Comparison of growth properties of carrot hairy root in various bioreactors. *Appl Microbiol Biotechnol* 32: 291–294.

Kuwabara S (1959). The forces experienced by randomly distributed parallel circular cylinders or spheres in a viscous flow at small Reynolds numbers. *J Phys Soc Japan* 14: 527–532.

Lang RJ (1962). Ultrasonic atomization of liquids. *J Acoust Soc Am* 34: 6–8.

McKelvey SA, JA Gehrig, KA Hollar and WR Curtis (1993). Growth of plant root cultures in liquid- and gas-dispersed reactor environments. *Biotechnol Prog* 9: 317–322.

Ramakrishnan D and WR Curtis (1994). Fluid dynamic studies on plant root cultures for application to bioreactor design. In *Advances in Plant Biotechnology*, edited by DDY Ryu and S Furusaki, pp. 281–305. New York: Elsevier.

Tarr MA, G Zhu and RF Browner (1991). Fundamental aerosol studies with an ultrasonic nebulizer. *Appl Spectroscopy* 45: 1424–1432.

Weathers PJ and KL Giles (1988). Regeneration of plants using nutrient mists. *In Vitro* 24: 727–732.

Weathers PJ and RD Zobel (1992). Aeroponics for the culture of organisms, tissues, and cells. *Biotechnol Adv* 10: 93–115.

Whipple M (1995). *Deposition of Nutrient Mist onto Hairy Root Cultures*. MS thesis, Worcester Polytechnic Institute, USA.

Whitney PJ (1992). Novel bioreactors for the growth of roots transformed by *Agrobacterium rhizogenes*. *Enzyme Microb Technol* 14: 13–17.

Yu S and PM Doran (1994). Oxygen requirements and mass transfer in hairy-root culture. *Biotechnol Bioeng* 44: 880–887.

Zobel RW (1989). Steady state control and investigation of root system morphology. In *Applications of Continuous and Steady-State Methods to Root Biology*, edited by JG Torrey and LJ Winthrop, pp. 165–182. Amsterdam: Kluwer.

Drip-Tube Technology for Continuous Culture of Hairy Roots with Integrated Alkaloid Extraction

18

Paul Holmes[1], Shun-Lai Li[1], Kenneth D. Green[1],
Brian V. Ford-Lloyd[2] and Neale H. Thomas[1]

[1]Fluid And Surface Transport (FAST) Team, School of Chemical Engineering, and [2]School of
Biological Sciences, University of Birmingham, Edgbaston, Birmingham, B15 2TT, UK;
email: N.H.Thomas@bham.ac.uk

Abstract

Throughout the eighties much time and money
was spent endeavouring to encourage plant cell suspension cultures to yield up valuable secondary alkaloid products in sufficient quantities as to afford
serious prospects for commercially viable processes,
sadly to little effect. In the late eighties hairy roots
appeared on the scene as potentially serious con-
tenders for viable operation, certainly at laboratory
scale but prospectively even at scales of industrial
interest. Having spent several frustratingly fruitless
years on various technology options for cell culture
(briefly recapitulated below to set the scene), we
eagerly seized on this radical alternative opportunity and have spent the past seven years endeavouring to work-up appropriate technologies to fully
exploit it. We describe the emergence and evolution
of 'drip-tube' fermenters as a moist or film-flow
environment much more suited to the fibrous morphology of hairy roots than established stirred tank
or airlift reactors. Much of our own activity has concentrated on process enhancement in bench-top (1-
L) units, especially on nutritional and environmental stressing by (for example) multi-stage medium
feed, temperature, pH and detergent, also by continuous product extraction. In round terms, we have
shown that productivity gains of up to a factor of
five or so can currently be delivered by imposing
various combinations of these process parameters,
although there remains perhaps a shortfall factor of
two or so to achieve commercial viability with higher
value alkaloids. Exploratory work with a pilot (8-L)
system has indicated no major obstacles in scale-up,
and we envisage industrial operations might be
conducted by bundling assemblages of these drip-
tubes. In sum, the prospects are rosier now than
at any time during the earlier era when ample fund-
ing was provided for cell suspension strategies but
the necessary resourcing to complete the exercise is
nowadays scarce indeed.

Introduction

It is over ten years since we first embarked on the
search for bioprocess strategies affording realistic
prospects of commercially viable alkaloid production in the laboratory. Following established wis-
dom at that time, we initially concentrated on free
cells as suspension culture in airlift units, endeavouring to secure and sustain monocell dispersions

for the perceived advantages of biological homogeneity. In common with others, we encountered and failed to overcome (Thomas and Janes, 1987a) operational problems of resistant rheology at the high biomass densities regarded as commercial prerequisite at that time.

The biological unreliability of free suspension cultures in airlift units led us to evaluate cell carriers (Snape *et al.*, 1989) as a protective microenvironment, not least against the ravages of disengager bubble-bursting (Kowalski and Thomas, 1991, 1994). As an alternative strategy to eliminate bubble hazards, we also explored (Thomas and Janes, 1987b) a rotary membrane device utilising Taylor vortices, thereby delivering not only bubble-free conditions but also eliminating the shear turbulence of stirrer blades. At that time stirrer 'shear' was widely perceived to be a main source of cell damage, although this notion seems to be less fashionable nowadays.

Whilst the rotary membrane unit held some promise as a bench-top bioreactor (Janes *et al.*, 1988), its constructional and operational complexity has probably precluded much interest among practitioners as a sensible proposition for adoption at usefully large scale. However, our interest in membrane reactors has recently been revived (Brough, 1995) in connection with utilising pulsatile streams to augment gas transfer and simultaneously control the motions of suspended carrier particles. On the other hand, our interest in airlift reactors has survived only with regard to bubbly flow transport dynamics as physical measurements (Hatton and Thomas, 1992) and computational simulations (Snape and Thomas, 1992), the latter hybridising expert system methodology for equipment specification and Monte Carlo tracking for bioparticle transit records and oxygen exposure.

In 1988 we awoke to the sea-change of emerging prospects offered by hairy roots. At the time we were working with *Nicotiana* species in suspension culture (Green, 1988), both as a convenient model for evaluating process performance and because of industrial interest in nicotine precursors for tobacco-related products (Lyons *et al.*, 1988). We quickly confirmed for ourselves (Green *et al.*, 1992a) that hairy root cultures are more rugged, reliable and repeatable than anything achieved with suspension or immobilised cell cultures. Soon afterwards we encountered (Green, 1991) the notorious problems associated with filamentous morphology in seeking to scale-up from shake flask to pilot plant, in our case using airlift units. Abandoning any serious attempt to adapt submerged culture equipment, we transferred our efforts (Green *et al.*, 1992b) to the notion of drip-feeding the nutrient stream as a 'moist' or film-flow tubular culture with oxygen supplied via upflowing gas, later incorporating product ex-

traction from the external loop of nutrient exiting from the bottom of the reactor. Subsequent refinements (Li *et al.*, 1994a) included side-porting for multiple inoculation onto mesh supports, aseptic introduction by peristaltic pumping, and drainage assurance via a perforated pipe along the axis. So far as scale-up from pilot to production is concerned, we have always felt (Green *et al.*, 1992b) that bundling of individual tubular elements would not only avoid scale-up uncertainties but also afford operational robustness insofar as individual tubes can be renewed with minimal disturbance to the operation, akin to the on-line refuelling strategy of nuclear power plants.

Far and away most of our effort since 1990 has been given to exploring options for boosting biomass and product performance by combinations of nutritional and environmental stressing factors (e.g. Green, 1991; Holmes, 1993; Li, 1994), including staging and cycling of medium composition, elicitation by temperature and toxic shock, permeabilisation by detergent, and pH switch. The most effective combinations of these factors together with continuous product extraction in a resin absorption column provided for gains in productivity of up to five fold compared with shake flask culture; that is, for nicotine and anabasine from *Nicotiana* (Green *et al.*, 1992a). More recently we have increasingly turned our attention to a wider variety of species and commercially more significant products, achieving for example (Li *et al.*, 1994b) gains of five fold in productivity of hyoscyamine from *Datura stramonium*.

Our account below begins with a perspective on conventional technologies for hairy root culture, arguing the case for drip-tube configurations with reference to illustrative examples drawn mainly but by no means exclusively from our own experience. Our account of performance evaluations, including recent studies of tropane alkaloids from *Datura stramonium*, concentrates on bench scale (1-L) work-up of process refinements including multi-stage medium feed and integrated product extraction, but we mention its implementation at pilot (8-L) scale. Our discussion offers an honest appraisal of current difficulties (e.g. inoculation) as well as painting a rosy picture of the prospects, potentially of course extending to commercial viability for higher value products. We conclude by noting that the biggest dividends in future optimisation will likely come from continuing to pay detailed attention to nutritional and environmental stressing factors.

Current Technology

The unique morphology of hairy roots makes conventional fermenters mostly unsuitable as culture vessels. Impellers in stirred tanks damage the fragile interconnected root material and induce callus formation with low productivities characteristic of undifferentiated cells (Hilton and Rhodes, 1990; Kim and Yoo, 1993; Nuutila *et al.*, 1994). Wire meshes have been used to separate roots from the impellers (Hilton and Rhodes, 1990; Uozumi *et al.*, 1991; Kinooka *et al.*, 1992; Kim and Yoo, 1993; Muranaka *et al.*, 1993; Nuutila *et al.*, 1994) but biomass production can still be inhibited compared with unstirred vessels (Kim and Yoo, 1993; Nuutila *et al.*, 1994), apparently due to high shear stresses induced by the stirrer.

Air sparged reactors have also been used with varying degrees of success (Rodríguez-Mendiola *et al.*, 1991; Shimomura *et al.*, 1991; Muranaka *et al.*, 1992, 1993; Kim and Yoo, 1993; Yoshikawa *et al.*, 1993; Nuutila *et al.*, 1994), though we have found that with airlift fermenters the roots float at low biomass densities and entrap gas at higher densities (Green, 1991). Similar difficulties have been reported elsewhere (Muranaka *et al.*, 1993), where inclusion of polyurethane foam in the vessel helped reduce this problem and improved biomass and alkaloid productivity by approximately 50% and 100%, respectively. Reactors in which sparged air is kept separate from the roots, either by using separate tanks or a partitioned tank, have been developed and recorded in the patent literature (Masao *et al.*, 1990). Agitation is provided by two pumps in the former and by sparged air in the latter. Though both of these systems can produce equivalent dissolved oxygen concentrations (40 ppm), the two-tank, pump-agitated system proved more suitable for root culture as it provided better biomass and alkaloid accumulation (20 compared to 16 g L^{-1} dry weight, and 1.3 compared to 0.8 wt%, respectively).

Drip-Tube Technology

An alternative approach has been to move away from submerged culture towards moist culture systems. Here two main strategies have evolved: film-flow (as in the drip-tube described below) and nutrient mist (DiIorio *et al.*, 1992). With nutrient mist fermenters, sonication of the medium induces a mist which is then introduced into the fermenter. Though this method allows for more homogeneous distribution of media, the drip-tube's simpler design not only has advantages of simplicity and low power demand, but also avoids excessive heat generation so there is no need for cooling to suppress heat-induced impairment of medium composition. Our approach (Green, 1991; Green *et al.*, 1992b) has been to adopt a trickling film strategy (Figure 18.1) whereby liquid medium is dripped into the fermenter and then returned to a medium reservoir for recirculation. Aeration is by counter-current flow, using filter sterilised, sparged saturated air to avoid evaporative drying of the film.

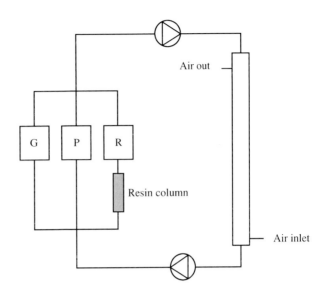

Figure 18.1. Schematic diagram of the 'FAST' drip-tube (adapted from Li *et al.*, 1994a) with medium reservoirs G, P and R for growth, production and release.

Our experimental fermenters (1-L bench-top and 8-L pilot; Figure 18.2) are of glass construction (500 × 50 mm and 1453 × 85 mm, respectively) having 5–7 inoculation ports distributed uniformly to encourage uniform growth of the inoculum throughout the tube. Root inoculum introduced through these ports is supported on stainless steel mesh trays located just below each port. The dripped medium is fed by peristaltic pumps and returned via a reservoir. Extracellular product extraction is achieved in a packed resin column located between the tube outlet and the medium reservoir (see Figure 18.1). Product extraction and concentration is a crucial consideration as downstream processing costs have been estimated at 50–90% of total production costs (Dwyer, 1984). Overall, the drip-tube's simple design is intended not only to be appropriate for the biological needs of roots, but is also a recognition of the overriding constraint of cost competitiveness for commercial operation.

Performance Evaluation

Materials and Methods

In our work we have considered tropane alkaloids as products of interest. Tropane alkaloids are pharmaceutically active (anticholinergic) and are used as antispasmodics (Hashimoto and Yamada, 1983) for treatment of nerve agent poisoning. They are synthesised at high levels in hairy root cultures compared to only very low levels in cell and callus cultures, a phenomenon attributed to the differentiated

Figure 18.2. (a) 1-L prototype 44 d after inoculation (5 sites). The dark areas correspond to the sites of inoculation of *Datura stramonium*. (b) *Nicotiana glauca* and (c) *Datura stramonium* hairy roots cultured in the 8-L drip tube (18 and 10 d post inoculation, respectively) showing the lack of geotropism exhibited by hairy roots and also the suitability of the drip tube as a bioreactor for root morphology.

morphology of root tissue (Hashimoto and Yamada, 1983; Yamada and Hashimoto, 1988). Another advantage hairy roots have over cell suspension cultures is lack of somaclonal variation, which gives root cultures stability in terms of both growth and productivity for periods in excess of five years (Maldonado-Mendoza *et al.*, 1993).

We used LBA 9402 transformed *Datura stramonium* root line C28 for the work described below (Li, 1994). This clone was selected after being screened for both high growth rate (doubling time: 2.9 d) and productivity (600 μg hyoscyamine per gram fresh weight, intracellular). Culture conditions were initially optimised in shake flask cultures (50 mL medium in 250 mL flasks) at 85 rpm and 28°C in the dark using Gamborg's B5 growth medium with 3% sucrose. For production, we switched to double strength White's medium supplemented with 5% sucrose and 0.3 M mannitol. To promote product release, a further switch was made after 10 d to permeabilisation medium ($^1/_2$ B5 supplemented with 100 ppm Triton-X 100, pH 5). Production and permeabilisation media can be cycled repeatedly in a ratio of 10 d to 2 h over a total of seven cycles before the roots lose productivity. Alkaloids were quantified by HPLC using the method described by Christen *et al.* (1989).

Bench-Top Reactor

The 1-L drip-tube was inoculated at five points along the tube's length with 5 cm root tips from 10 d old shake flask cultures. Excellent growth followed a lag phase of approximately 3 d. Depletion of sucrose, generation of fructose and glucose and pH change were monitored as shown in Figure 18.3. Taking sucrose depletion as a sensitive reflection of

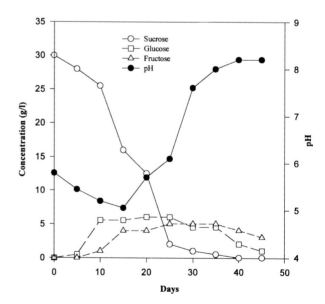

Figure 18.3. Typical sucrose, glucose, fructose and pH levels during culture of hairy roots in a 1-L drip-tube reactor (Li, 1994).

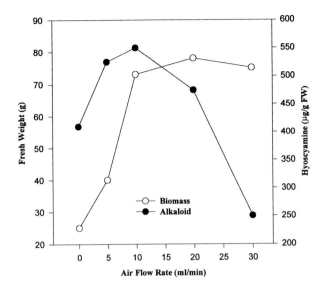

Figure 18.4. Effect of air flow rate on biomass and alkaloid production (Li *et al.*, 1994a) in the 1-L drip-tube reactor.

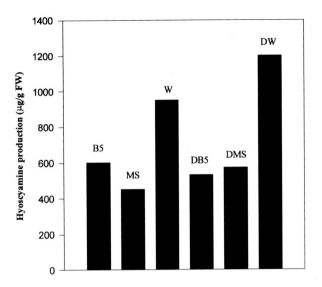

Figure 18.5. Effect of medium type on hyoscyamine production in shake flask culture (Li, 1994). Growth media: B5 = Gamborg's B5; MS = Murashige and Skoog; W = White's; D = double strength medium; all with 3% (w/v) sucrose.

root growth (Snape *et al.*, 1989; Green, 1991), we assessed the biomass doubling time to be similar to shake flask grown cultures, that is, approximately 3 d. pH shift also mirrored that seen in shake flasks, falling from 5.8 to 5 over a 15 d period then rising to 7.8 as the roots reached stationary phase (Day 30). At Day 45 the roots superficially occupied approximately 80% of the fermenter volume and were harvested giving a yield of 96 g fresh weight (growth index: 788). These results confirmed previous work showing excellent growth properties of *Nicotiana glauca* roots in the 1-L drip-tube (Green, 1991).

It has been reported (DiIorio *et al.*, 1992) that nutrient mist cultured *Beta vulgaris* failed to perform (in terms of biomass production) as well as in shake flask controls. However, with our 1-L drip-tube system, *Datura stramonium* hairy roots produced almost identical growth rates to shake flask cultures (Li, 1994), and at 8 L the growth rate was only slightly lower (doubling time 3.2 compared to 2.9 d).

Air flow rates of approximately 10 mL min^{-1} were required to achieve good growth and productivity (Figure 18.4). Above 10 mL min^{-1} growth rate was largely independent of the air supply but below 5 mL min^{-1} growth rates were significantly reduced. Hyoscyamine production was also greatest at 10 mL min^{-1} but dropped off rapidly at air flow rates in excess of 20 mL min^{-1}. Similar to shake flask cultures (Snape *et al.*, 1989), oxygen starvation during stationary phase can be used to promote product release, up to 5% of total hyoscyamine for 10 h starvation, which had little effect on root via-

bility measured in terms of regrowth on subculture. However, 16 h O$_2$ starvation increased the lag phase and doubling times by $^1/_2$ and $^1/_4$, respectively. pH shift from 8 to 5 increased the percentage of total hyoscyamine released, by about 18% compared with 24% in shake flasks.

Process Refinements

Multi-Stage Media

Using shake flask cultures we have demonstrated (Li, 1994; Figure 18.5) the benefit of switching from B5 medium to double strength White's medium (DW). Here hyoscyamine production was increased from 600 to 1200 µg g^{-1} fresh weight. A supplement of 5% sucrose and 0.3 M mannitol to the production medium may deliver comparable enhancement (Li, 1994). With roots grown in the 1-L fermenter on B5 medium for 26 d then switched to DW, 118.2 g fresh weight and 69 mg of hyoscyamine (580 µg g^{-1} fresh weight) were produced after 40 d. However this result does not match the two-fold increase obtained from shake flask experiments with and without medium switch.

To encourage hyoscyamine release we introduced a third medium, $^1/_2$ B5 supplemented with 100 ppm Triton X-100 ($^1/_2$ B5). In shake flask cultures these conditions produced 63% release of hyoscyamine over a 2 h period with only a small decrease in viability (lag phase increased by less than 1 d, total

Table 18.1. Enhancements in productivity during the development of three-stage drip-tube culture at the 1-L scale (Li, 1994).

System	Time (d)	Fresh weight (g)	Hyoscyamine (mg)			
			Medium	Resin	Roots	Total
One stage	35	124	0	–	52	52
Two stages	35	118	0	–	69	69
Three stages	45	130	47	–	63	110
Three stages with resin	50	135	33	174	11	218

biomass produced over a 30-d period was reduced by less than 20%). DW and $^1/_2$ B5 were cycled three times in the ratio of 10 d to 2 h and, by the end of 55 d cultivation, a total of 110 mg of hyoscyamine was produced of which 43% was extracellular.

Integrated Product Extraction

Using the same growth, production and release conditions as above but now incorporating an amberlite XAD-7 column directly between the fermenter outlet and medium reservoirs (the column only being switched in line during the release phase; see Figure 18.1), we were able to conveniently localise, concentrate and recover alkaloids over a 7-week culture period. A total of 218 mg (1.61 mg g^{-1} fresh weight; Table 18.1) of hyoscyamine was produced, with 15, 5 and 80% recovered from roots, medium and resin, respectively. Compared to single-stage culture, evolution of drip-tube technology through two stages, three stages, and three stages with recovery has given increased hyoscyamine recovery by factors of 1.35, 2.2 and 4.3, respectively.

Our results are similar to other published data by Muranaka *et al.* (1993) who found that integration of an XAD-2 column into an external loop of an air-sparged reactor (2 L) enhanced five-fold the alkaloid production from scopolamine-secreting hairy roots of *Duboisia leichhardtii*. We suggest that their slightly better enhancement in performance (5 compared to our 4.3) is because the scopolamine was naturally secreted whereas for our system the eliciting chemicals are inevitably deleterious to the well-being of the roots.

Pilot Scale Assessment

So far we have only evaluated single-stage (growth) cultures at 8 L. The saccharide depletion rate was similar to observations in shake flasks, though the doubling time was slightly longer compared to 1-L and shake flask cultures (3.25 compared to 3 d and 2.85 d, respectively). Ultimate biomass also fell short of an 8-fold gain (800 g fresh weight was expected

and 546 g fresh weight was achieved), a shortfall attributed to the lower fractional volume of inoculum which will be remedied by incorporating a larger number of mesh supports. In addition, specific alkaloid productivity was reduced in the 8-L compared to the 1-L drip-tube (372 and 460 µg g^{-1} fresh weight, respectively), again attributed (Li, 1994) to the smaller fractional volume of inoculum and the greater age of the roots at stationary phase.

Discussion

Bench-top scale (1-L) drip-tube cultures exhibit virtually the same growth characteristics as shake flask cultures. These rates are only slightly reduced in the 8-L pilot fermenter, indicating good prospects for satisfactory scale-up of this technology. We envisage that the existing simple design and construction can be retained for tube volumes up to about 100 L with further upsizing achieved by bundling of the tubes (Green, 1991; Green *et al.*, 1992b). Compared with capital costs for other reactors, the drip-tube represents a substantial cost saving (only $^1/_4$ or less the cost of a 50 m^3 stirred tank; Holmes *et al.*, unpublished). At substantially larger scales than envisaged here (more than 100 m^3, say), conventional fermenters would be cheaper (Drapeau *et al.*, 1986), but then their unsuitability for hairy root morphology would count against them.

Other than medium initial pH, we have not considered it important to control pH during the growth and production phases of hairy root culture. Indeed, it seems the roots themselves compensate for unfavourable imposed pHs, achieving this adjustment in less than 3 d or so (Holmes *et al.*, unpublished data). Only for release does medium pH become important, and here the roots are exposed for no more than 2 h or so (Li *et al.*, 1994b). Sugar concentrations are routinely monitored as an indicator of biomass production and, to the extent that medium switching from growth to production is introduced on this basis, they represent a control on the process.

Optimisation of the inoculation strategy remains an uncompleted task and is almost certainly responsible for the existing shortfall in performance seen at the 8-L scale. We have evaluated several methods, including pumping short sections of roots or root segments bundled in agar, also manual introduction. Pumping short sections is unattractive due to the extended lag phase. Preparing and pumping bundles or manual inoculation is impracticable for other than very small-scale operations. For example, a 1 m^3 culture conducted in 10×100 L tubes would require approximately 10 kg of inoculum to be introduced via at least 50 inoculation ports whilst still maintaining sterile conditions. We recently re-designed our drip-tube to incorporate wider diameter ports (35 and 45 mm, respectively, for 1- and 8-L tubes) and this has considerably eased the inoculation process.

External resin column extraction has delivered in excess of four-fold enhanced alkaloid production. However the present target product (with the clone used here) was intracellular, and the release promoters we introduced certainly did not favour root well-being whilst cycled production and permeabilisation conditions reduced the overall toxic shock loading on the roots. Clearly a better strategy would be to screen for extracellularly secreting root lines. We are presently addressing this option.

Conclusions

Drip-tube reactor technology favours growth and productivity as compared with the known short-comings of conventional units like stirred tanks and airlift devices. The system we have developed is simple in design and construction with prospect of low-cost scale-up achieved by bundling of tubes. Inoculation remains an operational challenge for larger tube units, although with our tube-bundling strategy there is the prospect of, say, daily renewal of one tube in 100, each with a useful lifetime of 100 d. Supposing 100 L is the maximum practicable tube size, then this strategy would provide for an effective 10 m^3 working volume maintained indefinitely with a daily demand of about 1 kg of inoculum. Such a nearly continuous operation would have substantial advantages over a single large tank process, not least for its minimisation of commercial exposure both as set-up risk and as ongoing risk of batch failure. Specifically, then, the full-scale plant can be verified for each 100-L tube on its own and the loss of a few 100-L tubes represents only a few percent impact on the total operation. We look forward to reporting the success of the strategy in less than the 10 years we have spent so far pursuing the goal of commercially viable alkaloid production in bioreactors.

References

Brough M (1995). *Augmentation of a Novel Membrane Reactor at High Biomass Loadings*. MSc thesis, University of Birmingham, UK.

Christen P, MF Roberts, JD Phillipson and WC Evans (1989). High-yield production of tropane alkaloids by hairy root cultures of a *Datura candida* hybrid. *Plant Cell Rep* 8: 75–78.

DiIorio AA, RD Cheetham and PJ Weathers (1992). Growth of transformed roots in a nutrient mist bioreactor: reactor performance and evaluation. *Appl Microbiol Biotechnol* 37: 457–462.

Drapeau D, HW Blanch and CR Wilke (1986). Economic assessment of plant cell culture for the production of ajmalicine. *Biotechnol Bioeng* 30: 946–953.

Dwyer JL (1984). Scaling up of bioproduct separation with HPLC. *Bio/Technol.* 2: 975–984.

Green KD (1988). *Plant Cell/Root Culture: Evaluation of Plant Cell and Root Tissue Culture With Reference to Physiological and Metabolic Aspects*. MSc thesis, University of Birmingham, UK.

Green KD (1991). *Diagnostic Evaluations for Improved Plant Cell and Root Tissue Culture*. PhD thesis, University of Birmingham, UK.

Green KD, NH Thomas and JA Callow (1992a). Product enhancement and recovery from transformed root cultures of *Nicotiana glauca*. *Biotechnol Bioeng* 39: 195–202.

Green KD, NH Thomas and JA Callow (1992b). Development and optimisation of a novel large scale root-tube bioreactor-separator. *Proc IChemE Research Event*. IChemE, Rugby, England, pp. 278–280.

Hashimoto T and Y Yamada (1983). Scopolamine production in suspension cultures and redifferentiated roots of *Hyoscyamus niger*. *Planta Med* 47: 195–199.

Hatton P and NH Thomas (1992). Experimental facility for local dynamics of bubbly flow in airlift bioreactors. *Proc IChemE Research Event*. IChemE, Rugby, England, pp. 368–370.

Hilton MG and MJC Rhodes (1990). Growth and hyoscyamine production of 'hairy root' cultures of *Datura stramonium* in a modified stirred tank reactor. *Appl Microbiol Biotechnol* 33: 132–138.

Holmes P (1993). *Feasibility Evaluation and Process Implementation of Plant Hairy Roots: Prospect for Fungal Elicitation*. MSc thesis, University of Birmingham, UK.

Janes DA, NH Thomas and JA Callow (1988). Red beet batch culture demonstration of a bubble-free Taylor-Couette bioreactor. In *Manipulating Secondary Metabolism in Culture*, edited by RJ Robins and MJC Rhodes, pp. 257–262. Cambridge: Cambridge University Press.

Kim YH and YJ Yoo (1993). Development of a bioreactor for high density culture of hairy roots. *Biotechnol Lett* 7: 859–862.

Kino-oka M, Y Hongo, M Taya and S Tone (1992). Culture of red beet hairy root in bioreactor and recovery of pigment released from the cells by repeated treatment of oxygen starvation. *J Chem Eng Japan* 25: 490–495.

Kowalski AJ and NH Thomas (1991). Bursting bubbles and cell damage in bioreactors. *IChemE Research Event* IChemE, Rugby, England, pp. 195–196.

Kowalski AJ and NH Thomas (1994). Bubbles burst the myth of shear damage in cell suspension bioreactors. *Proc 2nd Conf Advances in Biochemical Engineering*. IChemE, Rugby, England, pp. 82–84.

Li S-L (1994). *Plant Hairy Roots: Environmental Factors for Enhanced Recovery of Secondary Metabolite Products*. PhD thesis, University of Birmingham, UK.

Li S-L, KD Green, NH Thomas and BV Ford-Lloyd (1994a). Ongoing optimisation of a drip-tube bioreactor with integrated product recovery from transformed root cultures. *Proc 2nd Conf Advances in Biochemical Engineering*. IChemE, Rugby, England, pp. 118–120.

Li S-L, NH Thomas and BV Ford-Lloyd (1994b). Recovery of alkaloid released from hairy roots of *Datura stramonium* by repeated treatment with pH gradient and chemical agents. *Proc 2nd Conf Advances in Biochemical Engineering*. IChemE, Rugby, England, pp. 136–138.

Lyons I, NH Thomas and JA Callow (1988). The effect of N-methyl putrescene on growth and alkaloid production by transformed root culture of *Nicotiana glauca*. Poster, PSE/SEB Internat. Symp, Kings College, University of London, UK.

Maldonado-Mendoza IE, T Ayora-Talavera and VM Loyola-Vargas (1993). Establishment of hairy root cultures of *Datura stramonium*. *Plant Cell Tiss Organ Cult* 33: 321–329.

Masao S, Y Yukimune and D Hiroshi (1990). Method for culturing plant tissue, apparatus therefor and method for producing metabolite. European Patent 0 387 065.

Muranaka T, H Ohkawa and Y Yamada (1992). Scopolamine release into media by *Duboisia leichhardtii* hairy root clones. *Appl Microbiol Biotechnol* 37: 554–559.

Muranaka T, H Ohkawa and Y Yamada (1993). Continuous production of scopolamine by a batch culture of *Duboisia leichhardtii* hairy root clone in a bioreactor system. *Appl Microbiol Biotechnol* 40: 219–223.

Nuutila AM, L Toivonen and V Kauppinen (1994). Bioreactor studies of *Catharanthus roseus*: comparison of three bioreactor types. *Biotechnol Lett* 8: 61–66.

Rodríguez-Mendiola MA, A Stafford, R Cresswell and C Arias-Castro (1991). Bioreactors for growth of plant roots. *Enzyme Microb Technol* 13: 697–702.

Shimomura K, H Sudo, H Saga and H Kamada (1991). Shikonin production and secretion by hairy root cultures of *Lithospermum erythrorhizon*. *Plant Cell Rep* 10: 282–285.

Snape JB and NH Thomas (1992). Modelling particle transport by bubbles for performance guidelines in airlift fermenters. *Biotechnol Bioeng* 40: 337–345.

Snape JB, NH Thomas and JA Callow (1989). How suspension cultures of *Catharanthus roseus* respond to oxygen limitation: small scale tests with applications to large-scale cultures. *Biotechnol Bioeng* 34: 1058–1062.

Thomas NH and DA Janes (1987a). Fluid dynamic considerations in airlift bioreactors. *Ann NY Acad Sci* 506: 171–189.

Thomas NH and DA Janes (1987b). Fluid dynamic considerations in airlift and annular vortex bioreactors for plant cell culture. In *Biotechnology Processes*, edited by CSH Oldshue and JY Oldshue, pp. 60–71. New York: AIChE.

Uozumi N, K Kohketsu, O Kondo, H Honda and T Kobayashi (1991). Fed-batch culture of hairy root using fructose as a carbon source. *J Ferment Bioeng* 72: 457–460.

Yamada Y and T Hashimoto (1988). Biosynthesis of tropane alkaloids. In *Applications of Cell and Tissue Culture*, edited by G Bock and J Marsh, pp. 199–212. New York: Wiley.

Yoshikawa T, Y Asada and T Furuya (1993). Continuous production of glycosides by a reactor using ginseng hairy root culture. *Appl Microbiol Biotechnol* 39: 460–464.

Application of Fungal Elicitation for Enhancing Production of Secondary Metabolites by Hairy Roots Cultured in Large-Scale Bioreactors

19

Gurmeet Singh

Department of Chemical Engineering, The Pennsylvania State University, 133 Fenske Laboratory, University Park, PA 16802, USA; email: GXS11@PSUVM.PSU.EDU

Introduction

"Hairy roots" possess the biosynthetic capabilities of differentiated tissue (Signs and Flores, 1990; Hamill *et al.*, 1987; Doran, 1989) and growth rates comparable to plant cell suspension cultures (Flores *et al.*, 1988). This unique combination offers limitless potential for production of chemicals in the controlled environment of a bioreactor. However, low levels of production in plant tissue have prevented use of this technology on a commercial scale.

Various procedures to enhance production of secondary metabolites in plant tissue are being explored. Previous research has identified strain selection, medium optimisation and the imposition of nutrient limitations as techniques which can influence the production of secondary metabolites from plant tissue culture (Zenk *et al.*, 1977; Tabata *et al.*, 1978; Heinstein, 1985; Scragg *et al.*, 1988; Takahasi and Fujita, 1991). These methods are based on a random approach and are very labour intensive. Chang and Sim (1994) suggest implementation of methods such as *in situ* extraction, elicitation and manipulation of signal transduction which are based on enzymatic control and have potential for commercial application. This article discusses the potential of elicitation for product enhancement and its application to large-scale bioreactors.

Two features that make elicitation particularly suitable for commercial production of chemicals from plant tissue culture are its ability to induce *de novo* synthesis, and its ability to act synergistically with other techniques of production enhancement. Induction of *de novo* synthesis of secondary metabolites is the only method available for producing many commercially important chemicals from tissue culture (Ingham, 1982; Singh *et al.*, 1994). Application of elicitation in conjunction with other traditional and novel techniques of production enhancement such as membrane permeabilisation (Chang and Sim, 1994), *in situ* extraction (Corry *et al.*, 1993) and nutrient limitation (Dunlop and Curtis, 1991) has led to a manifold increase in secondary metabolite production at the laboratory scale. This has spurred interest in elicitation.

Elicitation

In plants, elicitation refers to the synthesis of small molecular weight secondary metabolites called phytoalexins in response to physical, chemical, microbial or fungal stress. It is also possible to elicit plant tissue in culture. This is accomplished by contacting the tissue with substances known as elicitors. Elicitors alter the normal metabolism of the plant tissue and induce synthesis of enzymes that catalyse reactions in the defence-related pathways leading to secondary metabolites. At first, the word elicitor was restricted to crude extracts obtained from biological sources (such as crude fungal cell wall extracts), but its definition has been broadened in recent years to include all molecules (biotic or abiotic) and effects (heat, UV light, etc.) which stimulate secondary metabolite production in plants (DiCosmo and Misawa, 1985; Ebel, 1986). A significant amount of work has been done in identifying the structures of elicitor molecules in crude fungal extracts. The work of Ayers *et al.* (1976a, 1976b), Sharp *et al.* (1984), Cheong *et al.* (1991), Frey *et al.* (1993), Hahn *et al.* (1993) and many others has led to characterisation of the oligosaccharide elicitors of soybean cell suspensions obtained from crude fungal extracts of *Phytophthera megasperma*. Other biotic elicitor molecules, such as the endogenous elicitor molecule for *Lithospermum erythrorhizon* cell suspension cultures, have been determined to be oligogalacturonides (Tani *et al.*, 1992). Fragments derived from oligochitins and oligochitosans have been shown to elicit phytoalexins in suspension cultures of parsley (Conrath *et al.*, 1989), tomato (Grosskopf *et al.* 1991) and *Catharanthus roseus* (Kauss *et al.*, 1989).

It has been proposed that elicitor recognition involves the existence of cell surface receptors for specific elicitors (Darvill and Albersheim, 1984; Ebel, 1986). Elicitors initiate their physiological activity by interacting with receptors. The presence of receptors on the plant cell surface has been examined by direct binding assays using labelled ligands and isolated cell membranes (Yoshikawa *et al.*, 1983). Reports on the existence and characterisation of high-affinity binding sites for branched 1,3-1,6-β-glucans from *Phytophthora megasperma* cell walls provided the initial evidence that elicitor recognition may be a receptor mediated process (Schmidt and Ebel, 1987). Research continues to fully characterise the binding sites for elicitor molecules (Cosio *et al.*, 1988, 1990).

Elicitation of Hairy Root Cultures

Most of the initial work on elicitation was conducted on plant cell suspension cultures. More recently, elicitation has been shown to be successful for secondary metabolite induction in hairy root cultures (Table 19.1). Flores and co-workers (1988) reported that hairy roots of *Bidens sulphureus* responded to elicitation with fungal culture filtrates by increasing the production of specific polyacetylenes. Mukundan and Hjortso (1990) studied the effect of fungal elicitor obtained from mycelial extracts of *Fusarium con-*

glutinans on thiophene production in hairy root cultures of *Tagetes patula*. These studies were conducted by contacting root tissue with varying concentrations of elicitor (0–0.3 mg carbohydrate per mL of medium). The authors concluded that thiophene production increased with the addition of elicitors and suggested a need for identifying optimal conditions for elicitor treatment. Robbins *et al.* (1991) studied the production of isoflavan phytoalexins upon elicitation of *Lotus corniculatus* hairy root cultures with glutathione. A 10 mM solution of glutathione resulted in production of 160 μg of isoflavonoids per gram fresh weight of tissue, whereas the controls produced only trace amounts. Flores and Curtis (1992) reported the *de novo* production of sesquiterpenes from hairy root cultures of *Hyoscyamus muticus* upon elicitation with mycelial extracts of *Rhizoctonia solani*. Addition of 0.5 mL of crude elicitor to Erlenmeyer flasks containing 10 g fresh weight of root tissue resulted in production of 3–4 mg of solavetivone per litre of medium. Buitelaar *et al.* (1993) worked with elicitor obtained from *Aspergillus niger* for eliciting hairy root cultures of *Tagetes patula*. Elicitation of 8-d-old tissue with 0.2 mL of crude mycelial extract resulted in a 4-fold increase in thiophene production over that of the controls. Sim *et al.* (1994) studied the synergistic affect of fungal elicitation, membrane permeabilisation and *in situ* adsorption on production of indole alkaloids by hairy root cultures of *Catharanthus roseus*. They observed a 2–3.5 times increase in production of ajmalicine and catharanthine by combining these techniques. Whitehead and Threlfall (1992) report elicitation of thorn apple and jimson weed hairy root cultures with abiotic elicitors. These root cultures belong to the *Solanaceae* family (along with potato, tomato, and *Hyoscyamus muticus*) and produce lubimin and rishitin upon elicitation. The thorn apple was elicited with cadmium chloride and resulted in production of 140 nmol of lubimin per gram fresh weight, 24 h after contact. Elicitation of jimson weed with copper sulfate resulted in production of traces of rishitin. The authors argue strongly for continued research investment in elicitation due to its potential biotechnological value.

Application of Elicitation at a Large Scale

Production of phytoalexins is a function of the concentration of elicitor applications. The amount of phytoalexins produced increases with increase in the concentration of elicitor applications and eventually reaches a maximum value. Increasing the application concentrations of elicitor beyond this level does not result in a further enhancement in the production of phytoalexins. Use of elicitation on a large scale requires the ability to predict the optimal

Table 19.1. Enhancement of secondary metabolite production by elicitation of plant tissue in culture.

Product	Species	Elicitor	Elicitor concentration	Product concentration in control	Product concentration after elicitation	Reference
Ajmalicine	*Catharanthus roseus* (cells)	*Pythium* sp. (crude)	> 1 mL	0	400 µg L^{-1}	DiCosmo et al. (1987)
Anthraquinones	*Morinda citrifolia* (cells)	Chitin 50 (pure)	100–250 µg mL^{-1}	3 µg g^{-1} fresh weight	7 µg g^{-1} fresh weight	Dornenburg and Knorr (1994)
Alkaloids (indole)	*Catharanthus roseus* (cells)	*Pythium aphanidermatum* (crude)	10 mL per 50 mL culture	50 µmol L^{-1}	75 µmol L^{-1}	Moreno et al. (1993)
Alkaloids (tropane)	*Datura stramonium* (cells)	*Phytophthora megasperma* (crude)	55 mg L^{-1} culture	0.85 mg g^{-1} dry weight	4.27 mg g^{-1} dry weight	Ballica et al. (1993)
Berberine	*Thalictrum rugosum* (cells)	*Saccharomyces cerevisiae* (crude)	0.2 mg g^{-1} fresh weight	0.5% of dry weight	2% of dry weight	Funk et al. (1987)
Capsidiol	*Capsicum annuum* (cells)	*Trichoderma viride* (crude)	20 µg mL^{-1}	0	1 mg per flask	Brooks et al. (1986)
Capsidiol	*Nicotiana tabacum* (cells)	*Phytophthora cryptogea* (crude)	0.15 µg cryptogein mL^{-1}	0	25 µg mL^{-1}	Milat et al. (1991)
Diosgenin	*Dioscorea deltoidea* (cells)	*Rhizopus arrhizus* (crude)	2 mL	134 mg L^{-1}	230 mg L^{-1}	Rokem et al. (1984)
Dopamine	*Sanguinaria canadensis* (cells)	*Verticillium dahliae* (cells)	1 mL per 15 mL culture	3 mg g^{-1} fresh weight	15 mg g^{-1} fresh weight	Cline et al. (1993)
Kinobeon A	*Carthamus tinctorius* (cells)	Blue green algae (crude)	5 mg dry weight per 30 mL	0.6 mg L^{-1}	5.78 mg L^{-1}	Hanagata et al. (1994)
Sanguinarine	*Eschscholtzia californica* (cells)	Yeast extract (crude)	100 µg g^{-1} fresh weight	20 mg L^{-1}	60 mg L^{-1}	Byun and Pedersen (1994)
Sanguinarine	*Sanguinaria canadensis* (cells)	*Verticillium dahliae* (crude)	1 mL per 15 mL culture	3 µg g^{-1} fresh weight	12 µg g^{-1} fresh weight	Cline et al. (1993)
Sanguinarine	*Papaver bracteatum* (cells)	*Dendryphion* (crude)	1.5 mL	50 µg g^{-1} fresh weight	450 µg g^{-1} fresh weight	Cline and Coscia (1988)
Shikonin	*Lithospermum erythrorhizon* (cells)	Endogenous (crude)	1 mg mL^{-1}	0	28 µg per 10 mL	Fukui et al. (1990)

Table 19.1. Continued.

Product	Species	Elicitor	Elicitor concentration	Product concentration in control	Product concentration after elicitation	Reference
Alkaloids (indole)	*Catharanthus roseus* (hairy roots)	*Penicillium* sp. (crude)	0.01 g L^{-1} dry weight	3 mg g^{-1} dry weight	9 mg g^{-1} dry weight	Sim *et al.* (1994)
Isoflavonoids	*Lotus corniculatus* (hairy roots)	Glutathione (abiotic)	10 mM	0	160 µg g^{-1} fresh weight	Robbins *et al.* (1991)
Rishitin	Jimson weed (hairy roots)	Copper sulfate	NA	0	traces	Whitehead and Threlfall (1992)
Sesquiterpenes	*Nicotiana tabacum* (hairy roots)	Yeast extract	4 mg hexose equivalents per 10 mL	1 µg g^{-1} fresh weight	87 µg g^{-1} fresh weight	Wibberley *et al.* (1994)
Sesquiterpenes	*Hyoscyamus muticus* (hairy roots)	*Rhizoctonia solani* (crude)	100 µg glucose equivalents mL^{-1}	0	1 mg per 10 g fresh weight	Singh (1995)
Sesquiterpenes	Thorn apple (hairy roots)	Cadmium chloride	NA	0	140 nmol g^{-1}	Whitehead and Threlfall (1992)
Thiophene	*Tagetes patula* (hairy roots)	*Fusarium conglutinans* (crude)	0.2 mg per mL medium	0.2 g per 100 g dry weight	0.55 g per 100 g dry weight	Mukundan and Hjortso (1990)
Thiophene	*Tagetes patula* (hairy roots)	*Aspergillus niger* (crude)	0.2 mg per 20 mL medium	1.5 µmol g^{-1} dry weight	3.5 µmol g^{-1} dry weight	Buitelaar *et al.* (1993)

NA: not available.

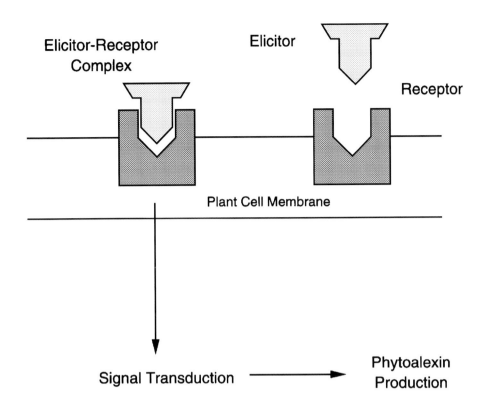

Figure 19.1. Schematic of the elicitor–receptor interaction. Elicitor molecules bind to specific receptor sites on the plant cell membrane. The binding triggers a signal transduction pathway which eventually results in a defence response, such as phytoalexin production.

elicitor concentration which will result in maximum production. Eliciting at sub-optimal concentrations of elicitor applications will under-utilise the production capacity of the hairy root bed. On the other hand, indiscriminate use of very high concentrations could result in excessive use of elicitor. Since the cost of crude fungal elicitors can be comparable to the operating cost of the plant bioreactor itself, the cost of the excessive elicitor used could offset the advantage gained by the enhancement in production of phytoalexins. Besides adding to the overall cost of the process, excess elicitor may also stress the tissue, resulting in lower production (Byun and Pedersen, 1994). Optimal use of elicitor will bring down production costs and make production of chemicals from plant tissue culture more competitive with other technologies.

Though the potential of elicitation and its applicability to enhancement of secondary metabolite production from plant tissue culture has been well studied, most of the work has been restricted to proving the ability of the technique to induce phytoalexins in various species. Little work has been done to obtain a quantitative understanding of the process and to apply the principle to secondary metabolite production in bioreactors. One reason for this is the limited knowledge about elicitation. Fortunately, physiological interpretation of elicitation is not a prerequisite to application of this technique to large-scale chemical production. All that is needed is a

physical interpretation which will permit calculation of the necessary amount of elicitor given the operating conditions within the bioreactor. This issue is considered in the subsequent sections.

Elicitor Quantification

To optimise elicitor dosage the elicitor must be quantified. Elicitor dosage can be reported in concentration terms for defined elicitor molecules. However, most of the elicitors in use are crude fungal extracts and chemical identity of the molecule responsible for triggering the elicitation response is not available. Representing such crude preparations with meaningful concentrations is a challenging problem. A wide variety of nomenclature has been used in the literature to specify elicitor dosage and concentration. Many papers simply refer to the volume of the crude fungal extract added to the plant tissue culture (Rokem *et al.*, 1984; Buitelaar *et al.*, 1993). The inherent variability between different batches of fungal extract does not facilitate comparison between experiments if elicitor dosages are reported on a simple volume basis. Some groups report elicitor dosage as the dry weight of mycelia per unit volume of elicitor preparation (Cline *et al.*, 1993). This approach suffers from the same variability problem as specifying elicitor dosage in terms of volume, and cannot be adopted as a reproducible basis for reporting elicitor concentration. Ayers *et al.* (1976a, 1976b) observed that elicitor activity was closely related to the amount of carbohydrate in the purified elicitor fraction. This permits characterisation of elicitor activity and dosage for carbohydrate or glycosylated elicitors whenever the exact identity of the elicitor molecule is unknown (Basse *et al.*, 1992; Hahn *et al.*, 1993; Singh *et al.*, 1994; Singh, 1995).

The Elicitation Model

Elicitation is a result of the interaction between the elicitor molecule and receptor sites on the cell membrane. The interaction triggers a signal transduction pathway which ultimately results in product formation. A schematic of this process is depicted in Figure 19.1. Preliminary models can be proposed for quantifying elicitation without knowledge of the exact structures of the elicitor and receptor sites. From a phenomenological standpoint, the level of phytoalexin response (or the amount of secondary metabolite produced) depends on the number of receptor sites bound to elicitor molecules, which in turn is a function of the strength of binding between the elicitor molecule and the receptor sites.

Qualitative comparisons of the strength of elicitor-receptor binding can be done by conducting dose response experiments. Known amounts of elicitor are added to specific amounts of tissue and the product levels calculated. A plot of the product level versus the elicitor concentration thus obtained is called the dose response curve. This curve typically follows saturation behaviour. It is linear at low elicitor dosages and saturates at higher elicitor levels. The dose response experiment has been the most common method for predicting the elicitor dosage required for maximum production (Basse *et al.*, 1992; Buitelaar *et al.*, 1993; Byun and Pedersen, 1994).

Dose response data can also be used to obtain the binding constant which characterises the interaction between the elicitor and the receptor. From a simplistic standpoint, the concentration of elicitor resulting in half the maximum response is considered to be equal to the inverse of the binding constant, K_B (Ayers *et al.*, 1976a). However, this estimate of the binding constant, and hence the corresponding elicitor dosage, cannot be used for scale-up because it assumes that the number of receptor sites is much greater than the number of elicitor molecules. To maximise production from root tissue, elicitation will need to be done at concentrations where the numbers of receptor molecules and elicitor molecules are comparable. Mathematical modelling of the elicitation process can be carried out to quantify phytoalexin production as a function of total elicitor and receptor concentrations (Singh *et al.*, 1994). Elicitor binding to membranes can be assumed to be saturable and reversible. This can be represented by a simple bimolecular reaction:

$$E + R \rightleftharpoons ER \qquad (1)$$

where ER is the elicitor–receptor complex, E is the free elicitor and R is the free receptor. At equilibrium, the binding constant K_B is related to the concentrations [ER], [E] and [R] by the following equation:

$$K_B = \frac{[ER]}{([R]_T - [ER])([E]_T - [ER])} \qquad (2)$$

where [E] and [R] have been expressed in terms of the total elicitor concentration $[E]_T$ and total receptor concentration $[R]_T$ by assuming conservation of elicitor and receptor (Singh *et al.*, 1994). The above development suggests that knowledge of the bound elicitor concentration is required to estimate the binding constant. For crude fungal elicitors, the concentration of bound elicitor or the elicitor–receptor complex, [ER], can be related to the phytoalexin response (Stoddart, 1986):

$$[ER] = k \, \gamma \, S \qquad (3)$$

where S is the specific productivity (mg product per g tissue), γ is the tissue concentration (g fresh weight per L medium) and k is the proportionality constant for production (mmoles of bound elicitor per mg of product). In this representation, production characterises the amount of elicitor bound to the tissue. Substituting for [ER] in Equation (2) and solving for the roots of the resulting quadratic equation gives the following expression for specific productivity in terms of the total elicitor added:

$$S = \frac{\left(\dfrac{1}{K_B} + [E]_T + [R]_T\right) - \sqrt{\left(\dfrac{1}{K_B} + [E]_T + [R]_T\right)^2 - 4[E]_T[R]_T}}{2 \, k \, \gamma}$$

$$(4)$$

The above equation can be fitted to the dose response data and the binding constant can be determined from least-squares regression. This approach provides a value of K_B which can be used to calculate the amount of elicitor to be added to a large-scale reactor.

The binding constant can be obtained in the above manner by conducting binding experiments on a variety of systems, e.g. purified receptors, membrane fractions, isolated cells and whole tissue. The bound and free elicitor concentrations can be measured, either directly by chemical assays if the elicitor concentration is known, or by bioassays and relating the phytoalexin response to the bound elicitor concentration. The binding data obtained from the various systems will differ from each other. Experiments conducted with purified receptors reveal the actual nature of elicitor–receptor binding as the system is free of regulatory controls and non-specific binding; however, application of the parameters thus obtained is limited for the same reasons. In the context of understanding and applying the results of elicitor binding for large-scale plant tissue culture, the ultimate goal must be a quantitative description of binding properties in tissue culture itself. Therefore, regulatory influences, diffusional barriers and non-specific binding will play a role in interpreting any binding data. An apparent binding constant which lumps these quantities together is, therefore, a more useful parameter for application to reactor dosage. In this respect, whole tissue presents the best system for conducting binding experiments and the results may be directly extrapolated to large-scale reactor systems. Therefore, the parameters obtained are "apparent" or "application" values.

The results from equilibrium experiments such as dose response measurements give estimates of the binding constants in the absence of mass transfer limitations. The effect of diffusional barriers can be studied by comparing parameters obtained from experimental systems in transient and equilibrium states. Binding experiments have been conducted in flow-through systems to obtain estimates of the binding constant under transient conditions (Singh, 1995). The binding constants obtained were found to be lower than those estimated from equilibrium studies, suggesting mass transfer limitations in the reactor. Therefore, it is important to study the flow profile of the medium and tissue surface–liquid contact patterns within the vessel for successful elicitation of root beds in large-scale reactors.

Effect of Reactor Configuration on Elicitation Strategy

The analysis presented in the previous section suggests that phytoalexin production can be enhanced by increasing the number of bound receptor sites within a reactor. The number of bound receptors will increase with an increase in the total receptor concentration at the same elicitor dosage due to a shift in equilibrium towards the bound state. The receptor concentration in a reactor can be increased by increasing the total mass of tissue within a reactor. Growth studies with tissue cultures in reactors indicate that concentrations of 200 g L^{-1} fresh weight are typical in conventional submerged reactors. However, attempting to increase production by increasing total tissue mass (and hence the total number of receptors) is a challenging problem due to the difficulties encountered in mass transfer and mixing at higher tissue concentrations (Singh and

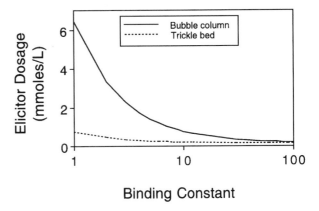

Figure 19.2. Comparison of elicitor dosage required for a bubble column (submerged bed) with an effective hairy root concentration of 200 g L^{-1} fresh weight, and a trickle bed with an effective hairy root concentration of 2000 g L^{-1} fresh weight. The total hairy root mass in both reactors is equal. Effective hairy root concentration (g fresh weight per L medium) is higher for the trickle bed due to less medium being present. The elicitor dosage required for the trickle bed is much less than the dosage required for the bubble column. The difference is larger for systems with weaker interactions and reduces as the binding strength increases.

Curtis, 1994). Even if high tissue density cultures can be obtained, channelling and dead zones in the reactor may reduce the effective surface available for contact with elicitor. An alternative approach is to operate a typical root reactor (200 g L^{-1} fresh weight) in a trickle bed configuration as opposed to a submerged configuration. This reduces the total amount of liquid medium while keeping the tissue mass constant. Thus, the "effective" tissue concentration per unit volume of medium can be increased by an order of magnitude without increasing the mass of tissue within the reactor. This implies that one can elicit at a higher region in the dose response curve by using less elicitor. The above theoretical development explains the experimental results reported by Flores and Curtis (1992). In a study comparing the production of sesquiterpenes by elicitation of hairy root cultures of *Hyoscyamus muticus* grown in a conventional submerged bed and in a trickle bed, these authors reported a 3.5-fold higher production in the latter. However, they could not identify the physiological basis for the enhanced production. It is clear from the binding analysis that eliciting in trickle bed mode results in a shift in the binding equilibrium towards the bound state of the elicitor–receptor, explaining the higher production levels. Figure 19.2 compares the elicitor dosage required for the same level of phytoalexin production for identical hairy root beds operated under

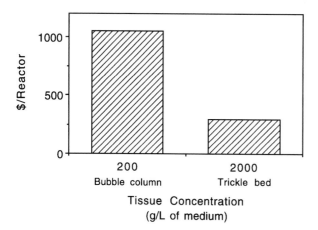

Figure 19.3. Hypothetical comparison of elicitor costs for production of sesquiterpenes from *Hyoscyamus muticus* hairy root cultures in bubble and trickle bed reactors. The costs have been calculated for a hairy root bed of 200 kg fresh weight using a binding constant of 12 L mmole^{-1} glucose equivalents. The bubble column is operated with 1000 L of medium and the trickle bed with 100 L of medium. The cost for *Rhizoctonia solani* elicitor has been assumed to be $1000 L^{-1}. The production level selected for the comparison is 0.1 mg of sesquiterpenes per g fresh weight of tissue. This corresponds to 80% of the maximum production capacity of the *Hyoscyamus muticus* hairy root cultures (Singh, 1995).

different modes: a conventional submerged reactor with tissue concentration of 200 g L^{-1} fresh weight, and a trickle bed reactor with an "effective" tissue concentration of 2000 g L^{-1} fresh weight. The elicitor dosage required for the trickle bed reactor is much lower than that required for a submerged bed reactor irrespective of the strength of binding of the elicitor–receptor pair. The difference is much larger for systems with weaker interactions and reduces as the interaction increases. Thus, eliciting a reactor in trickle bed mode could reduce elicitor dosage and save money. As an example, the theoretical cost of elicitor required to dose a bed of *Hyoscyamus muticus* hairy roots has been calculated for two operating configurations: bubble column and trickle bed (Singh, 1995). The results are presented in Figure 19.3. It is apparent that eliciting in bubble column mode costs approximately three times more than eliciting in trickle bed mode. Therefore, mode of reactor operation is critical to the economic success of production of chemicals from plant tissue in culture.

Conclusions

Elicitation of secondary metabolites in plant tissue culture involves exploiting the defence response of the plant to invasion by pathogens. In spite of the limited understanding of the molecular basis of elicitation (comprising the interaction between elicitor and host receptor and signal transduction), the technique can be used to induce production of phytoalexins in plant tissue. More importantly, it is possible to use elicitation synergistically with other established methods to enhance production of phytoalexins to still higher levels.

Hairy root cultures are amenable to elicitation. In fact, elicitation of root cultures can be done more effectively than that of cell suspensions because of their immobilised nature. The root matrix permits reactor operation in configurations such as the trickle bed which require a much lower elicitor dosage than conventional reactors operated under identical conditions. Cell suspensions on the other hand do not afford this opportunity unless grown in an immobilised state.

Although physiological interpretation of elicitation is not required for its application, it is desired for effective exploitation of the technique. Purification of elicitors and receptors and identification of their molecular structures is important for understanding the interaction that triggers the defence response, while an understanding of signal transduction pathways is vital for manipulation of secondary metabolism and to overcome feedback limitation, substrate limitation and other control mechanisms which limit production levels. Thus it is important to understand the molecular basis of elicitation. It is equally important to develop an understanding of the engineering aspects involved in application of the technology. It has already been shown that mode of reactor operation is a key factor determining the elicitor dosage to a bed of root tissue. Other important aspects include fluid flow, mixing patterns and mass transfer limitations within the reactor, as they affect contact between the receptor and elicitor, and nutrient supply to tissue within the reactor. Ongoing research in these areas will shed new light on both the academic and applied aspects of elicitation and lead to its effective application on a large-scale.

References

Ayers AR, J Ebel, F Finelli, N Berger and P Albersheim (1976a). Host pathogen interactions. IX. Quantitative assays of elicitor activity and characterization of the elicitor present in the extra cellular medium of cultures of *Phytophthora megasperma* var. *sojae*. *Plant Physiol* 57: 751–759.

Ayers AR, J Ebel, B Valent and P Albersheim (1976b). Host pathogen interactions. X. Fractionation and biological activity of an elicitor isolated from the mycelial walls of *Phytophthora megasperma* var. *sojae*. *Plant Physiol* 57: 760–765.

Ballica R, DDY Ryu and CI Kado (1993). Tropane alkaloid production in *Datura stramonium* suspension cultures: elicitor and precursor effects. *Biotechnol Bioeng* 41: 1075–1081.

Basse CW, B Klaus and T Boller (1992). Elicitors and suppressors of the defense response in tomato cells. *J Biol Chem* 267: 10258–10265.

Brooks CJW, DG Watson and IM Freer (1986). Elicitation of capsidiol accumulation in suspended callus cultures of *Capsicum annum*. *Phytochemistry* 25: 1089–1092.

Buitelaar RM, EJTM Leenen, G Geurtsen, E de Groot and J Tramper (1993). Effects of addition of XAD-7 and of elicitor treatment on growth, thiophene production, and excretion by hairy roots of *Tagetes patula*. *Enzyme Microb Technol* 15: 670–676.

Byun SK and H Pedersen (1994). Two-phase airlift fermentor operation with elicitation for the enhanced production of benzophenanthridine alkaloids in cell suspensions of *Escherichia californica*. *Biotechnol Bioeng* 44: 14–20.

Chang HN and SJ Sim (1994). Production of plant secondary metabolites by extractive cultivation. In *Advances in Plant Biotechnology*, edited by DDY Ryu and S Furusaki, pp. 355–369. Amsterdam: Elsevier.

Cheong, J, W Birberg, P Fugedi, A Pilotti, PJ Garegg, N Hong, T Ogawa and MG Hahn (1991). Structure-activity relationships of oligo-β-glucoside elicitors of phytoalexin accumulation in soybean. *Plant Cell* 3: 127–136.

Cline SD and CJ Coscia (1988). Stimulation of sanguinarine production by combined fungal elicitation and hormonal deprivation in cell suspension cultures of *Papaver bracteatum*. *Plant Physiol* 86: 161–165.

Cline SD, RJ McHale and CJ Coscia (1993). Differential enhancement of benzophenanthridine alkaloid content in cell suspension cultures of *Sanguinaria canadensis* under conditions of combined hormonal deprivation and fungal elicitation. *J Nat Prod* 56: 1219–1228.

Conrath U, A Domrad and H Kauss (1989). Chitosan-elicited synthesis of callose and of coumarine derivatives in parsley cell suspension cultures. *Plant Cell Rep* 8: 152–155.

Corry JP, WL Reed and WR Curtis (1993). Enhanced recovery of solavetivone from hairy root cultures of *Hyoscyamus muticus* using integrated product recovery. *Biotechnol Bioeng* 42: 1–6.

Cosio EG, H Popperl, WE Schmidt and J Ebel (1988). High-affinity binding of fungal β-glucan fragments to soybean microsomal fractions and protoplasts. *Eur J Biochem* 175: 309–315.

Cosio EG, T Frey, R Verduyn, JV Boom and J Ebel (1990). High affinity binding of a synthetic heptaglucoside and fungal glucan phytoalexin elicitors to soybean membrane. *FEBS Lett* 271: 223–226.

Darvill AG and P Albersheim (1984). Phytoalexins and their elicitors — a defense against microbial infection in plants. *Ann Rev Plant Physiol* 35: 243–275.

DiCosmo F and M Misawa (1985). Eliciting secondary metabolism in plant cell cultures. *Trends in Biotechnol* 3: 318–322.

DiCosmo F, A Quesnel, M Misawa and SG Tallevi (1987). Increased synthesis of ajmalicine and catharanthine by cell suspension cultures of *Catharanthus roseus* in response to fungal culture-filtrates. *Appl Biochem Biotechnol* 14: 101–106.

Doran PM (1989). Prospects for production of plant chemicals from genetically transformed roots. *Aust J Biotechnol* 3: 270–277.

Dornenburg, H and D Knorr (1994). Elicitation of anthraquinones in *Morinda citrifolia* cell cultures. *Food Biotechnol* 8: 57–65.

Dunlop DS and WR Curtis (1991). Synergistic response of plant hairy-root cultures to phosphate limitation and fungal elicitation. *Biotechnol Prog* 7: 434–438.

Ebel J (1986). Phytoalexin synthesis: the biochemical analysis of the induction process. *Ann Rev Phytopathol* 24: 235–264.

Flores HE and WR Curtis (1992). Approaches to understanding and manipulating the biosynthetic potential of plant roots. *Ann NY Acad Sci* 665: 188–209.

Flores HE, JJ Pickard and MW Hoy (1988). Production of polyacetylenes and thiophenes in heterotrophic and photosynthetic root cultures of *Asteraceae*. *Bioactive Molec* 7: 233–254.

Frey T, EG Cosio and J Ebel (1993). Affinity purification and characterization of a binding protein for a heptaglucoside phytoalexin elicitor in soybean. *Phytochemistry* 32: 543–550.

Fukui H, M Tani and M Tabata (1990). Induction of shikonin biosynthesis by endogenous polysaccharides in *Lithospermum erythrorhizon* cell suspension cultures. *Plant Cell Rep* 9: 73–76.

Funk C, K Gugler and P Brodelius (1987). Increased secondary product formation in plant cell suspension cultures after treatment with a yeast carbohydrate preparation (elicitor). *Phytochemistry* 26: 401–405.

Grosskopf DG, G Felix and T Boller (1991). A yeast derived glycopeptide elicitor and chitosan or digitonin differentially induce ethylene biosynthesis, phenylalanine-lyase and callose formation in suspension cultures of tomato. *J Plant Physiol* 138: 741–746.

Hahn MG, J Cheong, R Alba and F Cote (1993). Oligosaccharide elicitors: structures and signal transduction. In *Plant Signals in Interactions With Other Organisms*, edited by J Schultz and I Raskin. *Amer Soc Plant Physiologists* 11: 24–46.

Hamill JD, AJ Parr, MJC Rhodes, RJ Robins and NJ Walton (1987). New routes to plant secondary products. *Bio/Technol* 5: 800–806.

Hanagata N, H Uehara, A Ito, T Tekeuchi and I Karube (1994). Elicitor for red pigment formation in *Carthamus tinctorius* cultured cells. *J Biotechnol* 34: 71–77.

Heinstein PF (1985). Future approaches to the formation of secondary natural products in plant cell suspension cultures. *J Nat Prod* 48: 1–9.

Ingham J (1982). Phytoalexins from the *Leguminosae*. In *Phytoalexins*, edited by JA Bailey and W Mansfield, pp. 21–80. London: Blackie.

Kauss H, W Jeblick and A Domrad (1989). The degree of polymerization and N-acetylation of chitosan determine its ability to elicit callose formation in suspension cells and protoplasts of *Catharanthus roseus*. *Planta* 178: 385–392.

Milat M, P Ricci, P Bonnet and J Blein (1991). Capsidiol and ethylene production by tobacco cells in response to cryptogein, an elicitor from *Phytophthora cryptogea*. *Phytochemistry* 30: 2171–2173.

Moreno PRH, R van der Heijden and R Verpoorte (1993). Effect of terpenoid precursor feeding and elicitation on formation of indole alkaloids in cell suspension cultures of *Catharanthus roseus*. *Plant Cell Rep* 12: 702–705.

Mukundan U and MA Hjortso (1990). Effect of fungal elicitor on thiophene production in hairy root cultures of *Tagetes patula*. *Appl Microbiol Biotechnol* 33: 145–147.

Robbins MP, J Hartnoll and P Morris (1991). Phenylpropanoid defense response in transgenic *Lotus corniculatis* 1. Glutathione elicitation of isoflavan phytoalexins in transformed root cultures. *Plant Cell Rep* 10: 59–62.

Rokem JS, J Schwarzberg and I Goldberg (1984). Autoclaved fungal mycelia increase production in cell suspension cultures of *Dioscorea deltoidea*. *Plant Cell Rep* 3: 159–160.

Schmidt WE and J Ebel (1987). Specific binding of a fungal glucan phytoalexin elicitor to membrane fractions from soybean *Glycine max*. *Proc Nat Acad Sci USA* 84: 4117–4121.

Scragg AH, R Cresswell, S Ashton, A York, P Bond and MW Fowler (1988). Growth and secondary product formation of a selected *Catharanthus roseus* cell line. *Enzyme Microb Technol* 10: 532–536.

Sharp JK, B Valent and P Albersheim (1984). Purification and partial characterization of a β-glucan fragment that elicits phytoalexin accumulation in soybean. *J Biol Chem* 259: 11312–11320.

Signs MW and HE Flores (1990). The biosynthetic potential of plant roots. *Bioassays* 12: 7–13.

Sim SJ, HN Chang, JR Liu and KH Jung (1994). Production and secretion of indole alkaloids in hairy root cultures of *Catharanthus roseus*: effects of *in situ* adsorption, fungal elicitation and permeabilization. *J Ferment Bioeng* 78: 229–234.

Singh G (1995). *Fungal Elicitation of Plant Root Cultures — Application to Bioreactor Dosage*. PhD thesis, Pennsylvania State University, USA.

Singh G and WR Curtis (1994). Reactor design for plant root culture. In *Biotechnological Applications of Plant Cultures*, edited by PD Shargool and TT Ngo, pp. 185–206. Boca Raton: CRC Press.

Singh G, GR Reddy and WR Curtis (1994). Use of binding measurements to predict elicitor dosage requirements for secondary metabolite production from root cultures. *Biotechnol Prog* 10: 365–371.

Stoddart JL (1986). Gibberellin receptors. In *Hormones, Receptors, and Cellular Interactions in Plants*, edited by CM Chadwik and DR Garrod, pp. 91–114. Cambridge: Cambridge University Press.

Tabata M, T Ogino, K Yoshioka, N Yoshikawa and N Hiraoka (1978). Selection of cell lines with higher yield of secondary products. In *Frontiers of Plant Tissue Culture*, edited by TA Thorpe, pp. 213–222. Calgary: University of Calgary Press.

Takahasi S and Y Fujita (1991). Cosmetic materials — production of shikonin. In *Plant Cell Culture in Japan*, edited by A Komamine, M Misawa and F DiCosmo, pp. 72–78. Tokyo: CMC Co. Ltd.

Tani M, H Fukui, M Shimomura and M Tabata (1992). Structure of endogenous oligogalacturonides inducing shikonin biosynthesis in *Lithospermum erythrorhizon* *Phytochemistry* 31: 2719–2723.

Whitehead IM and DR Threlfall (1992). Production of phytoalexins by plant tissue cultures. *J Biotechnol* 26: 63–81.

Wibberley MS, JR Lenton and SJ Neill (1994). Sesquiterpenoid phytoalexins produced by hairy roots of *Nicotiana tabacum*. *Phytochemistry* 37: 349–351.

Yoshikawa M, NT Keen and MC Wang (1983). A receptor on soybean membranes for a fungal elicitor of phytoalexin accumulation. *Plant Physiol* 73: 497–506.

Zenk MH, H El-Shagi, H Arens, J Stöckigt, EW Weiler and B Deus (1977). Formation of indole alkaloids serpentine and ajmalicine in cell suspension cultures of *Catharanthus roseus*. In *Plant Tissue Culture and its Biotechnological Applications*, edited by W Barz, E Reinhard and MH Zenk, pp. 22–43. Berlin: Springer-Verlag.

Shikonin Production by Hairy Roots of *Lithospermum erythrorhizon* in Bioreactors with *In Situ* Separation

20

Sang Jun Sim and Ho Nam Chang

Bioprocess Engineering Research Center and Department of Chemical Engineering,
Korea Advanced Institute of Science and Technology, Taejon 305-701, Korea;
email: hnchang@sorak.kaist.ac.kr

Introduction

In situ adsorption and extraction have been widely applied in the field of biotechnology for the removal of inhibitory products, including from plant cell cultures. Adsorption and extraction can be performed by adding to the culture inert hydrophobic chemicals (liquid or solid) which have a high adsorption capacity for the hydrophobic plant products. Secondary metabolites of plant cells are stored within the cell vacuoles or are present as compounds outside of the cell wall. Some are secreted into the medium or may appear in the medium due to cell lysis. These compounds may inhibit cell growth and the further production of secondary metabolites by interfering with oxygen supply, nutrients or both. There are several reasons for the low product yield of plant cell culture, such as cellular regulation of the ratio between intracellular and extracellular product concentrations, enzymatic or non-enzymatic degradation of the product in the medium, and product volatility. For the above cases, it should be possible to increase net production by the addition of an artificial substance for accumulation of secondary substances in the culture medium. Employing either a lipophilic second phase or a polar second phase has been reported to be beneficial for the accumulation and detection of secondary metabolites.

Brodelius and Nilsson (1983) showed that some solvents were useful for extracting products from immobilised plant cells without affecting cell viability. Berlin *et al.* (1984) used adsorbents to retain volatile plant cell products from cultures of *Thuja occidentalis*. Becker *et al.* (1984) performed two-phase cultures to recover lipophilic secondary metabolites and to prevent product degradation. Nakajima *et al.* (1986) entrapped *Lavandula vera* cells in synthetic resin pre-polymers to enhance pigment production and to increase pigment release from the cells. Deno *et al.* (1987) studied *in situ* extraction with hydrocarbons in suspension cultures of *Lithospermum erythrorhizon*. Rhodes *et al.* (1986), Robins and Rhodes (1986), and Payne and Shuler (1988) used polymeric adsorbents to recover plant cell products. *In situ* product removal enhanced total plant secondary metabolite production, and products were selectively released from cells and dissolved in the solvents or adsorbents. Byun *et al.* (1990) showed an example of simultaneous use of *in situ* extraction and elicitation. Elicitation in combination with two-phase culture additionally increased net sanguinarine production, as well as the sanguinarine concen-

Table 20.1. Effect of *in situ* extraction by *n*-hexadecane on shikonin production. (Inoculum concentration: 0.1 g L^{-1} dry weight)

Volume of added solvent (mL)	Culture time[a] (d)	Maximum shikonin concentration (mg L^{-1})	Final dry cell weight (g L^{-1})	Cellular productivity[b] (mg g^{-1} h^{-1})	Volumetric productivity (mg L^{-1} d^{-1})
0	20	42.0	3.2	0.88	2.1
10	21	43.8	3.3	0.87	2.1
20	26	82.8	3.8	1.33	3.2
30	26	120.6	5.0	1.94	4.7
40	26	106.2	5.0	1.71	4.1

[a] Time for maximum shikonin concentration.
[b] Rate of biomass production per gram dry weight of inoculum.

tration in the accumulation phase. Kim and Chang (1990) were able to enhance shikonin production by employing *in situ* extraction and cell immobilisation. Shimomura *et al.* (1991) and Sim and Chang (1993) successfully applied this culture method to hairy roots of *L. erythrorhizon*.

Requirements for the selection of a suitable secondary phase depend on the morphology of the plant cells and the characteristics of the target compounds. There is no general rule for selection to ensure cell viability is preserved. Therefore, for every new plant cell culture, a suitable secondary phase has to be found. In general, selection of an extractive secondary phase should be directed towards maximal separation capacity and minimal toxic effect on the cells. Both solid–liquid and liquid–liquid systems are effective for a diverse range of plant cell cultures.

Most common bioreactor types like the stirred tank and bubble column can be used as two-phase bioreactors. In contrast with microbial or plant cell suspensions, organised tissues such as hairy roots present significant problems in reactor design. The biggest difficulty with large-scale culture of roots is the formation of root mats. These mats restrict internal mass transfer, can entrap gas and float, and make maintaining a uniform environment difficult. Stirred tank reactors are not suitable for hairy root cultures because root cells can be directly injured by impeller rotation. Taya *et al.* (1989) reported that an airlift reactor with cell immobilisation was a superior cultivation system for hairy roots.

Most plant secondary metabolites produced by hairy roots are not released into the extracellular phase. It would be desirable that products are released *in situ* from the roots into the medium to eliminate the need to harvest cells after full growth. Hairy root culture is a promising technique for im-

proving product yields. However, bioreactor designs that provide homogeneous environments for hairy root cultures are only now being developed.

In this chapter, we discuss two examples of shikonin production by hairy roots in which different reactors (bubble column and stirred tank) and different second phases (extraction by solvent and adsorption by resin) were employed.

Two-phase Hairy Root Culture with *In Situ* Solvent Extraction

Hairy roots of *Lithospermum erythrorhizon* were cultured in shake flasks and a bubble column bioreactor employing *in situ* extraction with *n*-hexadecane. The culture medium used in the experiments was modified SH basal medium (Sim and Chang, 1993).

Effect of *In Situ* Extraction on Shikonin Production in Flask Culture

To investigate the effects of *in situ* product removal from hairy roots in shake flask culture, various amounts of sterile *n*-hexadecane (10, 20, 30 and 40 mL) were added to 50 mL of medium. Hairy roots were cultivated for one month (Table 20.1). Most (more than 95%) of the shikonin produced was recovered in the *n*-hexadecane layer. Use of 30 mL *n*-hexadecane gave the best results, producing a final shikonin concentration three times higher than that obtained without extraction. As the volume of *n*-hexadecane was increased up to 30 mL, shikonin production was enhanced by efficiently removing the shikonin produced and other products which might be inhibitory to cell growth. However, using a volume of *n*-hexadecane greater than 30 mL resulted in a lower shikonin production probably due to oxygen limitation in the culture. It is expected, however, that shikonin production will be enhanced

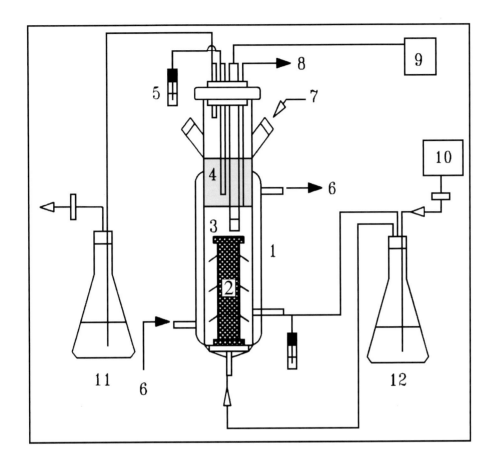

Figure 20.1. Schematic diagram of experimental set-up for two-phase bubble column reactor operation. 1. bubble column reactor, 2. stainless steel mesh structure for root immobilisation, 3. medium, 4. solvent, 5. sampling, 6. cooling water in and out, 7. inoculation port, 8. solvent out, 9. DO meter, 10. air generator, 11. humidifier air outlet, 12. humidifier air inlet.

in this system when oxygen can be supplied more effectively. It is also important to note that shikonin could be detected earlier when a higher volume of *n*-hexadecane was used. Final dry cell weight was higher in cultures containing larger volumes of solvent. This is due to the removal of shikonin and other metabolites that would inhibit growth if present.

Two-Phase Bubble Column Reactor Operation

A two-phase bubble column reactor was constructed to test procedures for improving shikonin production (Figure 20.1). A stainless steel mesh was installed inside the reactor to immobilise the hairy roots. Humidified air was introduced into the column through a sintered glass sparger at the bottom. Hairy roots were aseptically anchored to regular positions on the stainless steel mesh. The medium and solvent were replaced repeatedly through different inlet and outlet ports when the medium was depleted or the shikonin concentration in the solvent layer reached a high level. The culture broth in the reactor was aerated by supplying filtered air at a rate of 0.05–0.2 vvm. During bioreactor operation cell growth was indirectly estimated by conductivity measurement (Taya *et al.*, 1989). Approxi-

mately 100 mg of hairy root cells grown for 10 d in three-fold diluted SH medium were transferred to the reactor containing modified SH medium and *n*-hexadecane at pH 5.8. The volumes of culture medium and *n*-hexadecane were 600 and 300 mL, respectively.

The results are shown in Figure 20.2. Medium was replaced when the rate of sucrose consumption decreased as shown by down arrows. Shikonin levels increased continuously during the 54 d of the experiment. The final concentrations of cell mass and total shikonin produced were 15.6 g L^{-1} dry weight and 572.6 mg L^{-1}, respectively. Shikonin was produced at a constant volumetric rate of 10.6 mg L^{-1} d^{-1} during the whole period.

A problem that needs to be resolved for long-term operation with this reactor configuration is that channelling occurred in the vessel as hairy roots grew and aggregated, and this might have caused oxygen limitations in certain regions. For long-term operation this should be avoided.

The overall results suggest that hairy root culture with *in situ* extraction is useful for production and effective recovery of shikonin. Shikonin production can be further improved by using the two-phase bubble column reactor as shown in this study. This

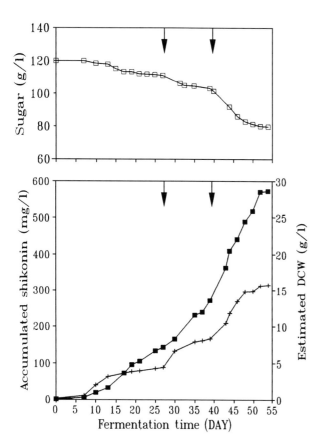

Figure 20.2. Time courses for sugar consumption, hairy root growth and shikonin production in a two-phase bubble column reactor. +: dry cell weight; ■: total shikonin; arrow: exchange of medium and solvent. (Reproduced from SJ Sim and HN Chang, 1993, Increased shikonin production by hairy roots of *Lithospermum erythrorhizon* in two phase bubble column reactor. *Biotechnol Lett* 15: 145–150, Figure 20.3; with permission from Science & Technology Letters, Northwood, England.)

technique may be useful for production of other secondary metabolites from hairy root culture and may also allow continuous operation.

Fed-batch Culture of Hairy Roots with *In Situ* Separation Using XAD-2 Resin

One of the most popular techniques for achieving high cell densities in microbial fermentations is fed-batch culture. This involves the control of nutrient feeding, which is often necessary for high productivity (Yamane and Shimizu, 1984). Fed-batch culture with intermittent feeding of substrate is usually used to overcome substrate inhibition or catabolite repression. However, as plant cell culture medium includes many different components (e.g. carbon source, salts, vitamins, and growth regulators), fed-

batch culture of plant cells can be hampered by the diversity of the medium components.

Another problem with fed-batch culture of plant cell is the difficulty of using monosaccharides as the carbon source. In fed-batch culture it is very difficult to control the concentration of a disaccharide such as sucrose without the accumulation of monosaccharides such as glucose and fructose. Although some authors have reported that the use of monosaccharides such as fructose improves plant cell density (Uozomi et al., 1991), monosaccharides are rarely used in plant cell culture.

During hairy root culture in bioreactors, one of the main problems is the impossibility of removing samples of biomass from the fermenter. Measuring the conductivity of the culture medium is a convenient tool for cell mass determination in plant cell or organ culture, as the effect on sugar concentration and other environmental variables is generally negligible and can be ignored. However, medium conductivity is greatly affected by the concentration of salt components such as nitrates. Therefore, it is expected that by controlling the medium conductivity, the salt level will also be controlled. Both sugar concentration and medium conductivity are suitable control parameters for fed-batch culture of plant cells.

In this study, fed-batch culture of *L. erythrorhizon* hairy roots was carried out by controlling sucrose concentration and medium conductivity in shake flasks and in a modified stirred tank reactor. For effective product recovery from the cultures, *in situ* adsorption by XAD-2 resin was also applied.

Effect of Carbon Source on Shikonin Production

Sucrose, glucose and fructose is the carbon source most often adopted in plant cell cultures. The effect of various carbon sources on shikonin production and hairy root growth was tested to obtain information about monosaccharide utilisation by *L. erythrorhizon* hairy roots (Figure 20.3). The highest shikonin production and hairy root growth were obtained when sucrose was used as the carbon source. On the other hand, when glucose or fructose was used instead, hairy root growth was severely inhibited. To compare the effect of sucrose hydrolysis on shikonin production, 20 g L^{-1} of glucose and 20 g L^{-1} of fructose which corresponded to 40 g L^{-1} of sucrose were added to the flask as carbon sources. But cell growth and shikonin production were also inhibited like other monosaccharide experiments. It seems that the hairy root line used in these experiments followed more hydrolytic process than non-hydrolytic ones. Although the detailed mechanism of sucrose uptake was not clarified, two types of process may be involved: hydrolytic and non-hydrolytic. The growth response of plant cells to carbon sources frequently depends on the plant

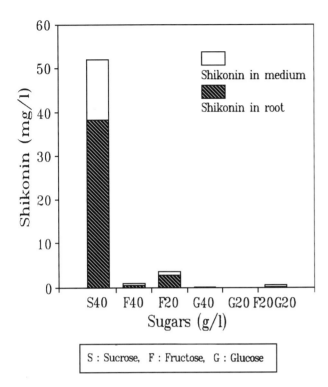

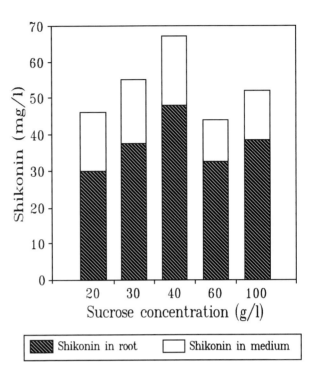

Figure 20.3. Effect of various carbon sources on growth and shikonin production by *L. erythrorhizon* hairy roots after 25 d culture in shake flasks.

Figure 20.4. Effect of sucrose concentration on growth and shikonin production by *L. erythrorhizon* hairy roots after 25 d culture in shake flasks.

species or clone of interest. For our hairy roots, it was difficult to perform fed-batch cultures using a monosaccharide such as fructose or glucose as the carbon source.

In order to examine the effect of osmotic strength on shikonin production, hairy roots were cultured in modified SH medium supplemented with different concentrations of sucrose (2, 3, 4, 6 or 10% w/v) (Figure 20.4). Total shikonin concentration reached a maximum of 67.3 mg L^{-1} at 4% sucrose. Further increase in the concentration of sucrose did not reduce the shikonin production so much, but at 6% sucrose shikonin production slightly decreased to 42 mg L^{-1}. The distribution of shikonin between cells and culture media was not affected by the osmotic stress of sucrose.

In Situ Adsorption by XAD-2 Resin

The Amberlite resin (XAD-2; 20–50 mesh) was used as the secondary phase in hairy root cultures in shake flasks. Prior to use, the resin was washed with methanol for 3 h followed by repeated rinsing with distilled water. Small bags of resin wrapped in nylon mesh cloth were autoclaved in culture medium for 15 min for equilibration with the culture medium. Resin was typically added to flasks 5 to 7 d after inoculation.

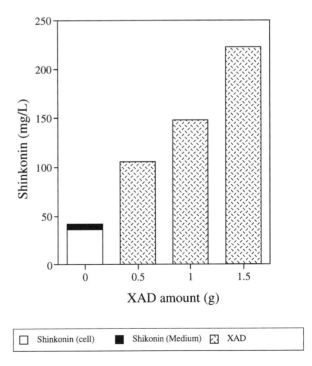

Figure 20.5. Effect of different XAD-2 concentrations on shikonin production.

The influence of XAD-2 concentration on shikonin production was investigated (Figure 20.5). Different amounts of sterilised adsorbent (0.5, 1.0 or 1.5 g per flask) were aseptically added to hairy roots in modified SH liquid medium (50 mL per 250 mL flask). Different numbers of resin bags containing 0.5 g XAD-2 were added to the hairy root culture flasks. Most of the shikonin produced was recovered in the XAD-2 resin. As the number of resin bags increased to three (equivalent to 1.5 g L^{-1}), shikonin production was enhanced by the efficient removal of shikonin and other products which might be inhibitory to cell growth.

Fed-batch Culture in a Modified Stirred Tank Reactor Coupled With an XAD-2 Column With Control of Sucrose Concentration and Medium Conductivity

The basic stirred tank reactor design used in this study was similar to the modified turbine blade reactor proposed by Kondo *et al.* (1990). Roots were grown in a 2-L glass vessel approximately 150 mm in diameter and 300 mm in height (Figure 20.6). The vessel was divided into two spaces: a hairy root growing space (1000 mL) and an impeller agitation space (300 mL), using a stainless steel mesh sheet which protected the roots from shear damage by the

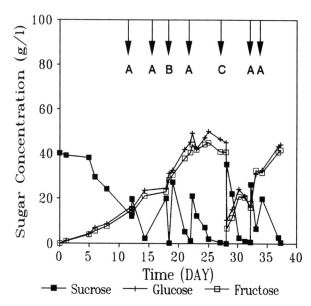

A: organic & salt media addition, B: organic media addition
C: media exchange

Figure 20.7. Time courses of sugar concentration of the fed-batch culture of *L. erythrorhizon* controlling by sucrose and medium conductivity in a modified stirred tank reactor.

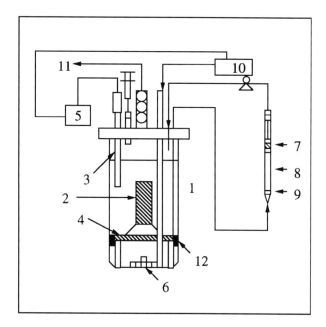

Figure 20.6. Schematic diagram of experimental set-up for a modified stirred tank bioreactor operation. 1. modified stirred tank bioreactor, 2. stainless steel mesh structure for root immobilisation, 3. DO probe, 4. stainless steel mesh, 5. DO meter, 6. turbine blade impeller, 7. glass wool, 8. XAD-2 resin, 9. sintered glass, 10. air generator, 11. condenser, 12. mesh frame.

floor-mounted impeller. A cylindrical stainless steel mesh structure was installed in the hairy root growing space for immobilisation of the roots. Using a control system for dissolved oxygen concentration (Celligen Cell Culture System, NBS Sci., USA.), air or a mixture of air and pure oxygen was introduced into the vessel intermittently. Temperature was maintained at 25°C by a water jacket. The modified stirred tank reactor was connected to a column containing 10 g XAD-2 resin via a peristaltic pump (circulation flow rate: 10 mL min^{-1}). In order to maintain the dissolved oxygen level constant, a mixture of air and pure oxygen was introduced into the reactor when the cell density became higher than 10 g L^{-1}. Severe oscillations of dissolved oxygen tension were observed.

Fed-batch culture of hairy roots was carried out by controlling both the medium conductivity and the sucrose concentration. Two solutions were used to feed concentrated nutrients to the culture: an organic medium containing sucrose, and a vitamins/salts medium containing modified SH medium salts. Figure 20.7 shows results for the evolution of sugar concentrations during the fed-batch culture. It was difficult to maintain the sucrose level constant due to the high rate of sucrose hydrolysis. By Day 28, glucose and fructose had accumulated at concentrations up to 50 g L^{-1}, which may have adversely

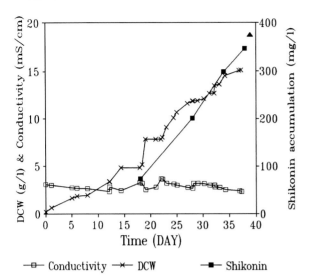

Figure 20.8. Time courses of estimated dry cell weight, medium conductivity and shikonin adsorbed by the XAD-2 column during fed-batch culture of *L. erythrorhizon* controlling by sucrose and medium conductivity in a modified stirred tank reactor. The triangle symbol represents the real dry cell weight measured after the fermentation.

affected cell growth. Therefore, at that time, the total volume of the culture broth was replaced with fresh medium. The conductivity of the culture medium was maintained by feeding concentrated salts medium. Figure 20.8 shows the results for conductivity, estimated dry cell mass and shikonin production. The XAD-2 resin in the adsorption column was replaced intermittently during the culture. The average shikonin productivity and hairy root growth rate were 9.34 mg L^{-1} d^{-1} and 0.49 g L^{-1} d^{-1} dry weight, respectively. By the end of the experiment, 355 mg L^{-1} of shikonin derivatives had been recovered from the adsorption column. However, at the end of the reactor operation, the roots inside were brown, had developed calli, and were fragmenting into the liquid. This was apparently due to the high osmotic stress produced by the accumulation of monosaccharides.

The results obtained in this study indicate that fed-batch culture coupled with XAD-2 adsorption could be utilised for high production of shikonin derivatives and hairy root growth rate when sucrose is used as the sole carbon source.

References

Becker H, JH Reichling, W Bisson and S Herold (1984). Two phase culture — a new method to yield lipophilic secondary products from plant suspension cultures. *Proc 3rd Eur Congr Biotechnol* 1: 209–213.

Berlin J, L Witte, W Schubert and V Wray (1984). Determination and quantification of monoterpenoids secreted into the medium of cell cultures of *Thuja occidentalis*. *Phytochemistry* 23: 1277–1279.

Brodelius, P and K Nilsson (1983). Permeabilization of immobilized plant cells resulting in release of intracellularly stored products with preserved cell viability. *Eur J Appl Microbiol Biotechnol* 17: 275–280.

Byun SY, H Pedersen and C Chin (1990). Two-phase culture for the enhanced production of benzophenandrine alkaloids in cell suspensions of *Eschscholtzia californica*. *Phytochemistry* 29: 3135–3139.

Deno H, C Suga, T Morimoto and Y Fujita (1987). Production of shikonin derivatives by cell suspension cultures of *Lithospermum erythrorhizon*. VI. Production of shikonin derivatives by a two-layer culture containing an organic solvent. *Plant Cell Rep* 6: 197–199.

Kim DJ and HN Chang (1990). Enhanced shikonin production from *L erythrorhizon* by in situ extraction and calcium alginate immobilization. *Biotechnol Bioeng* 36: 460–466.

Kondo O, H Honda, M Taya and T Kobayashi (1990). Comparison of growth properties of carrot hairy roots and their culture in bioreactors. *J Chem Eng Japan* 22: 84–89.

Nakajima H, K Sonomoto, H Morikawa, F Sato, K Ichimura, Y Yamada and A Tanaka (1986). Entrapment of *Lavandula vera* cells with synthetic resin prepolymers and its application to pigment production. *Appl Microbiol Biotechnol* 24: 266–270.

Payne GF and ML Shuler (1988). Selective adsorption of plant products. *Biotechnol Bioeng* 31: 922–928.

Rhodes MJC, M Hilton, AJ Parr, JD Hamill and RJ Robins (1986). Nicotine production by "hairy root" cultures of *Nicotiana rustica*: fermentation and product recovery. *Biotechnol Lett* 8: 415–420.

Robins RJ and MJC Rhodes (1986). The stimulation of anthraquinone production by *Cinchona ledgeriana* cultures with polymeric adsorbents. *Appl Microbiol Biotechnol* 24: 35–41.

Shimomura K, H Sudo, H Saga and H Kamada (1991). Shikonin production and secretion by hairy root cultures of *Lithospermum erythrorhizon*. *Plant Cell Rep* 10: 282–285.

Sim SJ and HN Chang (1993). Increased shikonin production by hairy roots of *Lithospermum erythrorhizon* in two phase bubble column reactor. *Biotechnol Lett* 15: 145–150.

Taya M, A Yoyoma, O Kondo, T Kobayashi and C Matsui (1989). Growth characteristics of plant hairy roots and their cultures in bioreactors. *J Chem Eng Japan* 22: 84–89.

Uozomi N, K Kohketsu, O Kondo, H Honda and T Kobayashi (1991). Fed-batch culture of hairy root using fructose as a carbon source. *J Ferment Bioeng* 72: 457–460.

Yamane T and S Shimizu (1984). Fed-batch techniques in microbial process. *Adv Biochem Eng /Biotechnol* 30: 147–194.

Hairy Roots of *Tagetes* in Two-Phase Systems

21

Lucilla Bassetti[1], Reinetta M. Buitelaar[2] and Johannes Tramper[1]

[1] *Department of Food Science and Technology, Food and Bioprocess Engineering Group, Wageningen Agricultural University, PO Box 8129, 6700 EV Wageningen, The Netherlands; email: Hans.Tramper@Algemeen.PK.WAU.NL*
[2] *Agrotechnological Research Institute, ATO-DLO, PO Box 17, 6700 AA Wageningen, The Netherlands*

Introduction

Hairy root cultures have received considerable attention for their capacity to grow indefinitely in hormone-free liquid media and produce valuable secondary metabolites at levels comparable to the original plants (Charlwood *et al.*, 1990; Curtis, 1993). Most of these products, however, are retained intracellularly or only partially released, hampering thus the use of hairy roots as catalytic systems for production of secondary metabolites. One of the strategies to facilitate metabolite release with possible increase of productivity is the use of two-phase systems and *in situ* extraction (Beiderbeck and Knoop, 1987).

We have applied two-phase system techniques to hairy root cultures of *Tagetes* (Asteraceae). In the past, these scented herbaceous plants were used as a source of insecticides, for medical purposes, and in a number of magical ceremonies. Thiophenes, secondary metabolites produced by *Tagetes*, are sulphurated, heterocyclic compounds mostly occurring in nature as substituted di- and terthienyls (Figure 21.1). They exhibit a strong biocidal activity. This finding was first reported in 1953 by a Dutch bulb breeder who noted the biological activity of mari-

golds (*Tagetes patula*) against root rot caused by free-living nematodes (Breteler and Ketel, 1993).

Most of the literature on *Tagetes* is about the three species *T. minuta*, *T. erecta* and *T. patula*. Their root cultures show stable patterns of growth and production, with the highest thiophene content in the root tips. Neither transformed nor untransformed roots release their secondary metabolites spontaneously, which makes them a suitable system for investigating how to trigger product release. Moreover, the biocidal activity of thiophenes motivates studies directed to their commercial production which, despite numerous efforts and some patents, has not yet been attained.

Two-Phase Systems

Any artificial site of accumulation and/or conservation of secondary substances in the culture medium is defined as a "second phase". Depending on the properties and the amount of the second phase material, an equilibrium should be approached between the concentrations of the selected compound inside the protoplasts (C_p), the culture medium (C_m) and the accumulation phase (C_a) (Beiderbeck, 1982):

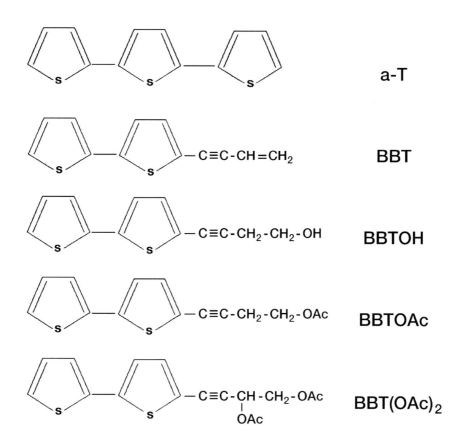

Figure 21.1. Molecular structure of the main thiophenes found in *Tagetes* species. a-T: α-terthienyl; BBT: butenylthiophene; BBTOH: hydroxybutenylthiophene; BBTOAc: acetoxybutenylthiophene; BBT(OAc)$_2$: diacetoxybutenylbithiophene. (Adapted from Croes *et al.*, 1989.)

$$C_p \rightleftharpoons C_m \rightleftharpoons C_a$$

As the culture medium ("first phase") is usually a liquid, we distinguish several types of two-phase systems:

(1) **Liquid–liquid**: the second liquid phase is provided by:
  (1.1) water-immiscible compounds, such as organic solvents or lipids (**organic–aqueous** two-phase systems);
  (1.2) water-miscible compounds, such as salt or polymer solutions (**aqueous** two-phase systems);
(2) **Solid–liquid**: the second phase is provided by a solid material, such as a resin or other type of adsorbent.

Apart from its state, the ideal component of a second phase should be chemically and physically stable, non-toxic, biocompatible, selective and specific for the desired product, without hampering down-

stream processing (Beiderbeck and Knoop, 1987). Clearly, the optimal second phase has to be tailored for each compound to be extracted and no general rules of thumb have been deduced so far.

We have used several two-phase systems to enhance excretion and/or thiophene production in root cultures of *Tagetes patula* transformed with *Agrobacterium rhizogenes* LBA 9402. Because aqueous two-phase systems do not usually offer advantages for downstream processing and because thiophenes are rather hydrophobic products, we have mostly investigated the effect of water-immiscible organic solvents.

Organic–Aqueous Two-Phase Systems

Biocompatibility studies

Strategies to select the proper solvent for *in situ* product removal have been described (Tramper *et al.*, 1987; Bruce and Daugulis, 1991). Among the desirable requirements, biocompatibility is absolutely necessary as the solvent is in contact with the biocatalyst. Various parameters have been used to as-

Table 21.1. Log P and density of the solvents used, and their effect on growth, oxygen consumption and thiophene production by *T. patula* hairy roots. Control roots were grown in one-phase systems without solvent. (Adapted from Buitelaar *et al.*, 1991.)

Solvent	Log P	Density (kg L^{-1})	Growth	Oxygen consumption	Thiophene production
Ethylacetate	0.7	0.90	−	−	−
Diethylphthalate	3.3	1.12	−	−	+
Hexane	3.5	0.66	−	−	−
Decanol	4.0	0.83	−	−	−
Dibutylphthalate	5.4	1.05	+	+	+
Decane	5.6	0.73	+	+	+
Hexadecane	8.8	0.77	+	+	+
Dioctylphthalate	9.6	0.99	+	+	+
FC 40	11.2	1.87	+	+	+

− Values much lower than control.
+ Values comparable to control.

sess solvents in terms of their biocompatibility: the Hildebrand solubility parameter (Brink and Tramper, 1985), the log P parameter (Laane *et al.*, 1985), the solvent partition coefficient in a membrane aqueous buffer (Osborne *et al.*, 1990), and the critical solvent concentration (Bassetti and Tramper, 1994). The last two parameters were introduced to explain the mechanism of solvent "molecular" toxicity, which is caused by solvent molecules dissolved in the aqueous phase. When the solvent concentration is above the saturating level, a second phase results and "phase" toxicity may occur (Bar, 1987).

Log P is the logarithm of the solvent partition coefficient in a standard 1-octanol/water two-phase system. Experiments with enzymes (Laane *et al.*, 1987) and microorganisms (Bar, 1988; Vermuë *et al.*, 1993) showed that biocompatibility of organic compounds increases with both decreased polarity and aqueous solubility. In particular, when log P is > 4 the solvent is generally not toxic to the biocatalyst, when log P is < 2 lethal toxicity is observed, and when log P is between 2 and 4 the activity retention cannot be predicted.

For experiments with transformed hairy roots of *T. patula* in organic–aqueous two-phase systems, log P was used to predict solvent biocompatibility. The influence of various solvents (log P range: 0.7–11.2) was tested by growing root cultures in shake flasks containing medium and 5% (v/v) solvent (Buitelaar *et al.*, 1991). After 10 d incubation in these two-phase systems, root growth, O$_2$ consumption and thiophene production were determined as indicators of the solvent toxic effect at both molecular

and phase levels (see Table 21.1). Results for growth and respiration activity were comparable. Only hydrophobic solvents with log P higher than 5 were not toxic, allowing cells to grow and produce secondary metabolites at levels similar to the controls. Analogous experiments with hairy roots immobilised in alginate gave the same trend. The results of Table 21.1 suggest that the log P threshold of 4 (Laane *et al.*, 1987) shifts for *Tagetes* to a value of about 5. This trend has been extensively confirmed by a study conducted with different classes of solvent and plant cells of *Morinda citrifolia* (Bassetti and Tramper, 1994), indicating that plant cells are more sensitive than other biocatalysts to the presence of organic solvents, with only the rather hydrophobic solvents not being harmful. If, however, an external extractor is introduced into the reactor configuration, volatile "toxic" solvents such as hexane and pentane can also be used for the recovery of plant cell secondary metabolites (Corry *et al.*, 1993), because the cells have no direct contact with the solvent.

Bioreactor studies

Physical parameters such as thiophene log P and partition coefficients, plus the phase separation behaviour of the medium/solvent two-phase system, must be considered for the rational choice of solvent for use in bioreactor experiments. From the biocompatible solvents listed in Table 21.1 (with log P values higher than 5), those — such as the phthalates — with density very close to water or medium were excluded because of the difficulty of

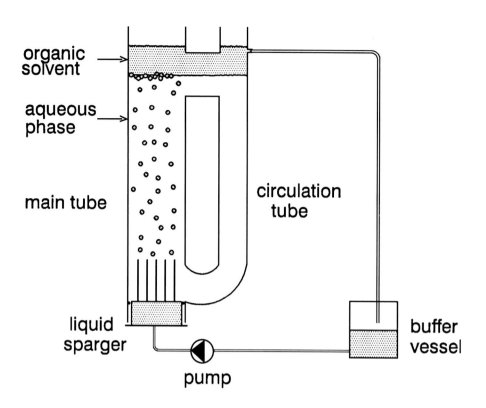

achieving phase separation during downstream recovery processes. Decane was also eliminated as it hampers procedures for thiophene determination (Buitelaar *et al.*, 1991). Therefore, hexadecane and FC 40 were selected for the experiments in bioreactors.

Three types of fermenter were tested: a liquid-impelled loop reactor, a stirred tank reactor, and a bubble column (Buitelaar *et al.*, 1991). The liquid-impelled loop reactor is depicted in Figure 21.2 and was specifically designed for organic–aqueous two-phase systems. It exploits the same operating principle as the airlift loop reactor, but instead of a gas phase, a dispersed liquid phase induces continuous circulation of the medium (liquid phase). The two phases should be immiscible and must have different densities (Tramper *et al.*, 1987). A standard fermenter of 2 L volume was used as the stirred tank reactor. The bubble columns tested were glass tubes with a sintered glass plate for air sparging in the lower part, and with an external pump to circulate the solvent and promote contact with the hairy roots.

A general problem associated with growing hairy roots in bioreactors is their growth as a single clump of interconnected roots. Alternative methods for biomass evaluation, such as medium conductivity measurement (Buitelaar *et al.*, 1991), are therefore required. In the liquid-impelled loop reactor with hexadecane as second phase, very fast growth of *T. patula* hairy roots occurred, and after only one week flow of liquid in the reactor was blocked. Immobilisation of hairy roots in calcium alginate beads could prolong the experiment for only one week, therefore not providing a practical alternative. Very fast growth did not occur in the other two reactor configurations, which were operated for about one month with hexadecane or FC 40 as the second phase. A higher volumetric growth rate was observed in the stirred tank than in the bubble column; better results were obtained with FC 40 than with hexadecane as dispersed phase. In contrast, the bubble column system with hexadecane gave the best performance with respect to thiophene production. Hairy roots of *Tagetes* do not excrete thiophenes under physiological conditions, but in the stirred tank reactor in the presence of hexadecane as dispersed phase, up to 70% of the intracellular thiophene content was released. FC 40 was able to extract 10–20% of the intracellular products. Organic–aqueous two-phase systems can thus be used for *in situ* extraction of thiophene metabolites with retention of cell viability.

Table 21.2 lists some non-lethal treatments recently used to accomplish secondary metabolite release from plant cells *in vitro*. A two-phase system containing *n*-hexadecane as extractant has been used effectively to recover up to 95% of shikonin from flask culture of *Lithospermum erythrorhizon* hairy roots. Hexadecane was observed in this case to have a peculiar effect on the morphology of *L. erythrorhizon*

Table 21.2. Treatments to achieve release of secondary metabolites from plant cells with retention of cell viability. Release is expressed as percentage of the total content.

Species	Type of system	Compound released	Treatment	Release (%)	Reference
Beta vulgaris	Suspended plant cells	Betanin	Ultrasound	5–10	Kilby and Hunter (1990)
	Hairy roots	Betanin	Mild heat	15	DiIorio et al. (1993)
	Hairy roots	Betanin	$(NH_4)_2SO_4$	15	DiIorio et al. (1993)
Rubia tinctorum	Hairy roots	Anthraquinones	O_2 starvation	5	Kino-oka et al. (1994)
Morinda citrifolia	Suspended plant cells	Anthraquinones	Lypozime®	7	Dörnenburg and Knorr (1992)
	Suspended plant cells	Anthraquinones	Hexadecane	37	Bassetti and Tramper (1995)
	Suspended plant cells	Anthraquinones	Pluronic F-68	55	Bassetti et al. (1995)
Lithospermum erythrorhizon	Hairy roots	Shikonin	Hexadecane	95	Sim and Chang (1993)
Coleus blumei	Suspended plant cells	Rosmarinic acid	DMSO[1]	66	Park and Martinez (1992)
Catharanthus roseus	Suspended plant cells	Indole alkaloids	DMSO	85–90	Brodelius and Nilsson (1983)
	Suspended plant cells	Inorganic phosphate	Triton X-100	50	Brodelius (1988)
	Hairy roots	Indole alkaloids	XAD-7 + DMSO[2]	20 and 70[3]	Sim et al. (1994)

[1] DMSO = dimethylsulfoxide.
[2] A fungal elicitor was also used.
[3] 20% for catharanthine, 70% for ajmalicine.

hairy roots; microscopic examination showed that roots treated with solvent appeared healthy and young, while those growing in the absence of solvent were damaged with some detached root hairs (Sim and Chang, 1993). Hexadecane, however, like other biocompatible solvents used in extractive biocatalysis, has a molecular weight and boiling point comparable to those of the secondary metabolites to be extracted (Corry et al., 1993). This can hinder downstream processing, especially product isolation from the organic phase. Different approaches to product release have therefore been investigated. *Beta vulgaris* hairy roots producing betanin as secondary metabolite have been forced to release up to 15% of the retained product by addition of ammonium sulphate or heat treatment (45 min at 42°C); higher levels of release were achieved but at the expense of cell viability (DiIorio et al., 1993). Kino-oka and co-workers (1994) exposed hairy roots of *Rubia tinctorum* to several cycles of oxygen starvation, obtaining a 5% release of anthraquinone pigments in a turbine-blade fermenter.

Aqueous Two-Phase Systems

Aqueous two-phase systems are mixtures of two aqueous solutions, either two polymers or a polymer and a salt. Originally developed for separating cells and macromolecules (Albertsson, 1986), these systems have been used also in bacterial, fungal and enzymatic bioconversions. For effective use of this technique, cells and macromolecules should partition into different phases according to their hydrophobicity and surface properties. Use of aqueous two-phase systems with plant cell cultures is at present very limited and only two reports have appeared in the last 5–6 years (Hooker and Lee, 1990; Buitelaar et al., 1992a). A reason could be the rather general hydrophobicity of plant cell secondary metabolites and the resulting problems of partitioning in such aqueous, therefore hydrophilic, systems.

Hairy roots of *Tagetes patula* were grown in two-phase systems containing several types of polyethyleneglycol (PEG) polymers as top phase and Dextran or Reppal as bottom phase (Buitelaar et al., 1992a). Hairy roots were found to be located completely in the bottom phase, while most of the released thiophenes were extracted by the more hydrophobic PEG phase. In both shake-flask and bioreactor (stirred tank) experiments, the growth rate of hairy roots in the two-phase systems was lower than in the control culture; thiophene production was apparently not affected. A slight increase was observed for product release, but this accounted for only 3% of the total thiophene content. Two-phase systems containing salts could not support hairy root growth.

Aqueous two-phase systems do not appear thus to be effective when applied to the production and extraction of thiophenes, for which organic–aqueous systems look more promising.

Solid–Liquid Two-Phase Systems

An attractive alternative to solvent extraction is the use of a solid extractant phase to recover the desired product from the culture medium and inhibit any feedback repression mechanisms, with the possible enhancement of secondary metabolite production. Because of their selective extractive power, solid phases can be used to extract from plant cells only some of many chemically-related compounds (Buitelaar and Tramper, 1992). Solid–liquid two-phase systems have been effectively employed to recover, for example, anthraquinones from *Cinchona ledgeriana* (Robins and Rhodes, 1986); alkaloids from *Catharanthus roseus* (Asada and Shuler, 1989), *Nicotiana rustica* (Rhodes *et al.*, 1986) and *Datura quercifolia* (Dupraz *et al.*, 1994); and coniferyl aldehyde from *Matricaria chamomilla* (Knoop and Beiderbeck, 1983). In most cases, resins with adsorbing properties, such as Amberlite XAD, were used.

In *Tagetes* cultures, increasing thiophene concentration in the growth medium causes inhibition of thiophene synthesis (Buitelaar *et al.*, 1993). It would therefore be advantageous to remove these compounds from the culture medium. *In situ* extraction with the ion-exchange resin XAD-7 was applied with this aim to hairy roots of *Tagetes patula*; XAD-7 affected neither hairy root growth nor thiophene production, although an increase in product excretion of up to 40% was observed. The simultaneous presence of ion-exchange resin and a fungal elicitor was also tested since the two treatments may have a synergistic effect (Asada and Shuler, 1989). Simultaneous treatment with XAD-7 and a fungal elicitor (extract of *Aspergillus niger*) increased product excretion to a even greater extent (> 50%) (Buitelaar *et al.*, 1993). A similar trend was observed in a study with hairy root cultures of *Catharanthus roseus*: the combination of *in situ* adsorption and fungal elicitation significantly increased excretion of the alkaloids ajmalicine and catharanthine (see Table 21.2) (Sim *et al.*, 1994). With *Tagetes patula*, presence of the resin played a key role in the above-mentioned synergistic effect as the elicitor alone was not able to trigger excretion (Buitelaar *et al.*, 1992b). XAD-7 selectively extracted the produced thiophenes: the predominant thiophenes found intracellularly were BBT and BBTOAc, with BBTOH not being detected at all; in contrast, BBTOH accounted for 70% of the total thiophene content adsorbed by the XAD-7 resin. A similar, marked difference in thiophene metabolite distribution was also found in hexadecane two-phase systems (Buitelaar *et al.*, 1991).

Conclusions

Of the several two-phase systems applied to hairy roots of *Tagetes patula*, the organic–aqueous combination gave the best results with respect to non-lethal thiophene recovery and *in situ* extraction. Despite the encouraging results achieved using this technique and other studies of hairy root biotechnology, we are still far from large-scale thiophene production and thus commercial exploitation. A deeper knowledge of hairy root physiology is an essential prerequisite for these technological developments (Fowler, 1988).

References

Albertsson PA (1986). *Partition of Cell Particles and Macromolecules*, 3rd ed. New York: Wiley.

Asada M and ML Shuler (1989). Stimulation of ajmalicine production and excretion from *Catharanthus roseus*: effects of adsorption *in situ*, elicitors and alginate immobilization. *Appl Microbiol Biotechnol* 30: 475–481.

Bar R (1987). Phase toxicity in a water–solvent two-liquid phase microbial system. In *Biocatalysis in Organic Media*, edited by C Laane, J Tramper and MD Lilly, pp. 147–153. Amsterdam: Elsevier.

Bar R (1988). Effect of interphase mixing on a water–organic solvent two-liquid phase microbial system: ethanol fermentation. *J Chem Tech Biotechnol* 43: 49–62.

Bassetti L and J Tramper (1994). Organic solvent toxicity in *Morinda citrifolia* cell suspensions. *Enzyme Microb Technol* 16: 642–648.

Bassetti L and J Tramper (1995). Increased anthraquinone production by *Morinda citrifolia* in a two-phase system with pluronic F-68. *Enzyme Microb Technol* 17: 353–358.

Bassetti L, M Hagendoorn and J Tramper (1995). Surfactant-induced non-lethal release of anthraquinones from suspension cultures of *Morinda citrifolia*. *J Biotechnol* 39: 149–155.

Beiderbeck R (1982). Zweiphasekultur. Ein weg zur isolierung lipophiler substanzen aus pflanzlichen suspensionskulturen. *Z Pflanzenphysiol* 108: 27–30.

Beiderbeck R and B Knoop (1987). Two-phase culture. In *Cell Culture and Somatic Cell Genetics of Plants*, edited by F Constabel and IK Vasil, pp. 255–266, vol 4. San Diego: Academic Press.

Breteler H and DH Ketel (1993). *Tagetes* spp. (marigolds): *in vitro* culture and the production of thiophenes. In *Biotechnology in Agriculture and Forestry*, edited by YPS Bajaj, pp. 377–412, vol 21. Medicinal and Aromatic Plants IV. Berlin: Springer-Verlag.

Brink LES and J Tramper (1985). Optimization of organic solvent in multiphase biocatalysis. *Biotechnol Bioeng* 27: 1258–1269.

Brodelius P (1988). Permeabilization of plant cells for release of intracellularly stored products: viability studies. *Appl Microb Biotechnol* 27: 561–566.

Brodelius P and K Nilsson (1983). Permeabilization with immobilized plant cells, resulting in release of intracellularly stored products with preserved cell viability. *Eur J Appl Microb Biotechnol* 17: 275–280.

Bruce LJ and AJ Daugulis (1991). Solvent selection strategies for extractive biocatalysis. *Biotechnol Prog* 7: 116–124.

Buitelaar RM and J Tramper (1992). Strategies to improve the production of secondary metabolites with plant cell cultures: a literature review. *J Biotechnol* 23: 111–141.

Buitelaar RM, AAM Langenhoff, R Heidstra and J Tramper (1991). Growth and thiophene production by hairy root cultures of *Tagetes patula* in various two-liquid-phase bioreactors. *Enzyme Microb Technol* 13: 487–494.

Buitelaar RM, EJTM Leenen and J Tramper (1992a). Growth and secondary metabolite production by hairy roots of *Tagetes patula* in aqueous two-phase systems. *Biocatalysis* 6: 73–80.

Buitelaar RM, MT Cesário and J Tramper (1992b). Elicitation of thiophene production by hairy roots of *Tagetes patula*. *Enzyme Microb Technol* 14: 2–7.

Buitelaar RM, EJTM Leenen, G Geurtsen, Æ de Groot and J Tramper (1993). Effects of the addition of XAD-7 and of elicitor treatment on growth, thiophene production, and excretion by hairy roots of *Tagetes patula*. *Enzyme Microb Technol* 15: 670–676.

Charlwood BV, KA Charlwood and J Molina-Torres (1990). Accumulation of secondary compounds by organized plant cultures. In *Secondary Products from Plant Tissue Culture*, edited by B Charlwood and MJC Rhodes, pp. 167–200. Oxford: Clarendon.

Corry JP, WL Reed and WR Curtis (1993). Enhanced recovery of solavetivone from *Agrobacterium* transformed root cultures of *Hyoscyamus muticus* using integrated product extraction. *Biotechnol Bioeng* 42: 503–508.

Croes AF, AJR van der Berg, M Boesveld, H Breteler and GJ Wullems (1989). Thiophene accumulation in relation to morphology in roots of *Tagetes patula*; effects of auxin and transformation by *Agrobacterium*. *Planta* 179: 43–50.

Curtis WR (1993). Cultivation of roots in bioreactor. *Curr Opinion Biotechnol* 4: 205–210.

DiIorio AA, PJ Weathers and RD Cheetham (1993). Nonlethal secondary product release from transformed root cultures of *Beta vulgaris*. *Appl Microbiol Biotechnol* 39: 174–180.

Dörnenburg H and H Knorr (1992). Release of intracellularly stored anthraquinones by enzymatic permeabilization of viable plant cells. *Proc Biochem* 27: 161–166.

Dupraz JM, P Christen and I Kapetanidis (1994). Tropane alkaloids in transformed roots of *Datura quercifolia*. *Planta Med* 60: 158–162.

Fowler MW (1988). Problems in commercial exploitation of plant cell cultures. In *Applications of Plant Cell and Tissue Culture*, edited by G Bock and J Marsh, pp. 239–253. Chichester: Wiley.

Hooker BS and JM Lee (1990). Cultivation of plant cells in aqueous two-phase polymer systems. *Plant Cell Rep* 8: 546–549.

Kilby NJ, and CS Hunter (1990). Repeated harvest of vacuole-located secondary product from *in vitro* grown plant cells using 1.02 MHz ultrasound. *Appl Microb Biotechnol* 33: 448–451.

Kino-oka M, K Mine, M Taya, S Tone and T Ichi (1994). Production and release of anthraquinone pigments by hairy roots of madder (*Rubia tinctorum* L.). under improved culture conditions. *J Ferment Bioeng* 77: 103–106.

Knoop B and R Beiderbeck (1983). Adsorbent culture method for the enhanced production of secondary substances in plant suspension culture. *Z Naturforsch* 38C: 484–486.

Laane C, S Boeren, K Vos and C Veeger (1985). On optimizing organic solvents in multi-liquid-phase biocatalysis. *Trends in Biotechnol* 3: 251–252.

Laane C, S Boeren, K Vos and C Veeger (1987). Rules for optimization of biocatalysis in organic solvents. *Biotechnol Bioeng* 30: 81–87.

Osborne SJ, J Leaver, MK Turner and P Dunnill (1990). Correlation of biocatalytic activity in an organic–aqueous two-phase system with solvent concentration in the cell membrane. *Enzyme Microb Technol* 12: 281–291.

Park CH and B Martinez (1992). Enhanced release of rosmarinic acid from *Coleus blumei* permeabilized by dimethyl-sulfoxide (DMSO) while preserving cell viability and growth. *Biotechnol Bioeng* 40: 459–464.

Rhodes MJC, M Hilton, AJ Parr, JD Hamill and RJ Robins (1986). Nicotine production by hairy root culture of *Nicotiana rustica*: fermentation and product recovery. *Biotechnol Lett* 8: 415–420.

Robins RJ and MJC Rhodes (1986). The stimulation of anthraquinone production by *Cinchona ledgeriana* cultures with polymeric adsorbents. *Appl Microbiol Biotechnol* 24: 35–41.

Sim SJ and HN Chang (1993). Increased shikonin production by hairy roots of *Lithospermum erythrorhizon* in two-phase bubble column reactor. *Biotechnol Lett* 15: 145–150.

Sim SJ, HN Chang, JR Liu and KH Jung (1994). Production and secretion of indole alkaloids in hairy root cultures of *Catharanthus roseus*: effect of *in situ* adsorption, fungal elicitation and permeabilization. *J Ferm Bioeng* 78: 229–234.

Tramper J, I Wolters and P Verlaan (1987). The liquid-impelled loop reactor: a new type of density-difference-mixed bioreactor. In *Biocatalysis in Organic Media*, edited by C Laane, J Tramper and MD Lilly, pp. 311–316. Amsterdam: Elsevier.

Vermuë MH, J Sikkema, A Verheul, R Bakker and J Tramper (1993). Toxicity of homologous series of organic solvents for the gram-positive bacteria *Arthrobacter* and *Nocardia* sp and the gram-negative bacteria *Acinetobacter* and *Pseudomonas* sp. *Biotechnol Bioeng* 42: 747–758.

INDEX